海洋浮游微食物网生态学

张武昌 赵 苑 编著

科学出版社

北 京

内 容 简 介

海洋浮游微食物网生物包括病毒、细菌、蓝细菌、真核藻类、鞭毛虫、纤毛虫等，它们在海洋水体物质生产和能量传递中扮演着重要角色。

本书主要包括三个部分，共 14 章。第一部分为微食物网生物的状态，主要介绍海洋浮游微食物网的发现和研究历史、微食物网的结构及其影响因素、海洋浮游细菌的生存状态，以及微食物网生物的运动和趋化等。第二部分为微食物网生物的营养关系，分别介绍了微食物网生物的异养营养、混合营养，以及各类生物之间的摄食营养关系。第三部分为微食物网与海洋生物地球化学循环，包括微食物网生物与海洋颗粒之间的关系、微食物网生物对营养盐的再生作用、微食物网和经典食物链的相对重要性，以及全球变暖对微食物网的影响等。相关内容包括了现场调查和培养实验的结果，在空间上覆盖了全球各海区。

本书可为海洋生物学和海洋生态学专业的研究人员、教师和学生提供参考资料，也可供海洋环境监测人员、养殖人员阅读参考。

图书在版编目（CIP）数据

海洋浮游微食物网生态学 / 张武昌，赵苑编著. -- 北京 : 科学出版社，2024.11. -- ISBN 978-7-03-080135-7

Ⅰ. Q938.8

中国国家版本馆 CIP 数据核字第 20243FX390 号

责任编辑：韩学哲　薛　丽 / 责任校对：严　娜
责任印制：吴兆东 / 封面设计：刘新新

科学出版社 出版
北京东黄城根北街 16 号
邮政编码：100717
http://www.sciencep.com

三河市春园印刷有限公司印刷
科学出版社发行　各地新华书店经销

*

2024 年 11 月第 一 版　开本：720×1000　1/16
2025 年 2 月第二次印刷　印张：15 3/4
字数：320 000
定价：158.00 元
（如有印装质量问题，我社负责调换）

自　　序

1996 年，我在中国科学院海洋研究所王荣研究员的指导下进行博士研究生阶段的学习，开展微型浮游动物生态学研究。在随后近 30 年的时间里，我与肖天研究员、黄凌风教授、孙军教授等学者进行了密切的合作，共同开展海洋浮游微食物网生态学研究。2012 年科学出版社出版《砂壳纤毛虫图谱》之后，我与赵苑副研究员及历届研究生共同总结、整理了海洋浮游微食物网生态学各个方面的研究成果，开始着手《海洋浮游微食物网生态学》的准备工作。经过多年的积累，终于整理成书呈献给读者，期望本书能成为海洋生态学研究生的入门参考资料。海洋水体是地球上最大的生境，浮游微食物网是这一生境中最基本的生产者和分解者，其重要性不言而喻。希望本书能激发大家对这些微小生物的兴趣，了解它们在微小尺度生境内与众不同的生活方式。

于莹、张翠霞、陈雪、丰美萍、李海波、王超锋、梁晨、宣俊、赵丽、董逸等参与了部分章节的资料整理和编写。焦念志院士对部分稿件内容进行了有益指导，在此深表谢意。本书在资料收集过程中得到了国内外学者的大力支持，张鸿雁博士、John Dolan 研究员和 Michel Denis 研究员帮助查找了很多资料，Farooq Azam 教授和 Tron Frede Thingstad 教授分享了其微食物环论文发表的背景故事，在此致以诚挚的感谢。此外，本书还获得了国家自然科学基金（42076139、42476143）、国家重点研发计划（2022YFC3105300）、中国科学院战略性先导专项（XDB42030402）等项目的资助。

知识的高塔是众人垒成的。虽然我们已竭尽心力，但限于业务水平，常觉知识浅薄。书中难免有疏漏和不妥之处，敬请各位读者不吝指教（wuchangzhang@qdio.ac.cn, yuanzhao@qdio.ca.cn），以便再版时补充、改正。

张武昌

2024 年 5 月于中国科学院海洋研究所

前　　言

微食物环是指由溶解有机碳与细菌、鞭毛虫、纤毛虫等微型生物构成的呈环状回路的营养关系。海水中的溶解有机碳被细菌利用，细菌被鞭毛虫、纤毛虫等原生动物摄食，原生动物被浮游动物摄食，所有这些生物又产生溶解有机碳，碳在这个环形的回路中被循环利用。1983 年，Farooq Azam 等学者提出了微食物环的概念。之后随着微微型自养生物和浮游病毒在海洋生态系统中的重要性被了解，这些微型生物也被加入，使微食物环扩展为微食物网。

微食物环/网概念被科学界广泛认同。以微食物环/网概念为代表和核心的海洋浮游微型生物研究被美国国家科学基金会（National Science Foundation，NSF）评为 1950～2000 年海洋科学的里程碑式成就（Landmark Achievements of Ocean Sciences）之一，被点评为是其中意义最深远的成就（the most far reaching achievement）。

微食物环/网概念的兴起大大促进了各国科学家对海洋微型生物生态学的研究。本书主要汇集整理了这些已有的成果，以期为海洋生态学及相关领域的学者提供参考。微食物网在淡水和海洋生态系统、底栖和浮游生态系统中都有存在，本书主要介绍海洋浮游微食物网的内容，并将重点放在生态学领域，对微食物网生物的分类学内容从略。微食物网生物的分子生态学和生物地理学已逐渐兴起，目前虽有一些研究，但除砂壳纤毛虫之外，微食物网生物在种水平的生物地理分布资料非常有限，分子手段主要分到大类，如科、属等，本书对此部分内容从略。腰鞭毛虫（甲藻）营异养或混合营养，是微型浮游动物的重要组成部分，在浮游植物生态学研究中已有很多介绍，因此本书不再赘述。微食物网的核心是细菌生产，细菌对溶解有机碳的吸收、同化都有大量的研究，本书对此只做简单介绍，而将重点放在微食物网生物的存活状态和相互之间的摄食关系上。

张武昌
中国科学院海洋研究所
2024 年 4 月 28 日

目 录

第 一 部 分

第一章 海洋浮游微食物网的发现和研究历史 ················· 3
- 第一节 前微食物环时代 ················· 3
- 第二节 计数微型生物 ················· 8
- 第三节 微食物环概念的提出 ················· 14
- 第四节 后微食物环时代 ················· 17
- 第五节 西学东渐 ················· 20
- 第六节 小结 ················· 21
- 参考文献 ················· 21

第二章 微食物网的结构及其影响因素 ················· 25
- 第一节 微食物网的丰度和生物量结构 ················· 25
- 第二节 微食物网的粒级结构 ················· 27
- 第三节 微食物网结构的变化及其影响因素 ················· 29
- 参考文献 ················· 34

第三章 海洋浮游细菌的生存状态 ················· 40
- 第一节 细菌生存状态的研究指标 ················· 40
- 第二节 浮游细菌的细胞膜完整性 ················· 42
- 第三节 浮游细菌的呼吸活性 ················· 43
- 第四节 浮游细菌的核酸含量 ················· 46
- 第五节 浮游细菌生存状态研究指标间的关系 ················· 48
- 参考文献 ················· 48

第四章 微食物网生物的运动和趋化 ················· 55
- 第一节 运动和趋化的意义 ················· 55
- 第二节 运动原理和运动器官 ················· 57
- 第三节 微食物网生物的趋化性和其他趋性 ················· 61
- 第四节 微食物网生物的运动及其与细胞的营养吸收 ················· 66
- 第五节 细菌向热点的聚集 ················· 70
- 参考文献 ················· 72

第 二 部 分

第五章 微食物网生物的异养营养 ··· 79
第一节 异养细菌利用溶解有机碳 ··· 79
第二节 病毒对微食物网生物的下行控制 ··· 80
第三节 微型浮游动物的摄食行为 ··· 83
第四节 微型浮游动物的摄食速率 ··· 90
参考文献 ··· 95

第六章 自然海区的混合营养浮游鞭毛虫和纤毛虫 ··· 102
第一节 混合营养鞭毛虫 ··· 103
第二节 混合营养纤毛虫 ··· 105
参考文献 ··· 112

第七章 微型浮游动物和浮游病毒对细菌的下行控制 ··· 116
第一节 微型浮游动物对细菌的下行控制 ··· 116
第二节 浮游病毒对细菌的下行控制 ··· 123
参考文献 ··· 127

第八章 微型浮游动物对自养微食物网生物的摄食 ··· 132
第一节 微型浮游动物对聚球藻的摄食 ··· 132
第二节 微型浮游动物对浮游植物的摄食 ··· 139
第三节 微型浮游动物的 $\delta^{15}N$ 营养级 ··· 143
参考文献 ··· 145

第九章 海洋浮游纤毛虫的摄食 ··· 151
第一节 纤毛虫的食性 ··· 151
第二节 纤毛虫的摄食强度和对饵料的选择性 ··· 153
第三节 影响纤毛虫摄食的因素 ··· 157
第四节 同化和排遗 ··· 160
参考文献 ··· 160

第十章 海洋桡足类对纤毛虫的摄食 ··· 164
第一节 桡足类摄食纤毛虫的早期证据 ··· 164
第二节 桡足类摄食纤毛虫的室内培养研究 ··· 165
第三节 自然海区桡足类对纤毛虫的摄食 ··· 169
参考文献 ··· 172

第 三 部 分

第十一章 微食物网生物与海洋颗粒 ··· 179
第一节 何为海洋颗粒 ··· 179

第二节　颗粒上的微食物网生物 ··· 181
　　第三节　微食物网生物促进海洋颗粒形成 ····································· 184
　　第四节　微食物网生物对颗粒的转化 ·· 186
　　第五节　微食物网生物与海洋碳输出 ·· 188
　　参考文献 ·· 190
第十二章　微食物网生物对营养盐的再生作用 ·································· 195
　　第一节　异养细菌对营养盐的再生作用 ·· 195
　　第二节　病毒对营养盐的再生作用 ·· 196
　　第三节　微型浮游动物对营养盐的再生作用 ································ 197
　　参考文献 ·· 203
第十三章　微食物网和经典食物链的相对重要性 ······························ 207
　　第一节　自养微食物网生物的相对重要性 ···································· 207
　　第二节　异养微食物网生物的相对重要性 ···································· 212
　　第三节　自然海区浮游食物网的结构 ·· 216
　　第四节　微食物网和经典食物链的能量传递效率 ························ 220
　　参考文献 ·· 222
第十四章　全球变暖对微食物网的影响 ·· 225
　　第一节　生态学代谢理论 ·· 225
　　第二节　低温对微食物网生物代谢的影响 ···································· 226
　　第三节　全球变暖对微食物网生物代谢的影响 ···························· 229
　　第四节　温度升高对微食物网生物地理分布的影响 ···················· 235
　　第五节　全球变暖对微食物网重要性的潜在影响 ························ 236
　　参考文献 ·· 236

第一部分

第一章　海洋浮游微食物网的发现和研究历史

海洋浮游微食物网（以下简称微食物网）的研究始于 20 世纪 80 年代初。1983 年美国斯克利普斯海洋研究所（Scripps Institution of Oceanography）的 Farooq Azam 博士等发表了论文 *The ecological role of water-column microbes in the sea*，阐述了海洋细菌与其他微型生物的营养关系，并提出了微食物环（microbial loop）概念（Azam et al.，1983）。海洋微型生物泛指体长或粒径小于 200μm 的生物，包括病毒、细菌、蓝细菌、鞭毛虫、纤毛虫、浮游植物等。海洋细菌被鞭毛虫、纤毛虫等依次摄食，而这些摄食者释放到海水中的溶解有机碳又被细菌吸收再利用，形成细菌-鞭毛虫-纤毛虫的营养闭环，因此被形象地称为"microbial loop"，字面意思是微型生物环，国内普遍翻译为微食物环。

微食物环概念一经提出即得到科学界广泛认同，截至 2024 年 8 月，Azam 等（1983）发表的这篇论文已被引用了 7200 余次。Azam 博士因为提出微食物环的概念而被美国湖沼与海洋学会（American Society of Limnology and Oceanography，ASLO）先后授予 G. Evelyn Hutchinson 奖（G. Evelyn Hutchinson Award，1995）和 John Martin 奖（John Martin Award，2006），并入选 2002 年度汤森路透高被引学者（Thomson Reuters Highly Cited Researchers）名单。以微食物环概念为代表和核心的海洋微型生物研究被美国国家科学基金会评为 1950～2000 年海洋科学领域意义最深远的里程碑式成就之一。

具有广泛和深远影响的微食物环概念于 20 世纪 80 年代初形成，具有其历史的必然性，本章将介绍微食物环概念的历史背景及其演变为微食物网的过程。

第一节　前微食物环时代

一、各浮游生物类群的定量

浮游生物（plankton）一词由德国海洋生物学家 Victor Hensen 于 1887 年提出（Hensen，1887）。传统的浮游生物研究中，需要使用专门的浮游生物网具采集样品。尽管学者意识到有大量的浮游生物会透过网孔，但这些生物个体微小、容易破损、难以固定，因此，直到 20 世纪中期仍没有合适的方法观察其形态、统计其丰度和生物量。这一阶段浮游生物定量研究的对象主要是网采浮游植物（网孔径 70μm）和网采浮游动物（网孔径 200μm），通过显微镜观察计数。

20 世纪中后期随着研究技术进步，对其他浮游生物类群的研究得以开展。1966 年，联合国教科文组织（United Nations Educational, Scientific and Cultural Organization, UNESCO）制定了测定水体叶绿素浓度的标准方法（SCOR-UNESCO, 1966），使采用叶绿素 a 浓度评估该海区浮游植物的生物量成为可能。1967 年，有研究收集了粒径小于 35μm 的微型浮游动物样品，成功观察并计数了以纤毛虫和桡足类无节幼体为主的小型浮游动物丰度（Beers and Stewart, 1967）。

20 世纪初即有研究者用手动离心的方法在浮游生物网过滤后的海水中发现了鞭毛虫（Lohmann, 1911; Hentschel, 1936），随后有研究者在近岸海水培养液中观察到鞭毛虫，并提出这些鞭毛虫可能摄食细菌（Hardin, 1944; Haas and Webb, 1979），但尚无计数海水中鞭毛虫丰度的方法。

计数水体浮游细菌丰度的传统方法是使用琼脂平板培养，将海水适当稀释后涂布在平板上，经过一段时间的培养后统计平板上的菌落数，这样得出的结果是每毫升海水中有几百到几千个细菌。培养计数法步骤烦琐、耗时较长，而且并非所有细菌都能在平板培养基中生长，因此只是一种间接计数（indirect count）的方法，难以获得海洋细菌的准确丰度。

海洋微生物生态学者致力于寻找能够直接计数（direct count）细菌的方法。滤膜和显微镜技术的发展改变了计数海水中细菌的方法，将细菌过滤到滤膜上，使 1%赤藓红（erythrosine）溶解到 5%苯酚（phenol）溶液中制成染色剂，用其染色细菌，显微镜观察计数得到的细菌丰度比培养法高一个数量级（Jannasch and Jones, 1959）。20 世纪 70 年代前，这种赤藓红染色是最主要的细菌直接计数法，但是这一方法较难分辨球状的细菌与海水中的其他颗粒，如要获得较高的准确性，需要进行很多检验，得到的海洋细菌丰度结果并未被广泛接受（Sorokin and Overbeck, 1972）。

除了直接计数, 20 世纪 60 年代还出现了库尔特颗粒计数仪（Coulter counter），可以进行生物粒径谱研究。库尔特颗粒计数仪的原理是监测悬浮于电解液中的颗粒流经小孔时电导率的变化值，以快速统计电解液中颗粒的大小和数目。库尔特颗粒计数仪的最初研发目的是对血细胞（粒径 6~10μm）进行计数和体积测量，因此早期库尔特颗粒计数仪对海洋细菌、鞭毛虫等微型生物的计数欠精准。尽管如此，有学者用库尔特颗粒计数仪计数大洋海水中粒级为 1~100μm 的颗粒，结合已有的浮游生物观测资料，发现海水中从 1μm（细菌）到 10^6μm（鲸）各个粒级生物的体积浓度大致相同（Sheldon et al., 1972）。

总的来说，在前微食物环时代，学者可以对个体较大的网采浮游动物、浮游植物进行准确计数，而对海洋细菌、鞭毛虫等微型生物的研究尚无法定量。

二、浮游生物的生态效率

在对微型生物定量计数研究比较滞后的时候，对浮游生物生态效率的测定却走到了前面。

20世纪50年代初，丹麦海洋生态学家E. Steemann Nielsen在调查航次中成功使用放射性 ^{14}C 标记法对海洋初级生产力进行了测定，其原理为浮游植物通过光合作用吸收 CO_2 合成有机物，若在浮游植物培养体系内加入一定量 $H^{14}CO_3^-$，经过一定时间的培养后用滤膜收集浮游植物并测定其 ^{14}C 放射性强度，即可估算浮游植物的初级生产力（Nielsen, 1952）。该技术灵敏度高，可在野外进行原位培养，适用于测定寡营养大洋区的初级生产力，因此自出现后便迅速成为标准技术，被全球海洋生态学者广泛使用，并发现在近海和大洋，微型浮游植物对初级生产力的贡献均远超大中型浮游植物。

浮游生物呼吸率的检测技术在20世纪60年代中期成熟，Pomeroy和Johannes（1966）使用配备Ag-Pt电极的呼吸计，分别用孔径366μm的浮游生物网和0.8μm的滤膜过滤水样，成功测定了大中型浮游生物与微型生物的群落呼吸率。这一技术随后也被广泛使用，学者发现各个海区微型生物的呼吸作用普遍超过大中型浮游生物的呼吸作用，通常是大中型浮游生物的10倍，表明微型生物消耗了初级生产所提供的大部分能量。

上述研究表明，海洋初级生产和呼吸大部分都是由微型生物完成的，微型生物在海洋生态系统中有着重要的生态意义。

三、浮游生物利用溶解有机物的研究

人们很早就认识到溶解有机碳（dissolved organic carbon，DOC）是海洋中碳的重要存在形式，并且一直在思考溶解有机碳和生物的关系。Pütter（1907, 1909）认为DOC在海洋生物营养中起核心作用，许多生物可以DOC为食，而Krogh（1931）和Bond（1933）则认为大型生物（macroscopic organism）和大的微型生物（larger microscopic form）无法利用DOC存活。而由于一直没有好的方法来定量检测DOC，上述观点均未得到实验验证。

1934年，出现了测定海水中溶解有机物（dissolved organic matter，DOM）含量的技术（Krogh and Keys, 1934），人们才得以知晓海水DOM的准确浓度，即每立方米海水约含200μg溶解有机氮（dissolved organic nitrogen，DON）和6~10倍于该浓度的DOC。Keys等（1935）认为，由于细菌可被原生动物摄食，而且已知细菌可以从一些有机物的极稀溶液中摄取营养，探讨细菌在多大程度上可以利用海水中DOM是必要的。因此他们尝试过滤掉海水中其他生物，仅保留细菌粒级的生物进行培养实验，发现这些细菌粒级生物的代谢可消耗海水中 10%～15%的 DOM（Keys et al.，1935）。有其他研究也进行了类似实验，如在英吉利海峡，约 50%的浮游植物合成的 DOM 被异养过程所消耗（Andrews and Williams，1971）。在地中海、大西洋和英国近海，粒径小于 8μm 的生物利用了总 DOM 底物的 80%，粒径小于 1.2μm 的生物利用了总 DOM 底物的 50%（Williams，1970）。

总的来说，在前微食物环时代，培养实验证明小粒级的微型生物消耗了海水中大量的 DOM，但尚无法确定具体是哪些生物利用了这些 DOM。

四、对微型生物和大型生物在食物网中作用的认识

在前微食物环时代，学界对海洋食物网的认识主要是经典食物链的营养级结构，即浮游植物-浮游动物-鱼类，微型生物仅作为分解者参与海洋食物网的能流传递。

英国海洋生态学家 John H. Steele（1926～2013）在他的著作《海洋生态系统的结构》（*The Structure of Marine Ecosystems*）中这样写道："开阔海域的浮游植物几乎刚生产出来就被摄食掉，因此所有的浮游植物生产都经植食者传递。生活在海底的动物依靠这些植食者的粪便生活，而不是直接依靠沉降下来的浮游植物"（Steele，1974）。

美国浮游生物学家 Lawrence R. Pomeroy（1925～2020）总结道："在经典的海洋食物网模式中，初级消费者是网采浮游动物，如桡足类、糠虾和磷虾，次级消费者和三级消费者是游泳生物，包括鱼、头足类和鲸，微生物只是分解者"（Pomeroy，1974）。

五、对浮游生态系统的思考：Pomeroy 的故事

Lawrence R. Pomeroy 博士是 20 世纪中后期非常有影响力的生态学家之一，他的研究改变了我们对海洋磷循环的认识，促进和推广了食物网模型的研究与应用，解析了温度如何成为北冰洋食物网的限制因子。然而 Pomeroy 博士对海洋生态学领域最具影响力的贡献是提出了"微生物是海洋生态系统运作的核心"的新范式。1974 年，他在《生物科学》（*BioScience*）上发表的综述论文《海洋食物网：正在改变的范式》（*The ocean's food web: a changing paradigm*）（Pomeroy，1974）完全颠覆了学界对海洋生物营养级关系和动态的既定理解。他认为，微型生物而非大型生物，才是海洋食物网的驱动力。这一观点在当时是相当激进的，直到 20 世纪 80 年代才开始被接受，随后在陆地土壤生态学中也产生了很大影响。Pomeroy 博士对浮游生

态系统营养级关系的思考和发表论文提出新范式的过程简要概括如下。

美国生物学家 Eugene P. Odum（1913~2002）在美国佐治亚大学成立了生态学研究所，并在佐治亚州海岸一个仅能通过乘船到达的障壁岛（Barrier Island）——萨佩洛岛（Sapelo Island）上建立了佐治亚大学海洋研究所（University of Georgia Marine Institute），研究萨佩洛岛周边河口盐沼的生态特点。生态学家 John M. Teal（1929~2024）在此开展了一项关于盐沼生态系统的能量流动的研究（Teal，1962）。盐沼的东南部有大片耐盐的互花米草（*Spartina alterniflora*）。这种植物在佐治亚州沿岸的高温条件下生长迅速且生物量巨大，每平方米盐沼的互花米草生物量与艾奥瓦州玉米地的生物量相当。秋季互花米草变黄并枯萎，每天两次的潮汐淹没盐沼，把许多草叶和茎带到潮沟中。海洋真菌和细菌定植在互花米草的草叶上，并将草叶分解成碎屑，分散在盐沼表面和河口的水体中。在这个过程中，海洋微生物增加了碎屑中有机氮、磷的含量，为盐沼动物提供了更富营养的植物碎屑作为食物。Odum、Teal 和该海洋研究所其他学者的研究表明，河口盐沼食物网主要是以这种富含微生物的碎屑为基础。这个"碎屑食物网"（detritus food web）概念后来对 Pomeroy 量化和评估微生物降解作用在海洋浮游生态系统 DOM 循环中的重要性很有启发。

1954 年，年轻的 Pomeroy 博士受一个同样对水生生态系统的元素循环有浓厚兴趣的同事 Robert E. Johannes（1936~2002）博士邀请加入佐治亚大学海洋研究所。Johannes 培养了河口水体中采集的鞭毛虫，发现其单位生物量释放的磷（biomass-specific release rate of phosphorus）远高于桡足类和其他浮游动物，有力证明了微型浮游动物的生长率高于大中型浮游动物。该发现在一定程度上促进了 Pomeroy 和 Johannes 的合作。二人共同进行了海水黑白瓶对照培养实验，对比原海水的呼吸率和经 2 号浮游生物网（孔径 366μm）过滤的海水的呼吸率，他们的实验证明黑瓶的呼吸率主要是由于微型生物的活动，而不是后生动物如桡足类等造成的，单位体积海水中微型生物（主要是细菌等微生物）的呼吸率约是较大个体浮游动物的 10 倍（Pomeroy and Johannes，1966）。

到 20 世纪 70 年代初期，Pomeroy 已经意识到当时主流认可的海洋食物网的结构和功能的概念是不完整的，异养微生物在浮游生物食物网中的地位比 Steele（1974）的模型显示的更重要。Pomeroy 将传统生态系统理论与"碎屑食物网"概念结合起来，强调了微生物在海洋物质和能量流动中的贡献，形成了关于微生物在海洋生态系统中具有重要作用的独特观点。

大约在这个时候，*BioScience* 的编辑 John Bardach 正在寻求具有广泛关注度的综述论文以吸引读者。他向他的朋友 Johannes 征求潜在作者，而 Johannes 推荐了 Pomeroy。Pomeroy 总结了当时浮游生物的生态效率研究、生物粒径谱研究、浮游生物利用 DOC 的研究成果，汇总了支持他新观点的证据，向 *BioScience* 投了一篇题为《海洋食物网：正在改变的范式》（*The ocean's food web: a changing paradigm*）的综述。在文中 Pomeroy 表明以往研究的网采浮游生物只是海洋浮游生态系统生物量的一部分，网具采集不到的小粒级微型生物的生物量与网采浮游生物量相当，其生态功能在整个生态系统中的贡献很大，并最终提出微生物是海洋生态系统运作的核心这一新范式（Pomeroy，1974）。

据 Pomeroy 所说，这篇综述的原稿被送给了两位外审专家。其中一位从未回应，另外一位认为 Pomeroy 的综述是胡说八道（nonsense），应该拒稿。然而编辑 Bardach 很喜欢这篇综述，还是将其发表了。该综述一开始并未引起太多关注，但到 20 世纪 80 年代末时，Pomeroy 的观点已被广泛接受。至 2024 年 8 月该综述已被引用了 1800 余次。

Pomeroy 在文中虽然意识到了粒级较小的生物的重要性，但也承认这些微型生物与硅藻、桡足类之间的营养关系目前并不清楚（unseen strand in the food web）。他在新范式的图示中给出了一些假想的营养关系，如细菌被原生动物摄食，微型浮游生物（包括动物和植物）只能被海樽等过滤摄食的动物摄食（图 1-1）。

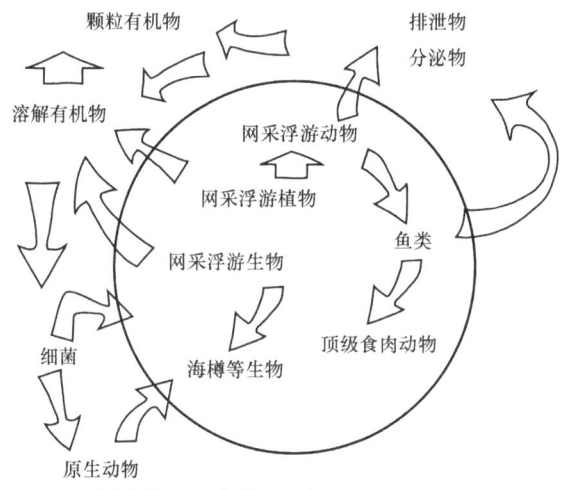

图 1-1　Pomeroy 提出的海洋食物网结构范式[改绘自 Pomeroy（1974）]
圆圈内是简化形式的经典海洋食物网，Pomeroy 新添加的微型生物途径在圆圈外

鉴于 Pomeroy 的开创性贡献，他于 1983～1984 年当选美国湖沼与海洋学会主席，1987 年获美国湖沼与海洋学会颁发的 G. Evelyn Hutchinson 奖，2001 年获海岸与河口研究联合会（Coastal and Estuarine Research Federation）颁发的 Odum 终身成就奖（Odum Lifetime Achievement Award）。

第二节　计数微型生物

Pomeroy（1974）提出应该加强对海洋细菌的研究："如果微生物是海洋中主

要的消费者，我们就应该知道哪些种类的微生物具有重要的代谢作用，以及它们是怎样整合入食物网的。然而我们目前（指发表文章的 1974 年）甚至不确定这些微生物的相对丰度。"

这个呼吁得到了科学界的赞同，海洋生态学家急需一种能准确计数微生物的方法，这个方法最终由 Hobbie 等（1977）完成。

一、滤膜技术的发展

（一）深度膜

深度膜（depth filter）一般由纤维交织而成，具有一定的厚度，液体中的颗粒进入到滤膜的内部时，大颗粒无法通过纤维之间的空隙而被滤膜截留，小颗粒则可以通过，从而达到过滤的目的（图 1-2a）。深度膜的孔径是一个范围，很难做到整齐划一。颗粒不是停留在滤膜的表面，而是位于滤膜内部交织的网状结构中，因此不适用于显微镜计数颗粒。最常见的深度膜是玻璃纤维滤膜（glass fiber filter，GF 膜）。

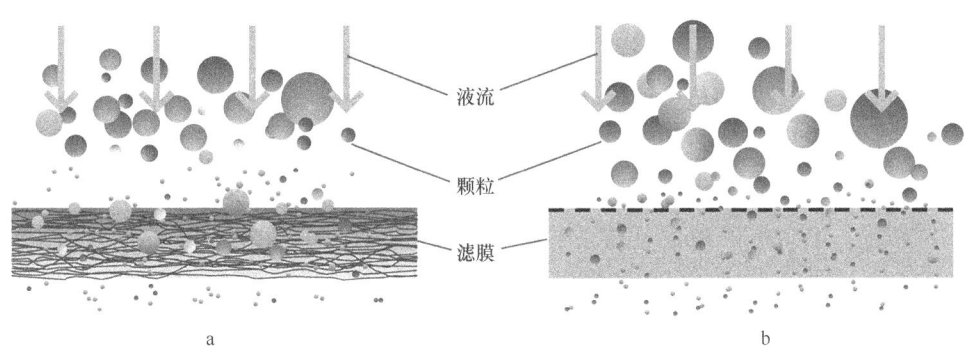

图 1-2 深度膜和表面膜的工作原理
a. 深度膜；b. 表面膜

（二）表面膜

与深度膜相对应的是表面膜（surface filter，又称 membrane filter），液体中的超过膜孔径的颗粒不能进入膜的内部，只能停留在膜的表面（图 1-2b）。表面膜是 20 世纪初由德国哥廷根大学（University of Göttingen）的诺贝尔化学奖得主 Richard Adolf Zsigmondy 和他的助手 Wilhelm Bachmann 发明的。生产滤膜使用的主要原料是醋酸纤维或硝酸纤维聚合物，将这些物质溶解在溶剂中，在玻璃平板上铺上薄薄的一层，溶剂挥发后，这些聚合物固化为滤膜，从玻璃平板上取下滤膜进行进一步加工即可使用。

德国的赛多利斯（Sartorius）公司最先使用这项发明生产滤膜。在第二次世界大战之前，赛多利斯公司生产的滤膜主要用于科学研究，商业用途很少。之后美国波士顿的 Lovell Chemical Company 开始生产基于滤膜技术的过滤装置和过滤系统，主要目的是分离流体中的不同分子。1953 年滤膜技术解密，可以被用于商业用途，这时 Lovell Chemical Company 的雇员 Jack Bush 买下这项技术，成立了密理博过滤器公司（Millipore Filter Company）以大规模开发和生产表面膜，这就是后来鼎鼎大名的密理博公司（Millipore Corporation）的前身，其名称来自亚德里亚海的一种海绵。从此表面膜逐渐应用于美国的微生物学研究。

（三）核孔滤膜

1958 年，英国原子能科学研究院（Atomic Energy Research Establishment，AERE）的 D.A. Young 博士首次发现核径迹（nuclear track），即核粒子穿透绝缘固体时，会在原子尺度上产生一条狭窄的强辐射损伤路径。Young 发现若使用合适的化学物质进行处理，可扩大这些辐射损伤路径，他在氟化锂晶体和云母中蚀刻出核粒子的痕迹并进行了光学显微观察（Young, 1958），随后，同属 AERE 的 E.C.H. Silk 和 R.S. Barnes 用透射电子显微镜观察了辐射损伤路径（Silk and Barnes, 1959）。

1961 年 5 月，美国宣布了登月计划。在月球研究的热潮中，美国通用电气公司研究实验室（General Electric Research Laboratory，GERL）的 Robert M. Walker 博士提出若使用电子显微镜观察月球矿物样品，有可能观察到地外矿物中的宇宙射线损伤痕迹。Walker 到 AERE 拜访了 Silk 和 Barnes，回到 GERL 后，Walker 邀请 P. Buford Price 博士和他一起，在陨石和其他地外物体中寻找古代高能带电粒子的痕迹。Walker 与 Price 用裂变碎片（fission fragment）作为能量源，以云母作为目标开展实验，他们发现辐射损伤路径可以被化学蚀刻并留下细孔，孔径尺寸可以由蚀刻条件控制（Price and Walker, 1962）。在这一系列研究中他们并未提及 Young 的工作。很快，GERL 研究团队为自己的成果提交了专利技术披露书并申请专利。

1962 年下半年，Robert L. Fleischer 博士加入了 Walker 和 Price 的团队。1963 年初，Fleischer 发现辐射损伤路径的化学蚀刻作用原理可用来在其他各种材料上"钻"出精准直径的孔。在此之前，GERL 团队仅在云母上进行化学蚀刻，而 Fleischer 成功将辐射损伤和化学蚀刻应用于通用电气公司生产的塑料材料 Lexan™ 聚碳酸酯（Fleischer et al., 1964），相关论文 *Novel filter for biological materials* 发表在《科学》（*Science*）期刊上。这一工艺制作的滤膜被称为塑料-辐射-蚀刻膜（plastics-irradiated-etched membrane，PIE 膜）（Karl, 2007）。1964 年，通用电气公司开始使用商标 Nuclepore®（nuclear-track pores 的缩写，也常被误写为 Nucleopore）生产和销售核孔滤膜（Nuclepore® membrane filter）。核孔滤膜的主要优点是表面更加平滑、孔径更加均一（图 1-3b），非常适用于后来的荧光显微镜观察计数细菌丰度。

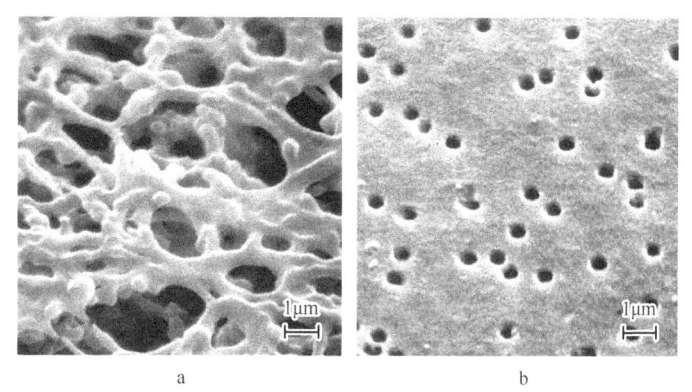

图 1-3 扫描电镜下的纤维素表面膜和核孔滤膜（Karl，2007）
a. 纤维素表面膜；b. 核孔滤膜

二、荧光显微技术的发展

1902~1903 年，德国卡尔蔡司公司（Carl Zeiss AG）的物理学家 Henry Siedentopf 与发明表面膜的 Zsigmondy 教授合作，研发出了超显微镜（ultramicroscope）。超显微镜是一种带有侧向照明的暗场显微镜（dark field microscope），适用于微小颗粒的观察。Siedentopf 在暗视野下观察紫外线照射的样品，发现样品发出荧光，这是首次在显微镜下观察到荧光。

1904 年，耶拿蔡司工厂（Jena Zeiss factory）的 August Köhler 使用石英单色物镜观察氰化钡铂晶体，发现该晶体可发射荧光。随后 Köhler 与 Siedentopf 共同发明了紫外显微镜（ultraviolet microscope），使用火花激发出波长为 275nm 的紫外光作为光源，这是荧光显微镜的前身。他们用紫外显微镜成功观察并拍摄了细胞核内的染色质（chromatin）的照片。

1903 年，美国约翰斯·霍普金斯大学（Johns Hopkins University）的 Robert Wood 使用亚硝基二甲基苯胺溶液从弧光灯中分离出长波紫外线辐射带（300~400nm），其可为荧光显微镜提供稳定可靠的光源。1910 年，耶拿蔡司工厂的德国物理学家 Heinrich Lehmann 根据 Wood 的研究制作出了紫外滤光片。参考 Lehmann 滤光片的原理，1911 年，奥地利卡尔·雷切特光学公司（Carl Reichert Optische Werke AG）的物理学家 Oskar Heimstädt 制作了一台配备暗场石英聚光器的荧光显微镜，使用透镜将弧光灯的光线通过 Wood 溶液滤器聚焦到石英聚光器。1912 年初，大约在雷切特荧光显微镜上市的同时，Lehmann 也发布了全新的蔡司明场荧光显微镜，并于 1913 年上市销售。蔡司荧光显微镜和雷切特荧光显微镜均使用透射紫外光作为光源，两者的区别为蔡司荧光显微镜使用明场聚光器，紫外激发光进入显微镜物镜以照亮样品，而雷切特荧光显微镜使用暗场聚光器，紫外光不会进入显微镜的物镜。

最初的荧光显微镜只能使用紫外光为光源来观察具有自发荧光的样品，而高能紫外线会对样品尤其是活细胞造成光化学损伤。1914 年，捷克生物学家 Stanislaus von Prowazek 发现荧光基团可以功能化并与活细胞结合，这意味着科学家可不依赖于细胞的自发荧光属性，人为设计特定激发和发射波长的荧光标记，从而观察细胞特定生理特征。这一技术使从紫外线到可见光的激发波长均可用于诱发荧光，激发能量大大减少，减少了样品的光损伤，可以保持样品尤其是活细胞的完整性。

落射荧光显微镜的原型是由德国药理学家 Philipp Ellinger 和解剖学家 August Hirt 在 1929 年设计的，他们称之为"活体荧光显微镜"（intravital fluorescence microscope），使用的光源是外焦型（epi-type）入射光（incident light）而非透射光（transmitted light），即激发和发射光学器件位于与样品相同的一侧。激发光经一系列的滤光片到达一个水浸物镜，以观察浸泡在生理溶液中的活器官或生物的荧光。在物镜和观察者眼睛之间有一个黄色的屏障过滤器，可防止大部分反射的激发光干扰观察。使用活体荧光显微镜，Ellinger 和 Hirt 对注射了荧光素（fluorescein）和吖啶黄（trypaflavin）的啮齿动物肾、肝组织进行了观察和成像。

1948 年，苏联学者 Evgenii Brumberg 发明了入射荧光显微镜（incident-light fluorescence microscope）。Brumberg 和他的同事 Krylova 使用干涉镜（interference mirror）引导入射光到样品的正上方，他把这项技术叫作"从上方照明"（"illumination from above"）。虽然 Brumberg 的这一改进是荧光显微镜发展史上的重大贡献，但是这一技术没有得到学界和显微镜生产企业（如蔡司等）的重视，原因之一可能是当时西方国家难以得知苏联的研究结果，另一个可能的原因是，当时明场侧射或暗场透射技术已很好地解决了紫外线激发和观察的问题，所以企业也无须开发其他入射结构的荧光显微镜。

1967 年，荷兰显微镜学家 Johan Sebastiaan Ploem 发明了二色分光元件或称分色镜（dichromatic mirror 或 dichroic mirror）。分色镜可反射约 90%的波长小于截止波长的光，并透射约 90%的波长大于截止波长的光（图 1-4）。Ploem 将分色镜与入射光呈 45°角放置，引导被反射的入射光垂直于样品到达物镜。物镜也起着聚光器的作用，确保样品被均匀激发。样品发出的荧光被物镜收集，并通过分色镜和滤光片回到目镜（图 1-4）。Ploem（1967）称这种入射照明方式为从上面照射，英语简化为 epi-illumination。至此，现代落射荧光显微镜（epifluorescence microscope）已成型。

可能是由于当时获取文献不便，Ploem 在首次发表介绍分色镜的论文时，并不知道苏联学者在 1948 年已有类似成果。但在随后的其他论文中，Ploem 对 Brumberg 和 Krylova 的工作进行了引用与致谢。

随着分色镜和落射荧光显微镜技术的成熟，蔡司和徕卡等厂家相继推出成熟的商品化落射荧光显微镜。荧光显微镜在医学诊断和生物学研究中成为常规手段，也为海洋细菌的精确计数提供了合适的工具。

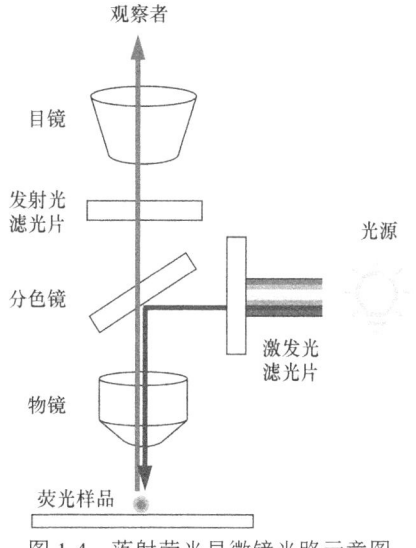

图 1-4 落射荧光显微镜光路示意图

三、海洋细菌的准确计数

如前所述，在 20 世纪 70 年代之前，海洋微生物学者主要通过培养的方法来估算海洋细菌丰度。由于一直没有准确的细菌丰度数据，海洋微生物生态学在很长一段时间内被传统海洋生态学研究者所忽视。

到 20 世纪 70 年代，落射荧光显微镜的性能更好，价格也变得便宜，在各大学术机构开始普及。1973 年，美国北卡罗来纳大学的研究生 Donald E. Francisco 首次报道了用落射荧光显微镜计数自然水体中的细菌丰度。Francisco 使用荧光染料吖啶橙（acridine orange）对水样进行染色，再将染色后的样品过滤至 0.45μm 孔径的黑色 Millipore®纤维素滤膜上，在荧光显微镜 40×和 100×物镜下观察，可清晰地分辨出实验室纯培养和自然淡水样品中的细菌，但他们在自然海水样品和潮间带沙样中的尝试未获成功（Francisco et al.，1973）。

不同类型滤膜在水体微生物计数方面的应用有一个比较和争论的过程。Todd 和 Kerr（1972）将细菌分别过滤到 Millipore®纤维素滤膜和 Nuclepore®核孔滤膜上，使用扫描电子显微镜观察后认为核孔滤膜的效果更理想。Zimmermann 和 Meyer-Reil（1974）认为，纤维素滤膜表面粗糙，易结合荧光染料导致较高的背景荧光，不利于区分细菌和杂质，而且许多细菌被截留在滤膜内部导致计数偏低，因此，推荐使用 Nuclepore®核孔滤膜过滤染色水样。Daley 和 Hobbie（1975）则认为纤维素滤膜的效果优于核孔滤膜。

1977年,美国海洋生物学实验室(Marine Biological Laboratory,MBL)的John E. Hobbie等针对海水样品的特点对前人的技术进行了改进,并发表论文《利用核孔滤膜和荧光显微镜计数细菌》(*Use of Nuclepore filters for counting bacteria by fluorescence microscopy*)(Hobbie et al., 1977)。Hobbie使用依加仑黑(irgalan black,又称酸性黑107)染料将Nuclepore®核孔滤膜染成黑色,以消除滤膜自身的荧光,并增加细菌与背景及杂质的对比。Hobbie在核孔滤膜下方垫上一片纤维素滤膜,以确保细菌在抽滤过程中均匀分布在核孔滤膜上。此外,考虑到核孔滤膜可能有一定的疏水性,还可以在滤膜表面滴上适量表面活性剂。采用这一改良方案,Hobbie等发现使用核孔滤膜得到的细菌丰度是纤维素滤膜的两倍。至此,精确计数海水中细菌丰度的方法已成熟。

Hobbie等(1977)的论文被认为是海洋微生物学历史上的里程碑之一,到1990年时就被引用超900次,被美国科学信息研究所(Institute for Scientific Information,ISI)授予经典引文奖(Citation Classic Award),截至2024年8月该论文已被引用5600余次。John E. Hobbie于1984~1986年当选美国湖沼与海洋学会主席,2007年当选美国艺术与科学院院士,并先后于1983年和2008年获美国湖沼与海洋学会颁发的G. Evelyn Hutchinson奖和A.C. Redfield终身成就奖(A.C. Redfield Lifetime Achievement Award),2005年获海岸与河口研究联合会颁发的Odum终身成就奖。

随着海洋细菌计数方法的成熟,落射荧光显微镜、荧光染料和核孔滤膜也被应用于其他海洋微型生物计数,如自养的聚球藻属蓝细菌(Waterbury et al., 1979)、混合营养和异养鞭毛虫(Davis and Sieburth, 1982; Haas, 1982),以及浮游病毒(Hara et al., 1991)等。各海洋微型生物类群的丰度已不再是谜。

第三节 微食物环概念的提出

Hobbie等(1977)的方法得到广泛应用后,学者发现在每毫升海水中普遍存在约10^6个细菌。这些细菌处于什么状态,其生长率如何,能否传递到上层营养级直至鱼类从而成为人类的食物呢?这些问题开始进入人们的视野。

20世纪80年代初,随着海洋微生物研究进展的加快,一系列国际研讨会的召开促进了海洋微生物学家的交流和思想碰撞。1980年9月底,在德国的基尔(Kiel)进行了第十五届欧洲海洋生物学研讨会(15th European marine biology symposium),并于次年出版会议论文集 *Lower Organisms and their Role in the Food Web*。在该论文集中,英国学者Peter J. leB. Williams在Pomeroy(1974)的基础上综述了最新研究结果,认为现有实验证据已不匹配海洋食物网的经典范式,至少有一半的海洋初级生产经异养细菌矿化，微生物二次生产的生物量，即细菌可提供给下一营养级的有机物，可能与草食浮游动物提供给初级肉食动物的数量相当甚至更多（Williams，1981）。Williams 延续使用 Pomeroy（1974）的海洋食物网范式图，但使用后生动物食物网（metazoan food web）和原生动物食物链（protozoan food chain）这两个概念来描述海洋食物网的结构。在综述的最后，Williams 进一步强调应重视海洋细菌的研究，只有确定了细菌在海洋生态系统中的作用，才能真正认识海洋食物网的物质和能量流动。

1981 年 11 月，北约高等研究所（NATO Advanced Research Institute）在葡萄牙卡斯凯什（Cascais）举办了一场关于海洋微生物代谢和有机物循环过程的学术会议，30 余名海洋微生物学者出席，包括 Azam、Hobbie、Pomeroy、Williams 等。会议主题为海洋异养营养（marine heterotrophy），即海洋光合作用固定的有机碳被转化和最终被呼吸分解的过程。尽管所有的动物和许多微生物均参与了海洋异养活动，并且异养活动在水体和沉积物中均有出现，但这次会议主题聚焦在水体微生物的异养过程（Hobbie and Williams，1984）。在针对"海洋异养过程在食物网中的作用是什么"这一问题的讨论中，与会学者提出了微型异养食物链（microheterotrophic food chain）和微生物食物链（microbial food chain）概念，即细菌将水体中的 DOM 转化为自身生物量，从而成为微型异养食物链的基础，并为后生动物食物链（metazoan food chain）提供除浮游植物之外的营养。会议认为，鞭毛虫、纤毛虫和后生动物可能摄食细菌，但是鞭毛虫可能是细菌的主要摄食者。Ducklow（1983）也提出了微生物食物链（microbial food chain）的概念。

1982 年 5 月，北约科学委员会海洋科学小组（Marine Sciences Panel of the NATO Science Committee）举办了"海洋生态系统中的能量和物质流动：理论与实践"（Flows of energy and materials in marine ecosystems: theory and practice）研讨会。促成会议召开的是海洋研究科学委员会（Scientific Committee on Oceanic Research，SCOR）的生物海洋学数学模型工作组（working group on mathematical models in biological oceanography），小组成员认为当时海洋生态系统的精简模型已经获得一定成功，但未来的前景在于发展海洋生态系统的整体模型。而无论是精简模型还是整体模型，都需要增加对各海洋生物类群生理速率的测量研究，以便计算生态通量。因此，工作组建议组织一次国际会议，设法召集理论生态学家和海洋生物学家，汇总已有和未来可测量的生态通量数据，并评估将这些数据纳入模型的可能性。

在北约和 SCOR 资助下，这次会议在法国卡尔康（Carcans）举办。据会议参加者 Tron Frede Thingstad 教授向笔者回忆，会场设在一个美丽的度假村，与会的各个讨论组围坐在室外松树下的桌子旁，旁边的餐厅里摆放着盛有 200L 波尔多红酒的橡木桶。海洋微生物学者 Farooq Azam、Lutz-Arend Meyer-Reil、Tom

Fenchel 和 Tron Frede Thingstad 给海洋生物学者 John S. Gray 和 John G. Field 讲解他们认为的科学界都应熟知的海洋微生物学研究进展。令他们意外的是，海洋微生物学者讲述的这些知识对海洋生物学者来说全然陌生，尤其是海洋浮游细菌与传统海洋食物网的营养关系。与会者意识到，此前海洋微生物生态学者的工作是独立于海洋生态学其他领域之外的，微生物生态学的进展并没有整合到海洋生态学中来。研究较高营养级浮游生物的学者并未认识到微生物在海洋生态系统中的中心地位，他们仅将微生物视为分解者，而不是生产者。因此，Gray 和 Field 提议应就此主题撰写一篇论文。1984 年，这篇题为 *The role of free bacteria and bacterivory* 的论文发表在以研讨会题目命名的专辑中（Gray et al., 1984），是该专辑的最后一篇，6 位作者在文章中提到了微食物环（microbial loop）这一名词。

也许是意识到海洋生物学者亟需了解海洋微生物的最新研究成果，在会议专辑付梓前的 1983 年，Azam、Fenchel、Field、Gray、Meyer-Reil 和 Thingstad 又在著名期刊《海洋生态学进展丛刊》（*Marine Ecology Progress Series*）上发表了内容相似的论文介绍微食物环（图 1-5），即后来鼎鼎大名的 *The ecological role of water-column microbes in the sea*（Azam et al., 1983）。论文作者之一的 Tom Fenchel 后来回忆道，6 位作者是按照姓氏的字母顺序排序的，至于是谁在探讨中首次提出了"microbial loop"这一名词，他们都记不清了。

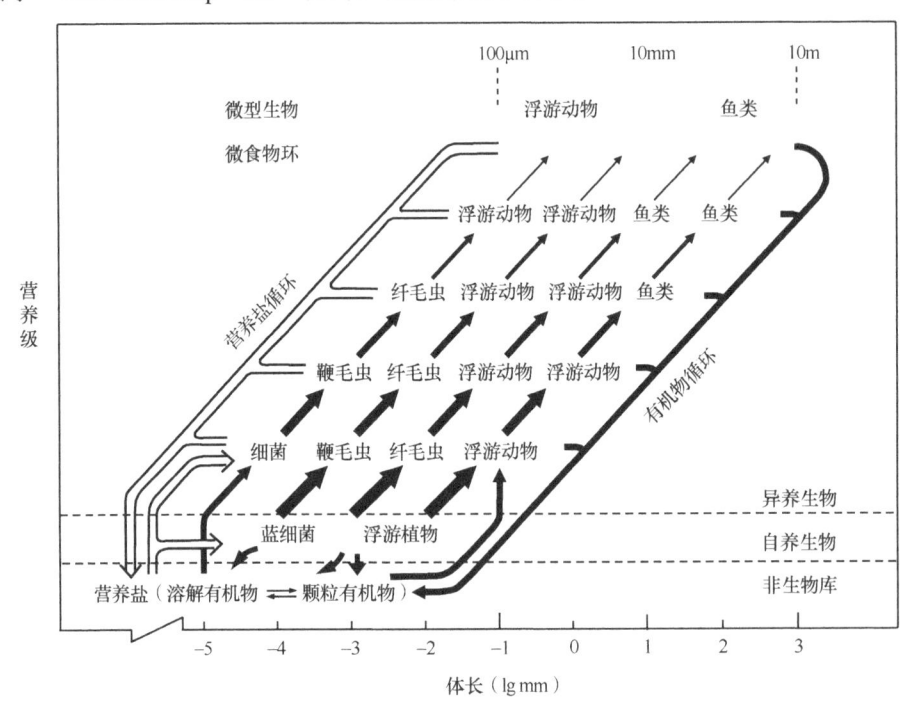

图 1-5　微食物环示意图[改绘自 Azam 等（1983）]

20世纪80年代初,为了将微型生物引入传统的海洋食物链,海洋微生物学者先后共提出了3个不同的名词,即原生动物食物链(protozoan food chain)、微生物食物链(microbial food chain)和微食物环(microbial loop)。从各位海洋微生物学者的原始论文和图示上看,这3个名词指的是同样的内容,但历史最终选择了"微食物环"这一名词。

第四节 后微食物环时代

一、微食物网替代微食物环

微食物环概念提出后,科学家发现海洋中还存在着大量与细菌粒径相近的自养生物,包括聚球藻属(*Synechococcus*)蓝细菌(Waterbury et al., 1979; Murphy and Haugen, 1985)、原绿球藻属(*Prochlorococcus*)蓝细菌(Chisholm et al., 1988; Chisholm et al., 1992),以及海洋微微型真核浮游生物(picoeukaryotes)(Johnson and Sieburth, 1982; Murphy and Haugen, 1985)。这些微微型(pico级,0.2~2μm)自养浮游生物的丰度与异养细菌相当,粒级仅比细菌稍大,通过光合作用进行自身生长,同样被鞭毛虫和纤毛虫所摄食而进入经典食物链。这表明海洋微食物环中的营养关系不是简单的线性环状,而是更复杂的网状,因此微食物环很快被微食物网(microbial food web)概念所替代。

1985年,法国生态学家Paul Bougis和Fereidoun Rassoulzadegan创办了期刊《海洋微食物网》(*Marine Microbial Food Webs*),最初的编委有Lawrence R. Pomeroy、Farooq Azam、Tom Fenchel和John McN. Sieburth等,集中发表海洋微食物网相关的最新研究成果。1986年8月,在纽约召开了题为"水生生态系统中微生物的作用"(The role of microorganisms in aquatic ecosystems)的学术会议,会议成果于1988年在期刊《水生生物学》(*Hydrobiologia*)上集成为专辑,其中,Pomeroy和Wiebe(1988)详细阐释了微食物网的营养级和能量流动。几乎是同时,Sherr E和Sherr B(1988)也对微食物环概念进行了修订,提出了更完善的海洋浮游微食物网(图1-6)。自此,微食物网概念已基本成熟,并取代了微食物环概念。在2001年出版的《海洋科学百科全书》(*Encyclopedia of Ocean Sciences*)中,尽管仍保留了微食物环这一词条,但其正文已基本是微食物网的相关内容(Landry, 2001)。

浮游病毒的计数方法成熟(Bergh et al., 1989)以后,海洋浮游病毒也很快被整合到微食物网中(Bratbak et al., 1992)。海洋浮游病毒作为分解者,裂解藻类和细菌向水中释放DOM,而这部分DOM又再次被细菌利用,形成微食物网的病毒回路(viral shunt)(Wilhelm and Suttle, 1999),进一步完善了微食物网的概念(图1-7)。

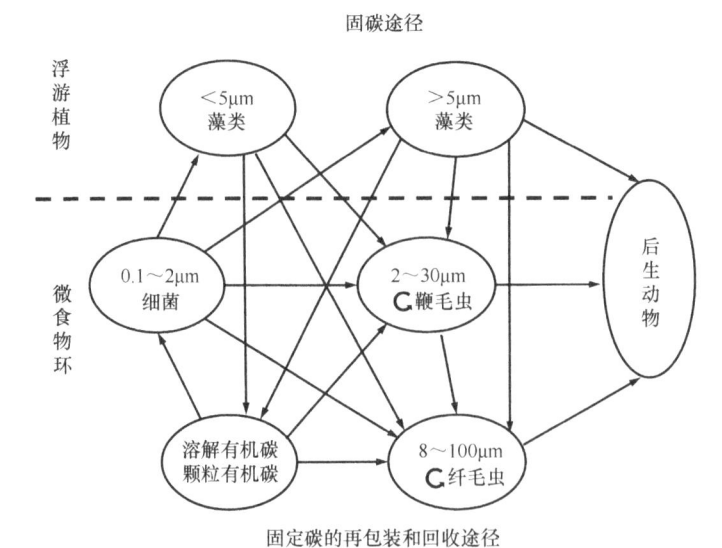

图 1-6 Sherr E 和 Sherr B（1988）提出的微食物网结构示意图

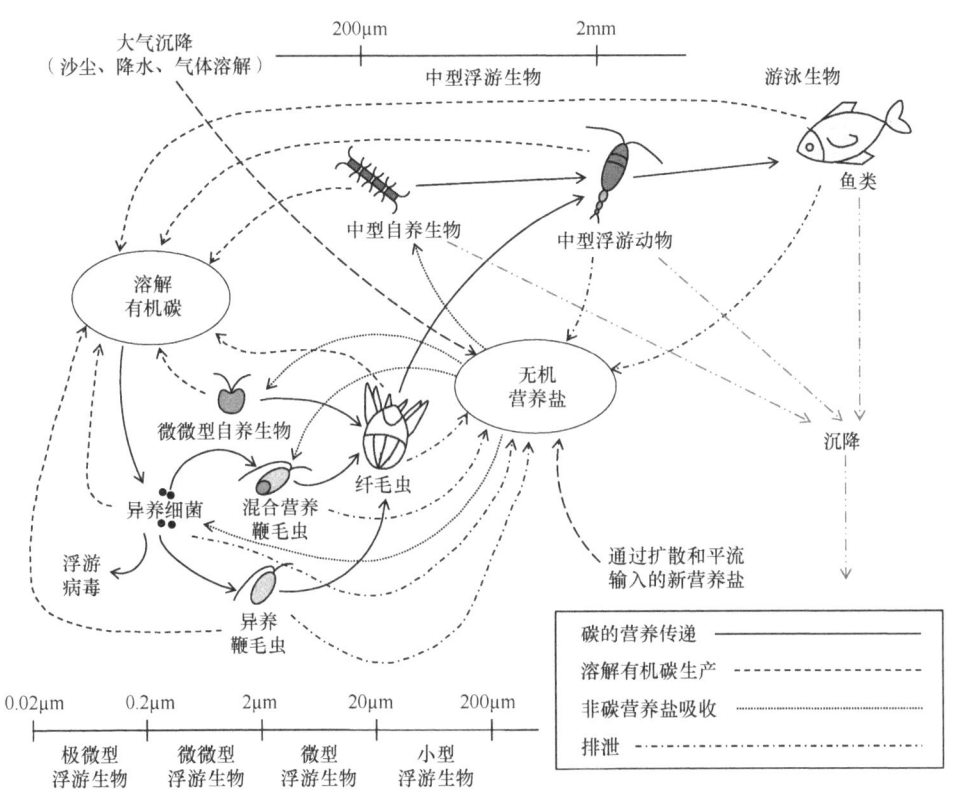

图 1-7 海洋微食物网的物质和能量流动[改绘自 Ocean Studies Board and National Research Council（2000）]

二、海洋浮游生态学的范式改变了吗

Pomeroy 认为海洋浮游生态学范式正在改变，他引用了美国科学哲学家托马斯·库恩（Thomas Kuhn）的著作《科学革命的结构》(*The Structure of Scientific Revolutions*)，写道"几十年来，海洋科学家一直在谨慎地对待这种食物网的观点，每当一个既定的范式受到质疑时，谨慎是可以预期的（Kuhn，1962）。现在有许多证据表明，一个新的海洋食物网范式确实正在出现"(Pomeroy，1974)。库恩认为"科学不是事实、理论和方法的简单堆砌，科学的发展也不是知识的简单积累，而是通过范式的不断转换所进行的不断革命的进程"(Kuhn，1962)。也就是说科学发展的方式不是知识由少到多逐渐发展的方式，而是旧的理论体系（范式）被新的理论体系推翻取代的过程，即范式转变（paradigm shift），最著名的范式转变是日心说取代地心说，以及相对论取代牛顿力学体系。Pomeroy 显然认为他在文中阐述的海洋食物链理论的发展是革命性的，很可能会导致海洋浮游生态学研究发生范式转变，以前的范式主张网采浮游动物是主要的摄食者，粒级较小的生物（1~4μm）仅是分解者，而现在的范式则认为粒级较小的生物是海洋生态系统代谢的主要执行者。

尽管 Pomeroy 持上述观点，但是就目前海洋生态学的知识框架来看，微食物环的提出是对以往观点的总结和补充，并没有导致范式转变。Pomeroy 的这篇论文虽然得到很多引用，但是此后再没有任何一篇论文提及海洋浮游生态学的范式发生转变。

三、对微食物环概念成功的思考

论文 *The ecological role of water-column microbes in the sea* 的作者都没有想到微食物环概念和这篇文章会如此成功，论文作者之一 Tom Fenchel 后来在回忆这段历史时说，这是一篇综述文章，总结了各位作者和其他学者的工作，微食物环概念在之前的许多论文中已被或明确或不明确地提出。早在 20 世纪 40 年代，美国学者 Keys 等在研究细菌对溶解有机物的利用时，就提出细菌可能被原生动物摄食，所以可以说 Keys 等（1935）的论文 *The organic metabolism of sea-water with special reference to the ultimate food cycle in the sea* 最早预测了微食物环的存在。论文 *The ecological role of water-column microbes in the sea*（Azam et al.，1983）的作者之一 Tron Frede Thingstad 对本书作者回忆道："我清晰记得，我当时真没看出有发表这篇文章的必要，因为对我来说没有任何新知识。这是我一生中为数不多的重大错误之一"。Thingstad 还提到，当时著名的海洋微生物学家 Peter J. leB. Williams 在"海洋生态系统中的能量和物质流动：理论与实践"研讨会上参加的是另一个讨论小组，但 Azam 等在文中多次引用 Williams 的成果，因此曾邀请 Williams 做论文的共同作者，但被他谢绝了，他很可能也是觉得没有新知识。

为什么 Azam 等（1983）发表的这篇论文能够大获成功呢？Tom Fenchel 认为，是因为"微食物环"这个名称起得好，仅两个词就清晰阐述了海洋和淡水环境中浮游食物链的复杂性，以及微生物和原生生物在其中发挥的巨大作用（Fenchel，2008）。此外，不得不说起个好名称也很重要。在"微食物环"出现之前，还有过"原生动物食物链""微生物食物链"这样的名称，"微食物环"之所以胜出，可能还有下述原因：第一，字数少，易记、易说、易写；第二，"食物链"（food chain）的提法太过专业化，而"环"（loop）属于大众语汇，易于被普通民众接受；第三，微食物环字面含义为营养物质循环利用，符合传统生态学者对微型生物的功能定位；第四，刊登 Azam 等（1983）论文的期刊《海洋生态学进展丛刊》（*Marine Ecology Progress Series*）易于获得，因而有较好的传播。

自 Sherr E 和 Sherr B（1988）的论文发表后，微食物网概念迅速被科学界接受，从此海洋生态学领域学者主要使用浮游微食物网这一名称，但其他领域学者仍普遍使用微食物环名称，引用 Azam 等（1983）的文献。实际上，在 Sherr E 和 Sherr B（1988）及以后的文献中提及的微食物环，在绝大多数情况下实际等同于微食物网概念。

第五节 西学东渐

20 世纪 90 年代初，焦念志（1994）率先将 microbial loop 概念引进中国，并译为微食物环。随后，宁修仁（1997）将 microbial loop 译为微型生物食物环，将 microbial food web 译为微型生物食物网，将 classical food chain 译为经典食物链。洪华生等（1997）则将上述名词分别译为微食物环、微型生物食物网和主食物链。目前，微食物环和微食物网的译法比较流行。焦念志（2006）的图示很好地展示了碳在微食物环中的传递路径（图 1-8）。

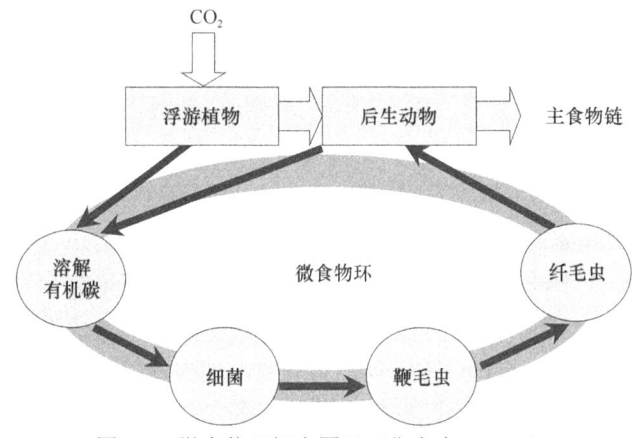

图 1-8 微食物环概念图示（焦念志，2006）

第六节 小 结

微食物环概念的提出是科研工作者共同努力的结果。在 1974 年到 1983 年，Lawrence R. Pomeroy 提出了呼吁，John E. Hobbie 等提供了关键研究方法，Farooq Azam 等进行了系统综述并提出"微食物环"这个术语。不同学者对微食物环概念诞生的时间认定不同，Berman 和 Stone（1994）认为 Pomeroy（1974）是微食物环的提出者，而 Fenchel（2008）则将 Azam 等（1983）发表的论文作为微食物环的发端。

微食物环/网概念的提出得益于新技术的发明和应用。这些新技术大多来自看似与海洋科学无关的领域，如光学、核物理学等，而海洋浮游生物的研究在一定程度上也受益于登月探索等重大历史事件（见本章第二节核孔滤膜的发明过程）。在同一时间段内，不同的作者有相像的发现，但有人的工作却被忽视，有作者的成果由于国家之间的封锁未能广泛流传。在微食物环概念的提出过程中，大众传播的规律发挥了作用，其中也不免有断章取义和夸大的成分。

微食物环和随后提出的微食物网概念大大推动了微型生物的生态学研究，本书后续章节将详细阐述这些研究成果。

参 考 文 献

洪华生, 阮五崎, 黄邦钦, 等. 1997. 台湾海峡初级生产力及其调控机制研究//洪华生. 中国海洋学文集 7. 北京: 海洋出版社: 1-15.

焦念志. 1994. 海洋生态学研究进展与发展趋势//韩晓鹏, 彭海青, 翟世奎. 海洋科学若干前沿领域发展趋势的分析与探讨. 北京: 海洋出版社: 107-119.

焦念志. 2006. 海洋微型生物生态学. 北京: 科学出版社: 525.

宁修仁. 1997. 微型生物食物环. 东海海洋, 15(1): 66-68.

Andrews P, Williams P J L. 1971. Heterotrophic utilization of dissolved organic compounds in the sea III. Measurement of the oxidation rates and concentrations of glucose and amino acids in sea water. Journal of the Marine Biological Association of the United Kingdom, 51(1): 111-125.

Azam F, Fenchel T, Field J, et al. 1983. The ecological role of water-column microbes in the sea. Marine Ecology Progress Series, 10: 257-263.

Beers J R, Stewart G L. 1967. Micro-zooplankton in the euphotic zone at five locations across the California current. Journal of the Fisheries Research Board of Canada, 24(10): 2053-2068.

Bergh Ø, Børsheim K Y, Bratbak G, et al. 1989. High abundance of viruses found in aquatic environments. Nature, 340(6233): 467-468.

Berman T, Stone L. 1994. Musings on the microbial loop: twenty years after. Microbial Ecology, 28(2): 251-253.

Bond R M. 1933. A contribution to the study of the natural food-cycle in aquatic environments: with particular consideration of micro-organisms and dissolved organic matter. Bulletin of the Bingham Oceanographic Collection, 4(4): 1-89.

Bratbak G, Heldal M, Thingstad T F, et al. 1992. Incorporation of viruses into the budget of microbial C-transfer. A first approach. Marine Ecology Progress Series, 83(2-3): 273-280.

Chisholm S W, Frankel S L, Goericke R, et al. 1992. *Prochlorococcus marinus* nov. gen. nov. sp.: an oxyphototrophic marine prokaryote containing divinyl chlorophyll a and b. Archives of Microbiology, 157(3): 297-300.

Chisholm S W, Olson R J, Zettler E R, et al. 1988. A novel free-living prochlorophyte abundant in the oceanic euphotic zone. Nature, 334(6180): 340-343.

Daley R J, Hobbie J E. 1975. Direct counts of aquatic bacteria by a modified epifluorescence technique. Limnology and Oceanography, 20(5): 875-882.

Davis P G, Sieburth J M. 1982. Differentiation of phototrophic and heterotrophic nanoplankton populations in marine waters by epifluorescence microscopy. Annales de l'Institut Oceanographique, Paris (Nouvelle Serie), 58(Suppl.): 249-260.

Ducklow H W. 1983. Production and fate of bacteria in the oceans. BioScience, 33(8): 494-501.

Fenchel T. 2008. The microbial loop-25 years later. Journal of Experimental Marine Biology and Ecology, 366(1): 99-103.

Fleischer R L, Price P B, Symes E M. 1964. Novel filter for biological materials. Science, 143(3603): 249-250.

Francisco D E, Mah R A, Rabin A C. 1973. Acridine orange-epifluorescence technique for counting bacteria in natural waters. Transactions of the American Microscopical Society, 92(3): 416-421.

Haas L W, Webb K L. 1979. Nutritional mode of several non-pigmented microflagellates from the York River estuary, Virginia. Journal of Experimental Marine Biology and Ecology, 39(2): 125-134.

Gray J S, Field J G, Azam F, et al. 1984. The role of free bacteria and bactivory//Fasham M J R. Flows of Energy and Materials in Marine Ecosystems: Theory and Practice. New York: Plenum Press: 707-723.

Haas L W. 1982. Improved epifluorescence microscopy for observing planktonic micro-organisms. Annales de l'Institut Oceanographique, Paris (Nouvelle Serie), 58(Suppl.): 261-266.

Hara S, Terauchi K, Koike I. 1991. Abundance of viruses in marine waters: assessment by epifluorescence and transmission electron microscopy. Applied and Environmental Microbiology, 57(9): 2731-2734.

Hardin G. 1944. Physiological observations and their ecological significance: a study of the protozoan, *Oikomonas Termo*. Ecology, 25(2): 192-201.

Hensen V. 1887. Über die bestimmung des planktons oder des im meere treibenden materials an pflanzen und thieren [On the determination of the plankton or the material floating in the sea on plants and animals]. Fünfter Bericht der Kommission zur wissenschaftlichen Untersuchung der deutschen Meere in Kiel für die Jahre 1882 bis 1886, 12(12-16): 1-108.

Hentschel E. 1936. Allgemeine Biologie des Südatlantischen Ozeans. Deutsche Atlantische Expedition METEOR, Band XI. Berlin, Germany: Walther de Gruyter: 344.

Hobbie J E, Daley R J, Jasper S. 1977. Use of nuclepore filters for counting bacteria by fluorescence microscopy. Applied and Environmental Microbiology, 33(5): 1225-1228.

Hobbie J E, Williams P J L. 1984. Heterotrophic Activity in the Sea. NATO Conference Series, IV. Marine Sciences. New York: Plenum Press: 569.

Jannasch H W, Jones G E. 1959. Bacterial populations in sea water as determined by different methods of enumeration. Limnology and Oceanography, 4(2): 128-139.

Johnson P W, Sieburth J M. 1982. *In-situ* morphology and occurrence of eucaryotic phototrophs of bacterial size in the picoplankton of estuarine and oceanic waters. Journal of Phycology, 18(3): 318-327.

Karl D M. 2007. Plastics-irradiated-etched: The Nuclepore® filter turns 45 years old. Limnology and Oceanography Bulletin, 16(3): 49-54.

Keys A, Christensen E H, Krogh A. 1935. The organic metabolism of sea-water with special reference to the ultimate food cycle in the sea. Journal of the Marine Biological Association of the United Kingdom, 20(2): 181-196.

Krogh A. 1931. Dissolved substances as food of aquatic organisms. Biological Reviews, 6(4): 412-442.

Krogh A, Keys A. 1934. Methods for the determination of dissolved organic carbon and nitrogen in sea water. The Biological Bulletin, 67(1): 132-144.

Kuhn T S. 1962. The Structure of Scientific Revolutions. Chicago: University of Chicago Press: 210.

Landry M. 2001. Microbial loops//Steele J H. Encyclopedia of Ocean Sciences. Oxford: Academic Press: 1763-1770.

Lohmann H. 1911. Über das nannoplankton und die zentrifugierung kleinster wasserproben zur gewinnung desselben in lebendem zustande. Internationale Revue der Gesamten Hydrobiologie und Hydrographie, 4(1-2): 1-38.

Murphy L, Haugen E. 1985. The distribution and abundance of phototrophic ultraplankton in the North Atlantic. Limnology and Oceanography, 30(1): 47-58.

Nielsen E S. 1952. The use of radio-active carbon (C^{14}) for measuring organic production in the sea. ICES Journal of Marine Science, 18(2): 117-140.

Ocean Studies Board, National Research Council. 2000. 50 Years of Ocean Discovery: National Science Foundation 1950-2000. Washington, DC: National Academies Press: 269.

Ploem J S. 1967. The use of a vertical illuminator with interchangeable dichroic mirrors for fluorescence microscopy with incidental light. Zeitschrift für Wissenschaftliche Mikroskopie Und Mikroskopische Technik, 68(3): 129-142.

Pomeroy L R. 1974. The ocean's food web: a changing paradigm. BioScience, 24(9): 499-504.

Pomeroy L R, Johannes R E. 1966. Total plankton respiration. Deep Sea Research and Oceanographic Abstracts, 13(5): 971-973.

Pomeroy L R, Wiebe W J. 1988. Energetics of microbial food webs. Hydrobiologia, 159(1): 7-18.

Price P B, Walker R M. 1962. Chemical etching of charged-particle tracks in solids. Journal of Applied Physics, 33(12): 3407-3412.

Pütter A F R. 1907. Der Stoffhaushalt des Meeres. Zeitschrift für Allgemeine Physiologie, 7: 321-368.

Pütter A F R. 1909. Die Ernährung der Wassertiere und der Stoffhaushalt der Gewässer. Jena: Gustav Fischer: 180.

SCOR-UNESCO. 1966. Determination of Photosynthetic Pigments in Sea-water. Monographs on Oceanographic Methodology. Paris: UNESCO: 69.

Sheldon R W, Prakash A, Sutcliffe J W H. 1972. The size distribution of particles in the ocean. Limnology and Oceanography, 17(3): 327-340.

Sherr E, Sherr B. 1988. Role of microbes in pelagic food webs: a revised concept. Limnology and Oceanography, 33(5): 1225-1227.

Silk E C H, Barnes R S. 1959. Examination of fission fragment tracks with an electron microscope. Philosophical Magazine, 4(44): 970-972.

Sorokin Y I, Overbeck J. 1972. Direct microscopic counting of microorganisms//Sorokin Y I, Kadota H. Techniques for the Assessment of Microbial Production and Decomposition in Fresh Waters, International Biological Programme Handbook 23. Oxford: Blackwell Scientific Publications: 44-47.

Steele J H. 1974. The Structure of Marine Ecosystems. Cambridge: Harvard University Press: 128.

Teal J M. 1962. Energy flow in the salt marsh ecosystem of Georgia. Ecology, 43(4): 614-624.

Todd R L, Kerr T J. 1972. Scanning electron microscopy of microbial cells on membrane filters. Applied Microbiology, 23(6): 1160.

Waterbury J B, Watson S W, Guillard R R, et al. 1979. Widespread occurrence of a unicellular, marine, planktonic, cyanobacterium. Nature, 277(5694): 293-294.

Wilhelm S W, Suttle C A. 1999. Viruses and nutrient cycles in the sea. Viruses play critical roles in the structure and function of aquatic food webs. BioScience, 49(10): 781-788.

Williams P J L. 1970. Heterotrophic utilization of dissolved organic compounds in the sea I. Size distribution of population and relationship between respiration and incorporation of growth substrates. Journal of the Marine Biological Association of the United Kingdom, 50(4): 859-870.

Williams P J L. 1981. Incorporation of microheterotrophic processes into the classical paradigm of the planktonic food web. Kieler Meeresforschungen Sonderheft, 5(Suppl.): 1-28.

Young D A. 1958. Etching of radiation damage in lithium fluoride. Nature, 182(4632): 375-377.

Zimmermann R, Meyer-Reil L A. 1974. A new method for fluorescence staining of bacterial populations on membrane filters. Kieler Meeresforschungen, 30(1): 24-27.

第二章 微食物网的结构及其影响因素

微食物环和微食物网概念的提出大大促进了微食物网生态学的研究，而学者首要的任务是了解组成微食物网的各个生物类群的丰度和生物量，而后随着资料的积累，学者开始思考各个类群丰度和生物量之间的关系。海洋微食物网的结构这一概念最早由 Garrison 等在 2000 年提出，但是他们并没有给出结构的定义，而是测定了微食物网不同生物类群的丰度和生物量，通过聚类分析的方法判断不同调查季节的微食物网是否相同（Garrison et al., 2000）。近年来，有学者从不同角度对海洋微食物网的结构进行了深入研究，本章将对微食物网的丰度结构、生物量结构和粒级结构，以及微食物网结构的变化和影响因素进行详述。

第一节 微食物网的丰度和生物量结构

一、丰度结构

尽管微食物网各个类群的丰度在不同海洋环境中有相对变化，但是这些变化都处于一定的范围之内。一般说来，每毫升海水中含有浮游病毒 10^7 个、细菌 10^6 个、蓝细菌 $10^4 \sim 10^5$ 个、真核藻类 10^4 个（Azam and Malfatti, 2007）、鞭毛虫 10^3 个（Fenchel, 1982）、纤毛虫（近岸海区）10 个（Leakey et al., 1992）。在典型的自然海水中，异养鞭毛虫、细菌、病毒的丰度比例通常为 $1:10^3:10^4$（Miki and Jacquet, 2008）。从空间分布上看，如果这些生物在空间均匀分布，则 1nl（100μm 的立方）中有一个细菌，两个细菌之间的距离为 100μm，其间分布有 2~3 个病毒；1μl（1000μm 的立方）中有一个鞭毛虫，两个鞭毛虫之间相隔 1000μm，其间分布有 10 个细菌。

在微食物网各类群中，异养鞭毛虫（heterotrophic nanoflagellate，HNF）丰度与细菌丰度（bacteria abundance，BA）的关系研究最多。Sanders 等（1992）集合了多个研究的数据，分析了在海洋、河口及淡水等各种水生环境中异养鞭毛虫丰度（HNF abundance，HNFA）与细菌丰度的数量关系（图 2-1a），发现二者呈显著正相关，可表示为 lgHNFA = –2.4+0.90×lgBA，r^2=0.50，P<0.001，n=600。Gasol 和 Vaque（1993）采用相同的方法，分析了不同水生环境中异养鞭毛虫丰度和细菌丰度的数量关系（图 2-1b），发现在海洋环境中二者呈显著正相关，可表示为 lgHNFA = –1.67+0.79×lgBA，r^2=0.31，P<0.001，n=41。这一结果整合了季节变化，因此，代表的是平均实际丰度（mean realized abundance，MRA）值。这两个研究的结果

表明，异养鞭毛虫丰度与细菌丰度之间存在显著的相关性，但相关关系较弱。此外，也有部分研究表明异养鞭毛虫丰度与细菌丰度之间并无相关性（McManus and Fuhrman，1990；van Duyl et al.，1990；Weinbauer and Peduzzi，1995）。

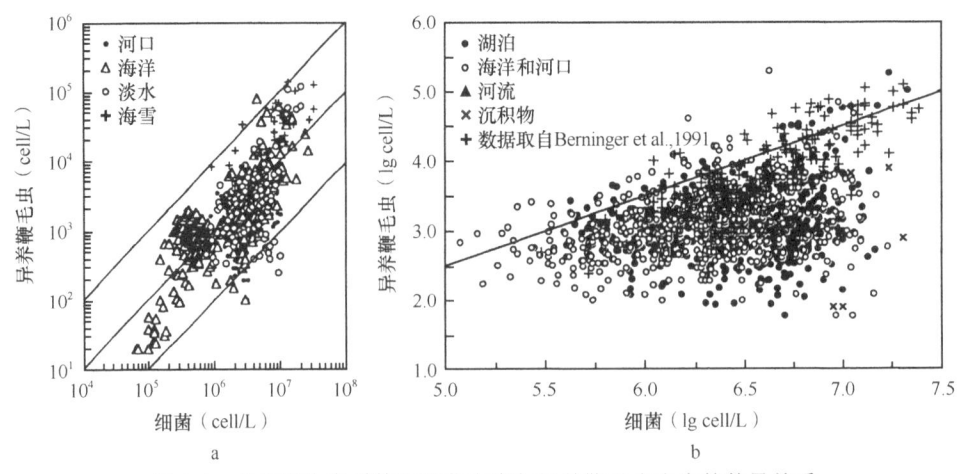

图 2-1　各浮游生态系统中细菌丰度与异养鞭毛虫丰度的数量关系
a. 改绘自 Sanders 等（1992），黑色实线自上至下分别代表细菌丰度：异养鞭毛虫丰度=$10^2:1$、$10^3:1$、$10^4:1$；
b. 改绘自 Gasol 和 Vaque（1993），细菌丰度和异养鞭毛虫丰度经对数化处理，黑色实线表示异养鞭毛虫丰度的理论最大值

在上述研究的基础上，Gasol（1994）提出了最大可实现丰度（maximum attainable abundance，MAA）值，即对于一个给定的细菌丰度，鞭毛虫所能达到的最大丰度。基于收集的数据，他认为这个最大值边界为：$\lg HNFA_{MAA} = -2.47 + 1.07 \times \lg BA$。但是在自然海区中实测的鞭毛虫丰度数据很难达到 MAA 值。此外，他还提出可用一个定性模型来分析下行控制（top-down control）和上行控制（bottom-up control）对鞭毛虫丰度的影响（图 2-2），即当鞭毛虫的丰度高于 MRA 时，上行控制为主要因素，反之则下行控制为主要因素（Gasol，1994）。

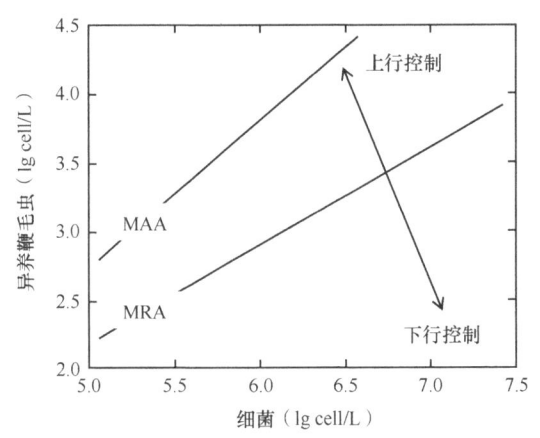

图 2-2　上行控制和下行控制影响异养鞭毛虫丰度的定性模型[改绘自 Gasol（1994）]
细菌丰度和异养鞭毛虫丰度经对数化处理

二、相对生物量的变化

目前微食物网研究多针对各生物类群各自的丰度和生物量变化,但对于各个类群相对生物量变化的研究非常少。对于如何判断不同时间、不同海区的微食物网的结构异同,Garrison 等(2000)使用多维尺度分析(multiple dimensional scaling)与聚类分析相结合的方法研究了阿拉伯海的微食物网生物量结构,发现不同季风季节的微食物网结构存在差异。但上述分析无法指出具体差异是什么,因此,Li 等(2020)提出了通过归一化的方法来寻找具体差异,即将异养细菌的生物量归为 1,其他类群生物量等比例变化,从而获得各微食物网类群的相对生物量。用归一化法所得的相对生物量可以反映微食物网的生物量结构,使得不同海区和不同季节的微食物网之间可以进行比较。

Li 等(2020)用归一化法研究了我国山东半岛东部桑沟湾海区微食物网的生物量结构,发现细菌生物量:异养鞭毛虫生物量:纤毛虫生物量为 1:0.44:0.13。聚类分析表明,桑沟湾微食物网结构存在季节变化和湾内外的地域性差异。进而使用归一化的方法分析发现,造成季节变化的主要原因是聚球藻、微微型真核浮游生物和鞭毛虫的相对生物量在夏秋季升高,而纤毛虫的相对生物量在冬季升高;湾内外微食物网结构的差异主要是由聚球藻、微微型真核浮游生物和自养鞭毛虫引起的。通常随着水体细菌丰度的增加,鞭毛虫和纤毛虫与细菌生物量的比值会增加(Thelaus et al.,2008),但在桑沟湾这个比值基本不变,若将含色素鞭毛虫归入摄食者,则该比值会增加(Li et al.,2020)。

第二节 微食物网的粒级结构

一、微食物网营养级的粒级结构

Azam 等根据 Sheldon 等(1972)的海洋浮游生物的粒径模型,推测海洋浮游食物链中捕食者和饵料生物粒径的最佳比值为 10:1,体积的最佳比值为 $10^3:1$,营养级间碳的输送效率为 10%(Azam et al.,1983)。而随着研究的深入,多位学者发现在微食物网中摄食者和饵料生物的粒径比值可能更小。例如,Goldman 和 Caron(1985)用实验证明,异养鞭毛虫 *Paraphysomonas imperforata* 的饵料粒级变化较大,以细菌为食时粒径比值为 7,以浮游植物为饵料时,粒径比值则降低至 2。Hansen 等(1994)总结了摄食实验中饵料和摄食者的粒径,给出了甲藻与其饵料的最佳粒径比值为 1:1,其他鞭毛虫(不含甲藻)为 3:1,纤毛虫为 8:1。在后来的研究中,有学者建议将鞭毛虫与其饵料的粒径比值控制为 3:1(Guillou et al.,2001)。

海洋浮游鞭毛虫的主要粒径范围为 1~20μm，按照 3∶1 的鞭毛虫与饵料粒径比值，在自然海区的鞭毛虫应该存在多个营养级，这一现象在自然海区海水多级过滤培养中得到了证实（Rassoulzadegan and Sheldon，1986；Wikner and Hagström，1988；Chen et al.，2009）。不同学者对微型浮游动物的营养级划分略有不同，如 Wikner 和 Hagström（1988）将鞭毛虫分为 1~5μm、5~8μm、8~10μm 和>10μm 4 个营养级，Chen 等（2009）则把微型浮游动物划分为<2μm、2~5μm、5~10μm、10~60μm 和 60~200μm 5 个营养级。

因为微型浮游动物存在多个营养级，所以微食物网内部也会存在营养级联摄食。Calbet 等（2001）在研究细菌与鞭毛虫之间的营养级联作用时发现，如果 2~3μm 粒级的异养鞭毛虫被去除，小于 2μm 粒级的异养鞭毛虫的丰度会增加，进而导致细菌丰度降低，当细菌有较高生物量（8~14μgC/L）的时候，这种营养级联现象会更加显著。分粒级稀释培养实验的结果也反映了营养级联现象，分别去除>10μm 和>3μm 的微型浮游动物后，微微型浮游生物（picoplankton，0.2~2μm）的内禀生长率和被摄食率都增加了，被摄食率的增加可能是因为微型浮游动物中存在营养级联作用，较大个体的微型浮游动物被过滤去除后，较小个体的微型浮游动物因为失去了摄食者而生长，进而增加了对微微型浮游生物的摄食率（Reckermann and Veldhuis，1997）。Chen 和 Liu（2010）通过分粒级稀释培养也发现，去除粒径 20μm 以上的浮游动物后，微微型浮游生物的被摄食率增加。

在大的空间尺度上，不同大洋环流中生物的大小有差异。从极地到热带海区，浮游纤毛虫的个体是逐渐变小的，符合伯格曼法则（Bergmann's rule）（Wang et al.，2020）。但是鞭毛虫与之相反，无论是表层还是深层，地中海的鞭毛虫的体积都比亚北极海区大 2 倍左右（Fukuda et al.，2007）。从垂直分布上看，表层鞭毛虫的体积（10.5μm^3）大于中层（7.11μm^3）和深层（6.57μm^3），而细菌的体积变化不大（表层 0.020μm^3、中层 0.017μm^3、深层 0.018μm^3）（Fukuda et al.，2007）。

二、微微型浮游生物和微型浮游生物的比例

微微型浮游生物和微型浮游生物（nanoplankton，2~20μm）是根据粒径来进行划分的（Sieburth et al.，1978）。微微型浮游植物（picophytoplankton）与微型浮游植物（nanophytoplankton）的丰度比（pico∶nano）是研究微食物网结构的参数之一，具有不受研究尺度影响的优点，可用于研究区域性和全球性微食物网结构（Li et al.，2013；López-Urrutia，2013）。海洋环境中 pico∶nano 值范围为 1~1000，海洋上层水体中的值高于底层，低纬度海区高于高纬度海区。pico∶nano 值常应用于"空间换时间"（space-for-time substitution）的研究方法中，用以预测海洋微

食物网的长期变化（López-Urrutia，2013）。

第三节　微食物网结构的变化及其影响因素

一、空间和季节的影响

空间和季节对微食物网的结构及变化有重要的影响，这些影响更多的是通过温度、盐度等环境因子对微食物网各不同类群产生影响而实现的。

在对不同空间微食物网的研究中，人们往往研究不同物理性质的水团中各类群生物丰度的异同，以此来表征微食物网结构的差异。在大的空间尺度上，不同环流中微食物网结构有区别，如聚球藻在亚热带环流和亚极地环流中都有分布，而原绿球藻主要分布在亚热带环流中，在高纬度海区丰度极低（Flombaum et al.，2013）。但随着全球变暖，聚球藻的分布区已出现北扩，甚至会进入北冰洋海区（Paulsen et al.，2016）。

在中尺度上，不同水团中的微食物网结构也有区别。Jacquet 等（2010）将地中海西南部 Alboran 海区分为 3 个水团：北部的地中海水团、南部的大西洋水团，以及二者之间的锋面，在低温高盐的地中海水团原绿球藻丰度较高，在高温低盐的大西洋水团聚球藻丰度较高，在锋面海区这两类蓝细菌的丰度都较高，此外，微微型真核浮游生物和异养细菌在大西洋水团和锋面区域丰度较高。Christaki 等（2011）研究了地中海中部 3 个涡内外微食物网的差异，发现在涡内部异养生物的丰度低于外部。在比斯开湾的 Urdaibai 河口，细菌、异养鞭毛虫和纤毛虫在受潮汐影响较大的站位比潮汐影响较小的站位生物量大（Iriarte et al.，2003）。

同一海区微食物网结构的季节变化也主要使用各个类群丰度和生物量的变化来表示。在微食物网生物中，微微型浮游生物的季节变化主要受水文环境因素影响，在温带海域其丰度的季节性变动通常为夏季和秋季最大、冬季最小（Agawin et al.，1998；Li，1998；Liu et al.，2002）。在比斯开湾，聚球藻和原绿球藻在夏、秋季丰度占优势，其他季节则是微微型真核浮游生物占优势（Calvo-Díaz and Morán，2006）。在南海，原绿球藻在夏季丰度最高，并且在各个季节均比聚球藻和微微型真核浮游生物丰度高 1~2 个数量级，聚球藻和微微型真核浮游生物的丰度虽然较低，但变化幅度很大，在冬季至早春达到丰度最大值，这可能与研究海域冬季混合层深度的变化有关（Liu et al.，2007）。鞭毛虫、纤毛虫等原生动物类群的丰度和生物量变化受自养生物类群的影响。Dupuy 等（2007）研究了大西洋近岸浮游生物群落结构，发现原生动物类群的生物量在夏季较高、冬季最低；但不论什么季节，鞭毛虫生物量所占比例都很高（53%~66%）。此外，季节变化会影响微食物网的粒级组成。冬、春季粒级较大（micro 级）的类群占优势，夏、秋季则以粒级较小的类群（pico 和

nano 级）为主（Savenkoff et al.，2000；Shinada et al.，2005）。

在垂直方向上，聚球藻、原绿球藻和微微型真核浮游生物等自养生物主要分布在海洋上层水体，在中层和深层中几乎不存在，这是垂直方向微食物网结构最重要的差异之一。此外，随深度增加，各微食物网类群丰度和生物量的总体变化趋势是降低，但不同类群降低的程度不一样，这也导致了微食物网结构的差异。通常鞭毛虫的丰度和生物量降低得更快，如在白令海和亚北极太平洋，异养鞭毛虫与细菌的生物量比值随深度增加而减小，在表层（0~100m）为 0.048，中层（100~1000m）为 0.032，深层（>1000m）仅为 0.0083（Fukuda et al.，2007）。异养鞭毛虫与细菌的丰度比在表层约为 1∶1000，随着深度的增加这一比值逐渐缩小为 1∶5000 甚至 1∶10 000（Yamaguchi et al.，2000）。海洋病毒与细菌的丰度比在表层约为 10，大洋深层则为 50~100（Yang et al.，2014）。

二、摄食者的影响

摄食者通过 3 个途径对微食物网各类生物产生影响：①中型浮游动物摄食；②中型浮游动物摄食微型浮游动物，通过营养级联效应影响低营养级生物；③中型浮游动物通过释放溶解有机物、营养盐影响细菌和其他低营养级生物。在摄食者对微食物网结构的影响研究中，研究最多的是桡足类和水母。

（一）桡足类

为研究桡足类对微食物网各类生物的影响，学者通常采用在自然海水中添加桡足类培养的方法。桡足类摄食饵料的最小粒级约为 5μm，这意味着桡足类能摄食微食物网中的纤毛虫、鞭毛虫，其中纤毛虫是桡足类的优先饵料，几乎所有的摄食培养实验中都显示桡足类对纤毛虫有很高的清滤率（张武昌等，2014），但添加桡足类对其他微食物网类群的响应并无一致的结果（表 2-1），添加中型浮游动物对小粒级微食物网生物影响很小（Calbet and Landry，1999）。

理论上说，按照微食物网的细菌-异养鞭毛虫-纤毛虫营养级结构，添加桡足类既然能导致纤毛虫减少，就应该会引起异养鞭毛虫增加，但是在部分实验中可观测到异养鞭毛虫增加，部分实验中异养鞭毛虫反而降低（表 2-1）。这一现象表明，细菌-异养鞭毛虫-纤毛虫的营养级联效应较弱，这可从 3 个角度进行解释：①异养鞭毛虫的摄食者，如桡足类、纤毛虫和摄食鞭毛虫的鞭毛虫等之间存在互相摄食关系，使理论上的线性摄食关系不成立；②异养鞭毛虫主要受到上行控制即饵料供给影响，下行控制即被摄食并非主要限制因素；③虽然在实验过程中已观察到纤毛虫丰度降低，但由此释放的对异养鞭毛虫的摄食压力在短期培养实验中来不及显现（Samuelsson and Andersson，2003；Sipura et al.，2003）。

表 2-1 桡足类添加培养实验结果

区域	培养时间	桡足类物种	叶绿素 a 浓度	纤毛虫丰度	异养鞭毛虫丰度	微型异养鞭毛虫丰度	含色素鞭毛虫丰度	微型含色素鞭毛虫丰度	聚球藻丰度	微微型真核浮游生物丰度	异养细菌丰度	参考文献
地中海西北部	24h				↓						↑	van Wambeke 等 (1996)
加利福尼亚圣佩德罗海峡	24h 和 72h	3月: Acartia spp.; 8月: Clausocalanus spp. 和 Oithona spp.	3月: —; 8月: ↑↑	3月: —; 8月: ↓↓	3月: —; 8月: ↑		3月: ↓ (72h); 8月: —		3月: ↓ (72h); 8月: ↓ (72h)		8月: ↓ (72h)	Schnetzer 和 Caron (2005)
挪威半封闭海洋潟湖	1周	Calanus finmarchicus		↓↑		↑↑	↑↑				↑	Zöllner 等 (2009)
阿拉斯加湾近海		Neocalamus cristatus	↑	↑	↑	↑↑			↑			Liu 等 (2005)
亚热带北太平洋寡营养开阔海域	12~18h		↑↑									Calbet 和 Landry (1999)
黄海寡营养海域	24h	Calanus sinicus	↑	↑	*				↑↑	*	*	Zhao 等 (2020)

注: —表示无变化, ↑表示提高, ↑↑表示显著提高, ↓表示降低, ↓↓表示显著降低, *表示无一致变化趋势

（二）水母

近年来水母暴发（jellyfish bloom）现象在全球近岸海区频发，其范围和严重程度有可能继续增加，水母对微食物网生物的影响得到了科学家的重视。培养实验证明添加水母后，水母释放的胶体态和溶解态有机物质可促进细菌丰度增加，并导致细菌的优势类群出现变化（Condon et al., 2011）。含虫黄藻的水母 *Catostylus mosaicus* 和不含虫黄藻的水母 *Phyllorhiza punctata* 都能摄食中型浮游动物，并通过营养级联作用导致异养甲藻 *Protoperidinium* sp.丰度升高（West et al., 2009）。在中型围隔（300L）进行水母 *Mnemiopsis leidyi* 添加培养实验，培养 168h 后发现水母造成中型浮游动物的生物量降低，甲藻生物量增加，小型浮游植物和纤毛虫丰度降低，而细菌丰度的变化很小，表明水母对中型浮游动物的影响较明显，而对细菌等低营养级生物的影响低微（Dinasquet et al., 2012）。

（三）贝类

双壳贝类对微食物网各生物类群有不同的截留效率（retention efficiency），可影响微食物网的结构（Dupuy et al., 2000a）。目前对贝类滤食微食物网组分的研究多集中在实验室内或海上原位实验，在自然海区的研究还未见报道。已有的贝类添加培养实验证明，双壳贝类，如贻贝、牡蛎和扇贝等可以滤食或截留（retain）异养细菌、蓝细菌、微微型浮游真核生物、鞭毛虫和纤毛虫等微食物网生物（Kreeger and Newell, 1996; Le Gall et al., 1997; Dupuy et al., 1999; Dupuy et al., 2000a; Dupuy et al., 2000b; Fournier et al., 2012），但对不同生物的截留效率不同，且截留效率随生物粒级的增加而增加。Kryger 和 Riisgård（1988）选用 6 种双壳贝类摄食不同粒级（2～9μm）的细菌和单胞藻饵料，发现有大型前侧纤毛的贝类可以完全截留粒径在 4μm 以上的饵料，对于 2μm 的饵料截留效率下降了 35%～70%；有小型或者没有前侧纤毛的贝类可以完全截留粒径 5～6μm 的饵料，但对 2μm 的饵料截留效率分别下降了 50%和 15%。牡蛎可摄食粒径超过 5μm 的鞭毛虫、微型浮游植物、甲藻、纤毛虫和大型浮游动物，对粒径小于 5μm 的鞭毛虫和微微型真核浮游生物的截留效率分别为 45%和 2%（Dupuy et al., 2000b）。尽管贝类可有效滤食硅藻和甲藻等微型藻类，以及鞭毛虫和纤毛虫等原生动物，但从碳贡献率角度看，异养或混合营养的原生动物的贡献率较高，其中主要为鞭毛虫，其次为纤毛虫，而硅藻和甲藻等微型藻类的碳贡献率很小，这意味着异养或混合营养的原生动物可有效将微食物网中的细菌生产力向更高营养级传递（Dupuy et al., 2000a）。

除了摄食截留，贝类还可通过释放无机营养盐影响微食物网。在桑沟湾进行的围隔培养实验表明，添加栉孔扇贝（*Chlamys farreri*）的围隔培养 7 天后，海水无机磷酸盐浓度从 0.04μmol/L 升高到 0.26μmol/L，显著高于对照组，异养细菌、聚球藻和微微型真核浮游生物的丰度也有显著升高，栉孔扇贝释放的磷酸盐对微食物网生物的生长有一定促进作用（Lu et al.，2015）。

（四）病毒

病毒在海洋中数量巨大，是微食物网的组分之一。平均每天有 20%～40%海洋细菌的死亡是由病毒介导的，与浮游动物摄食造成的影响相当（Suttle，2007）。在部分海区病毒的调节作用更为明显，如夏季在南海西部，有 26.44%～96.08%（平均值 77.82%）的细菌生产被病毒裂解（Chen et al.，2011）。通过病毒特异性侵染和裂解宿主，可以有效地调节异养细菌、蓝细菌、真核藻类等微食物网组分的丰度和群落结构，进而影响微食物网的结构。

三、浮游植物的影响

有些浮游植物会产生化感物质，这些化感物质对不同生物的作用有差异，可能对微食物网的结构产生影响。例如，甲藻塔玛亚历山大藻（*Alexandrium tamarense*）释放的化感物质对细菌影响不大，但会对异养鞭毛虫和纤毛虫产生抑制作用，导致异养鞭毛虫和纤毛虫出现自溶，向水体释放溶解有机物，从而增加水体中细菌的生物量（Weissbach et al.，2011）。

不同种类的浮游植物产生的溶解有机物组成不同，可能会影响异养细菌的群落结构，进而影响微食物网结构。在中型围隔（1000L）培养实验中，接种甲藻[球形棕囊藻（*Phaeocystis globosa*）]的围隔发生小规模水华，接种硅藻[并基角毛藻（*Chaetoceros decipiens*）与假微型海链藻（*Thalassiosira pseudonana*）]的围隔发生大规模水华，不额外接种生物的对照组未发生水华（Passow et al.，2007）。在不同围隔水华过程中，异养细菌丰度的变化趋势相似，但变化幅度不同，异养细菌的群落组成也有差异（Murray et al.，2007）。此外，发生水华的围隔中异养鞭毛虫丰度高于对照组，粒级结构也有差异，在接种球形棕囊藻的围隔中粒径超过 6μm 的异养鞭毛虫丰度较高（Passow et al.，2007）。

四、营养盐的影响

营养盐对微食物网生物有直接和间接的影响。直接影响指不同营养盐浓度和成分可导致细菌丰度有差异。间接影响则是营养盐浓度和成分不同，造成浮游植物的类群组成和释放的溶解有机物成分有区别，进而间接影响细菌的丰度和群落结构。然而这些影响交织在一起，很难辨别是哪个途径占主导。通常在高营养盐浓度下，浮游植物的生物量增加，可向水体释放更多的溶解有机物，高浓度的营养盐和溶解有机物进而促进细菌的生长。但是营养盐浓度升高能否导致异养鞭毛虫和纤毛虫的生物量也升高却有很大的不确定性。有营养盐添加培养实验发现细菌生物量升高，异养鞭毛虫和纤毛虫的生物量也升高（Andersson et al., 2006），也有研究发现随着营养盐浓度的降低，细菌丰度降低，但异养鞭毛虫的丰度反而增加（Šolić et al., 2010）。

大气沉降到海里的沙尘对微食物网的影响复杂多变。在地中海进行的两次中型围隔实验的结果迥异：第一次添加沙尘实验中，细菌群落组成发生了显著但短暂的变化，但病毒和异养鞭毛虫的丰度没有明显变化；在随后的第二次添加沙尘实验中，细菌群落没有变化，但是细菌的丰度降低,病毒的丰度增加（Pulido-Villena et al., 2014）。在地中海另一个沙尘添加实验中，细菌的丰度先升高，但在 48h 后丰度出现降低，此时异养鞭毛虫的丰度达到峰值（Romero et al., 2011）。在南海寡营养水体添加大气气溶胶过滤物后，聚球藻、原绿球藻和微微型真核浮游生物的丰度并没有显著增加，微型浮游植物的丰度则出现升高，表明高水平的气溶胶缓解了南海寡营养水体浮游植物的营养盐限制，增加了浮游植物总生物量，并将浮游植物群落从超微型浮游植物占主导转变为微型浮游植物为主（Guo et al., 2012）。

五、其他因素

除以上影响因素外，还有一些因素也会导致微食物网结构发生变化。例如，扰动可以使鞭毛虫和纤毛虫对细菌的摄食压力减小，转而摄食比细菌大的生物，从而导致细菌的丰度增加（Peters et al., 2002）。分散剂常被用来清除原油泄漏事故导致的海面漂浮原油，在自然海水中单独加入原油可导致纤毛虫生物量增加，若单独加入分散剂或同时加入分散剂与原油，可导致细菌生物量增加而纤毛虫生物量受到抑制（Ortmann et al., 2012）。缺氧是近年来备受关注的海洋灾害之一，在缺氧区浮游纤毛虫的群落结构会发生很大变化，超过 80%的纤毛虫是摄食细菌的尾丝虫（*Uronema* spp.）（Stauffer et al., 2013）。

参 考 文 献

张武昌, 陈雪, 李海波, 等. 2014. 海洋浮游桡足类摄食纤毛虫的研究. 海洋与湖沼, 45(4): 764-775.

Agawin N S, Duarte C M, Agusti S. 1998. Growth and abundance of *Synechococcus* sp. in a Mediterranean Bay: seasonality and relationship with temperature. Marine Ecology Progress Series, 170: 45-53.

Andersson A, Samuelsson K, Haecky P, et al. 2006. Changes in the pelagic microbial food web due to artificial eutrophication. Aquatic Ecology, 40(3): 299-313.

Azam F, Fenchel T, Field J, et al. 1983. The ecological role of water-column microbes in the sea. Marine Ecology Progress Series, 10: 257-263.

Azam F, Malfatti F. 2007. Microbial structuring of marine ecosystems. Nature Reviews Microbiology, 5(10): 782-791.

Berninger U G, Finlay B J, Kuuppo-Leinikki P. 1991. Protozoan control of bacterial abundances in freshwater. Limnology and Oceanography, 36(1): 139-147.

Calbet A, Landry M R, Nunnery S. 2001. Bacteria-flagellate interactions in the microbial food web of the oligotrophic subtropical North Pacific. Aquatic Microbial Ecology, 23: 283-292.

Calbet A, Landry M R. 1999. Mesozooplankton influences on the microbial food web: direct and indirect trophic interactions in the oligotrophic open ocean. Limnology and Oceanography, 44(6): 1370-1380.

Calvo-Díaz A, Morán X A G. 2006. Seasonal dynamics of picoplankton in shelf waters of the southern Bay of Biscay. Aquatic Microbial Ecology, 42(2): 159-174.

Chen B, Liu H, Wang Z. 2009. Trophic interactions within the microbial food web in the South China Sea revealed by size-fractionation method. Journal of Experimental Marine Biology and Ecology, 368(1): 59-66.

Chen B, Liu H. 2010. Trophic linkages between grazers and ultraplankton within the microbial food web in subtropical coastal waters. Marine Ecology Progress Series, 407: 43-53.

Chen X, Liu H, Weinbauer M, et al. 2011. Viral dynamics in the surface water of the western South China Sea in summer 2007. Aquatic Microbial Ecology, 63(2): 145-160.

Christaki U, Wambeke F V, Lefevre D, et al. 2011. Microbial food webs and metabolic state across oligotrophic waters of the Mediterranean Sea during summer. Biogeosciences, 8(7): 1839-1852.

Condon R H, Steinberg D K, del Giorgio P A, et al. 2011. Jellyfish blooms result in a major microbial respiratory sink of carbon in marine systems. Proceedings of the National Academy of Sciences of the United States of America, 108(25): 10225-10230.

Dinasquet J, Titelman J, Møller L F, et al. 2012. Cascading effects of the ctenophore *Mnemiopsis leidyi* on the planktonic food web in a nutrient-limited estuarine system. Marine Ecology Progress Series, 460: 49-61.

Dupuy C, Le Gall S, Hartmann H J, et al. 1999. Retention of ciliates and flagellates by the oyster *Crassostrea gigas* in French Atlantic coastal ponds: protists as a trophic link between bacterioplankton and benthic suspension-feeders. Marine Ecology Progress Series, 177: 165-175.

Dupuy C, Pastoureaud A, Ryckaert M, et al. 2000a. Impact of the oyster *Crassostrea gigas* on a microbial community in Atlantic coastal ponds near La Rochelle. Aquatic Microbial Ecology, 22(3): 227-242.

Dupuy C, Ryckaert M, Le Gall S, et al. 2007. Seasonal variations in planktonic community structure and production in an Atlantic coastal pond: the importance of nanoflagellates. Microbial Ecology, 53(4): 537-548.

Dupuy C, Vaquer A, Hoai T L, et al. 2000b. Feeding rate of the oyster *Crassostrea gigas* in a natural planktonic community of the Mediterranean Thau Lagoon. Marine Ecology Progress Series, 205: 171-184.

Fenchel T. 1982. Ecology of heterotrophic microflagellates. IV. Quantitative occurrence and importance as bacterial consumers. Marine Ecology Progress Series, 9(1): 35-42.

Flombaum P, Gallegos J L, Gordillo R A, et al. 2013. Present and future global distributions of the marine Cyanobacteria *Prochlorococcus* and *Synechococcus*. Proceedings of the National Academy of Sciences of the United States of America, 110(24): 9824-9829.

Fournier J, Dupuy C, Bouvy M, et al. 2012. Pearl oysters *Pinctada margaritifera* grazing on natural plankton in Ahe atoll lagoon (Tuamotu Archipelago, French Polynesia). Marine Pollution Bulletin, 65(10-12): 490-499.

Fukuda H, Sohrin R, Nagata T, et al. 2007. Size distribution and biomass of nanoflagellates in meso-and bathypelagic layers of the subarctic Pacific. Aquatic Microbial Ecology, 46(2): 203-207.

Garrison D L, Gowing M M, Hughes M P, et al. 2000. Microbial food web structure in the Arabian Sea: a US JGOFS study. Deep Sea Research PartII: Topical Studies in Oceanography, 47(7): 1387-1422.

Gasol J M, Vaque D. 1993. Lack of coupling between heterotrophic nanoflagellates and bacteria: a general phenomenon across aquatic systems? Limnology and Oceanography, 38(3): 657-665.

Gasol J M. 1994. A framework for the assessment of top-down vs bottom-up control of heterotrophic nanoflagellate abundance. Marine Ecology Progress Series, 113(3): 291-300.

Goldman J C, Caron D A. 1985. Experimental studies on an omnivorous microflagellate: implications for grazing and nutrient regeneration in the marine microbial food chain. Deep Sea Research Part A. Oceanographic Research Papers, 32(8): 899-915.

Guillou L, Jacquet S, Chrétiennot-Dinet M J, et al. 2001. Grazing impact of two small heterotrophic flagellates on *Prochlorococcus* and *Synechococcus*. Aquatic Microbial Ecology, 26(2): 201-207.

Guo C, Yu J, Ho T Y, et al. 2012. Dynamics of phytoplankton community structure in the South China Sea in response to the East Asian aerosol input. Biogeosciences, 9(4): 1519-1536.

Hansen B, Bjornsen P K, Hansen P J. 1994. The size ratio between planktonic predators and their prey. Limnology and Oceanography, 39(2): 395-403.

Iriarte A, Madariaga I, Revilla M, et al. 2003. Short-term variability in microbial food web dynamics in a shallow tidal estuary. Aquatic Microbial Ecology, 31(2): 145-161.

Jacquet S, Prieur L, Nival P, et al. 2010. Structure and variability of the microbial community associated to the Alboran Sea frontal system (Western Mediterranean) in winter. Journal of Oceanography, Research and Data, 3: 47-75.

Kreeger D A, Newell R. 1996. Ingestion and assimilation of carbon from cellulolytic bacteria and heterotrophic flagellates by the mussels *Geukensia demissa* and *Mytilus edulis* (Bivalvia, Mollusca). Aquatic Microbial Ecology, 11(3): 205-214.

Kryger J, Riisgård H U. 1988. Filtration rate capacities in 6 species of European freshwater bivalves. Oecologia, 77(1): 34-38.

Le Gall S, Bel Hassen M, Le Gall P. 1997. Ingestion of a bacterivorous ciliate by the oyster *Crassostrea gigas*: protozoa as a trophic link between picoplankton and benthic suspension-feeders. Marine Ecology Progress Series, 152(1-3): 301-306.

Leakey R, Burkill P, Sleigh M. 1992. Planktonic ciliates in Southampton Water: abundance, biomass, production, and role in pelagic carbon flow. Marine Biology, 114(1): 67-83.

Li H, Chen X, Denis M, et al. 2020. Seasonal and spatial variation of pelagic microbial food web structure in a semi-enclosed temperate bay. Frontiers in Marine Science, 7: 589566.

Li W K W, Carmack E C, McLaughlin F A, et al. 2013. Space-for-time substitution in predicting the state of picoplankton and nanoplankton in a changing Arctic Ocean. Journal of Geophysical Research: Oceans, 118(10): 5750-5759.

Li W K W. 1998. Annual average abundance of heterotrophic bacteria and *Synechococcus* in surface ocean waters. Limnology and Oceanography, 43(7): 1746-1753.

Liu H, Chang J, Tseng C M, et al. 2007. Seasonal variability of picoplankton in the Northern South China Sea at the SEATS station. Deep Sea Research Part II: Topical Studies in Oceanography, 54(14-15): 1602-1616.

Liu H, Dagg M J, Strom S L. 2005. Grazing by the calanoid copepod *Neocalanus cristatus* on the microbial food web in the coastal Gulf of Alaska. Journal of Plankton Research, 27(7): 647-662.

Liu H, Suzuki K, Minami C, et al. 2002. Picoplankton community structure in the subarctic Pacific Ocean and the Bering Sea during summer 1999. Marine Ecology Progress Series, 237: 1-14.

López-Urrutia A. 2013. Macroscopic patterns in marine plankton//Levin S A. Encyclopedia of Biodiversity. Amsterdam: Elsevier Science: 667-680.

Lu J, Huang L, Xiao T, et al. 2015. The effects of Zhikong scallop (*Chlamys farreri*) on the microbial food web in a phosphorus-deficient mariculture system in Sanggou Bay, China. Aquaculture, 448: 341-349.

McManus G B, Fuhrman J A. 1990. Mesoscale and seasonal variability of heterotrophic nanoflagellate abundance in an estuarine outflow plume. Marine Ecology Progress Series, 61(3): 207-213.

Miki T, Jacquet S. 2008. Complex interactions in the microbial world: underexplored key links between viruses, bacteria and protozoan grazers in aquatic environments. Aquatic Microbial Ecology, 51(2): 195-208.

Murray A E, Arnosti C, Christina L, et al. 2007. Microbial dynamics in autotrophic and heterotrophic seawater mesocosms. II. Bacterioplankton community structure and hydrolytic enzyme activities. Aquatic Microbial Ecology, 49(2): 123-141.

Ortmann A C, Anders J, Shelton N, et al. 2012. Dispersed oil disrupts microbial pathways in pelagic food webs. PLoS One, 7(7): e42548.

Passow U, Christina L, Arnosti C, et al. 2007. Microbial dynamics in autotrophic and heterotrophic seawater mesocosms. I. Effect of phytoplankton on the microbial loop. Aquatic Microbial Ecology, 49(2): 109-121.

Paulsen M L, Doré H, Garczarek L, et al. 2016. *Synechococcus* in the Atlantic Gateway to the Arctic Ocean. Frontiers in Marine Science, 3: 191.

Peters F, Marrasé C, Havskum H, et al. 2002. Turbulence and the microbial food web: effects on bacterial losses to predation and on community structure. Journal of Plankton Research, 24(4): 321-331.

Pulido-Villena E, Baudoux A C, Obernosterer I, et al. 2014. Microbial food web dynamics in response to a Saharan dust event: results from a mesocosm study in the oligotrophic Mediterranean Sea. Biogeosciences, 11: 5607-5619.

Rassoulzadegan F, Sheldon R. 1986. Predator-prey interactions of nanozooplankton and bacteria in an oligotrophic marine environment. Limnology and Oceanography, 31(5): 1010-1021.

Reckermann M, Veldhuis M J. 1997. Trophic interactions between picophytoplankton and micro-and nanozooplankton in the western Arabian Sea during the NE monsoon 1993. Aquatic Microbial Ecology, 12(3): 263-273.

Romero E, Peters F, Marrasé C, et al. 2011. Coastal Mediterranean plankton stimulation dynamics through a dust storm event: an experimental simulation. Estuarine, Coastal and Shelf Science, 93(1): 27-39.

Samuelsson K, Andersson A. 2003. Predation limitation in the pelagic microbial food web in an oligotrophic aquatic system. Aquatic Microbial Ecology, 30(3): 239-250.

Sanders R, Caron D, Berninger U G. 1992. Relationships between bacteria and heterotrophic nanoplankton in marine and fresh waters-an inter-ecosystem comparison. Marine Ecology Progress Series, 86(1): 1-14.

Savenkoff C, Vézina A, Roy S, et al. 2000. Export of biogenic carbon and structure and dynamics of the pelagic food web in the Gulf of St. Lawrence Part 1. Seasonal variations. Deep Sea Research Part II: Topical Studies in Oceanography, 47(3): 585-607.

Schnetzer A, Caron D A. 2005. Copepod grazing impact on the trophic structure of the microbial assemblage of the San Pedro Channel, California. Journal of Plankton Research, 27(10): 959-971.

Sheldon R W, Prakash A, Sutcliffe J W H. 1972. The size distribution of particles in the ocean. Limnology and Oceanography, 17(3): 327-340.

Shinada A, Ban S, Yamada Y, et al. 2005. Seasonal variations of plankton food web structure in the coastal water off Usujiri southwestern Hokkaido, Japan. Journal of Oceanography, 61(4): 645-654.

Sieburth J M, Smetacek V, Lenz J. 1978. Pelagic ecosystem structure: heterotrophic compartments of the plankton and their relationship to plankton size fractions. Limnology and Oceanography, 23(6): 1256-1263.

Sipura J, Lores E, Snyder R. 2003. Effect of copepods on estuarine microbial plankton in short-term microcosms. Aquatic Microbial Ecology, 33(2): 181-190.

Šolić M, Krstulović N, Kušpilić G, et al. 2010. Changes in microbial food web structure in response to changed environmental trophic status: a case study of the Vranjic Basin (Adriatic Sea). Marine Environmental Research, 70(2): 239-249.

Stauffer B A, Schnetzer A, Gellene A G, et al. 2013. Effects of an acute hypoxic event on microplankton community structure in a coastal harbor of southern California. Estuaries and Coasts, 36(1): 135-148.

Suttle C A. 2007. Marine viruses-major players in the global ecosystem. Nature Reviews Microbiology, 5(10): 801-812.

Thelaus J, Haecky P, Forsman M, et al. 2008. Predation pressure on bacteria increases along aquatic productivity gradients. Aquatic Microbial Ecology, 52(1): 45-55.

van Duyl F, Bak R, Kop A, et al. 1990. Bacteria, auto-and heterotrophic nanoflagellates, and their relations in mixed, frontal and stratified waters of the North Sea. Netherlands Journal of Sea Research, 26(1): 97-109.

van Wambeke F, Christaki U, Gaudy R. 1996. Carbon fluxes from the microbial food web to mesozooplankton. An approach in the surface layer of a pelagic area (NW Mediterranean Sea). Oceanologica Acta, 19(1): 57-66.

Wang C, Li H, Xu Z, et al. 2020. Difference of planktonic ciliate communities of the tropical western Pacific to the Bering Sea and the Arctic Ocean. Acta Oceanologica Sinica, 39(4): 9-17.

Weinbauer M G, Peduzzi P. 1995. Significance of viruses versus heterotrophic nanoflagellates for controlling bacterial abundance in the northern Adriatic Sea. Journal of Plankton Research, 17(9): 1851-1856.

Weissbach A, Rudström M, Olofsson M, et al. 2011. Phytoplankton allelochemical interactions change microbial food web dynamics. Limnology and Oceanography, 56(3): 899-909.

West E J, Pitt K A, Welsh D T, et al. 2009. Top-down and bottom-up influences of jellyfish on primary productivity and planktonic assemblages. Limnology and Oceanography, 54(6): 2058-2071.

Wikner J, Hagström Å. 1988. Evidence for a tightly coupled nanoplanktonic predator-prey link regulating the bacterivores in the marine environment. Marine Ecology Progress Series, 50(1): 137-145.

Yamaguchi A I, Ishizaka J, Watanabe Y. 2000. Vertical distribution of plankton community in the western North Pacific Ocean (WEST-COSMIC). Bulletin of the Plankton Society of Japan, 47(2): 144-156.

Yang Y, Yokokawa T, Motegi C, et al. 2014. Large-scale distribution of viruses in deep waters of the Pacific and Southern Oceans. Aquatic Microbial Ecology, 71(3): 193-202.

Zhao Y, Dong Y, Li H, et al. 2020. Grazing by microzooplankton and copepods on the microbial food web in spring in the southern Yellow Sea, China. Marine Life Science & Technology, 2: 442-455.

Zöllner E, Hoppe H G, Sommer U, et al. 2009. Effect of zooplankton-mediated trophic cascades on marine microbial food web components (bacteria, nanoflagellates, ciliates). Limnology and Oceanography, 54(1): 262-275.

第三章 海洋浮游细菌的生存状态

海洋浮游细菌（以下简称浮游细菌）在海洋有机物降解和溶解有机碳转化中起着重要作用，是影响海洋净生产力的重要生物因素。浮游细菌种类繁多，胞内含几千种化合物，代谢反应（途径）极其多样。因此，自然海水中的浮游细菌除种类的差别之外，也存在生理代谢和化学组成等生存状态的异质性。

细菌群落中的单个细菌的生存状态，包括生理状态、代谢活性、化学组成、基因转录翻译、行为等多个方面存在着差异，称为细菌的个体间异质性。区分细菌群落中不同个体特征的研究被称为单细胞微生物学。

对浮游细菌生存状态异质性及其影响因素的研究是理解细菌生产、代谢及其在生物地球化学循环中的重要作用的基础。本章主要介绍自然海区浮游细菌的细胞膜完整性、呼吸活性和核酸含量等生存状态的研究现状，以及不同海区（主要是近岸）浮游细菌的生存状态及其随环境的变化特征。

第一节 细菌生存状态的研究指标

一、存活状态

培养法可用来检测细菌的存活状态，即将细菌涂布在琼脂平板上能生长形成菌落的即为活的，不能生长的为死的，这曾经是判断细菌存活的金标准。但进一步研究发现，很多细菌在自然环境中处于"活的非可培养状态"（viable but non-culturable，VBNC）（Xu et al.，1982；Roszak et al.，1984），这些细菌无法在琼脂平板上生长，但仍保持一定代谢能力，在适宜环境条件下可复苏并繁殖。

使用落射荧光显微镜来计数细菌丰度的方法是海洋浮游细菌研究的重要突破（Hobbie et al.，1977）。在相当一段时间内人们认为这个方法得出的是存活细菌的丰度，然而，Zweifel 和 Hagström（1995）使用荧光染料 DAPI（4',6-二脒基-2-苯基吲哚，4',6-diamidino-2-phenylindole）染色波罗的海水样，用荧光显微镜观察发现仅有少数细菌（2%~32%）可看到拟核（nucleoid），而大多数细菌不含拟核，他们认为这些不含拟核的细菌可能是无活性（inactive）的，将其称为幽灵细菌（ghost）。

几乎与此同时，Heissenberger 等（1996）用电镜法观察到浮游细菌仅约 34% 具有完整的细胞结构，多数细菌细胞的内部结构受损（42%）或缺乏内部结构

（24%），这意味着这些细菌是无活性或者死亡的。

上述结果表明，可通过检验细胞结构的完整性来判断细菌的存活状态，主要检验指标有：①是否具有拟核；②细胞膜是否完整；③细胞膜是否具有极性和膜电位。

二、代谢活性

代谢活性是通过检验细菌对某种物质（底物）的吸收和转化能力来进行判断的，需要培养并检测细菌体内物质的积累情况，因此，对于底物的选择主要考虑可吸收和可检测两个指标。

许多物质由于分子量或极性等原因是不能被细菌所吸收的，因此并不是所有物质都能作为底物。根据可检测的要求，底物主要有两类：①放射性标记底物，即使用放射性标记的有机底物培养细菌并过滤到滤膜上，将滤膜和底片叠加放到玻片上，用荧光显微镜观察细胞周围是否有放射线造成的阴影，这项技术被称为显微放射自显影术（microautoradiography，MAR），这种放射性标记的有机底物除了被吸收进细菌细胞内以外，还可能部分吸附在细菌细胞外，也能出现所谓的"放射线阴影"，造成假阳性的误差，因此在过滤后对滤膜的清洗很重要；②可产生荧光物质的酶的底物，如加入呼吸作用电子传递链中某种酶的底物，具有呼吸作用的细菌胞内会积累可以产生荧光的物质，通过检测胞内有无荧光可判定是否有呼吸活性。

三、化学组成

细菌的化学组成异质性即测定细胞内化学成分的含量，目前常用指标包括核酸含量、核糖体含量和细胞内 pH。其中，核酸含量高、核糖体含量高的细菌通常被认为是活跃生存的，而核酸和核糖体含量低的细菌则可能是不活跃甚至死亡的。细菌胞内 pH 是决定细胞功能和酶活性的重要因素之一，只有在适宜的 pH 条件下细菌才能维持体内蛋白质的结构和功能，从而进行生理代谢活动。细菌通过质膜转运蛋白严格调控体内 pH，生理正常的细菌胞内 pH 通常在 7.0～7.4，pH 过高或过低将意味着细菌处于代谢异常状态。

四、对研究指标的评价

以上细菌生存状态指标的测定方法都有各自的不足，在浮游细菌研究中的应用受到多方面制约。有的方法技术复杂（如显微放射自显影术），有的方法依赖肉眼观察，而肉眼的敏感度和个体差异会对结果造成影响，所以，这些方法在浮游细菌研究中的应用普及程度不一样。

对于单个细菌，基本上在进行上述一种或两种生存状态指标测试后，细菌就会被测定过程杀死，无法再获得该细菌的其他指标。现有技术无法确定一个细菌的所有生存状态指标，这意味着很难确定同一细菌个体的各个指标之间的联系。代谢活性和存活状态相关联，但不完全代表存活状态，根据代谢活性得出的活跃细菌肯定是活的，但不活跃细菌可能是死的，也可能是活的，只是活跃程度低于该种方法的检测下限（Falcioni et al., 2008）。因此，不能通过测定其中的一个指标而推知其他指标。

上述各个生存状态指标给出的结果基本都是二分法，即阳性或阴性，但对于阳性指标，不能进一步分出强弱的等级。例如，对于细胞膜不完整的结果，这个细菌的细胞膜损坏 1/4 还是 1/8 就无法测定。除了阳性（或阴性）细菌的具体数目外，阳性（或阴性）细菌所占的比例也是重要的表达方式，一般以%表示。因为每个指标都是细菌生存状态或代谢活性的一个方面，因此，应尽可能同时测定多个指标来描述细菌群落的生存状态。

受技术手段所限，早期的研究大多只测定细菌一个或两个生存状态指标。近年来随着流式细胞术（flow cytometry，FCM）与荧光染料技术的发展和应用，越来越多的研究使用流式细胞术同时测定细菌 3 个生存状态指标，包括细胞膜完整性、呼吸活性和核酸含量。

第二节 浮游细菌的细胞膜完整性

一、细胞膜完整性检测原理

细胞膜完整性检测主要采用核酸双染色（nucleic acid double-staining，NADS）法，基于荧光染料对细胞膜的穿透性不同进行检测，可用落射荧光显微镜观察（Boulos et al., 1999）或者流式细胞术计数（Grégori et al., 2001）。NADS 的原理是使用两种荧光染料对细菌的核酸进行染色，绿色荧光染料 SYTO 9 或 SYBR Green I 可以透过完整和不完整的细胞膜，使细菌呈现绿色荧光；红色荧光染料碘化丙啶（propidium iodide，PI）由于分子较大，仅能穿过不完整的细胞膜，而且 PI 与核酸的结合能力强于 SYTO 9 或 SYBR Green I，可削弱细菌的绿色荧光，使具有不完整细胞膜的细菌呈红色荧光，计数具有不同颜色荧光的细菌即可分别得到样品中具完整细胞膜和不完整细胞膜的细菌数目。

NADS 法可以检测自然环境中细菌的死活比例（Falcioni et al., 2008），通常认为细胞膜完整的细菌是活的，细胞膜不完整的细菌是死的或受伤的。该方法操作简单，结果判定直观，已有完善成熟的实验流程（Nescerecka et al., 2016），被广泛应用于微生物学和生态学研究。NADS法的结果常用活细胞百分比来表示。

二、浮游细菌的存活状态

海洋中使用NADS法进行的细菌存活状态调查大多局限于近岸和浅水区域，大洋和深海的研究较少。根据这些调查，在表层海水中活细菌在总细菌丰度中占优势，其比例在50%~90%，大洋和近岸海区没有差异。随水深增加活细菌的丰度比例下降，大西洋加那利群岛附近海域的活细菌丰度比例在200m为40%~70%，到1000m则降低为10%~40%（Gasol et al., 2009），Baltar等（2010）在同一海区的研究发现，表层的活细菌丰度比例为70%~79%，到2000m降低为25%~60%。在一定温度范围（5~25℃）内，活细菌的丰度比例与温度呈负相关，随温度升高活细菌的丰度比例下降（Lasternas et al., 2010）。在比斯开湾的季节变化研究表明，夏季活细菌有较高的比例和生长率（Morán and Calvo-Díaz, 2009；Huete-Stauffer et al., 2015），而从昼夜变化上看，活细菌丰度比例没有明显的变化（Lefort and Gasol, 2014）。

水文现象和生物过程对活细菌比例有一定影响，如中尺度涡区域的活细菌比例高于涡的外部（Baltar et al., 2010）；在北冰洋喀拉海河口，活细菌比例随盐度（4.4~28.5）升高而增加（Mosharova et al., 2016）；在加那利群岛海域的调查表明，当浮游植物死亡时，浮游植物向细胞外释放的有机物增加，活细菌比例从60%上升到95%（Lasternas and Agustí, 2014）。在围隔实验中向水体添加营养盐诱发浮游植物水华，可提高水体活细菌的比例，原因可能是浮游植物水华释放大量溶解有机物，促进了细菌的生长（Lasternas and Agustí, 2014；Baltar et al., 2016）；培养实验表明，光照对活细菌比例的影响有季节性，夏季活细菌比例会降低，其他季节则影响不大（Alonso-Sáez et al., 2006；Ruiz-Gonzalez et al., 2012）。

第三节 浮游细菌的呼吸活性

一、呼吸活性的检测原理

细菌的呼吸需要电子传递系统（electron transport system，ETS），5-氰基-2,3-二（4-甲基苯基）四唑氯化物[5-cyano-2,3-di-(p-tolyl) tetrazolium chloride，CTC]是一种可以渗透细胞膜的荧光染料，在海水中加入CTC后，有活跃ETS的细菌会还原CTC，在细菌细胞内形成并积累具有红色荧光的CTC-甲臜(CTC-formazan，CTF)结晶（Rodriguez et al., 1992；Schaule et al., 1993），这些CTC阳性（CTC+）

细菌即为活跃呼吸的细菌（Rodriguez et al.，1992），可使用落射荧光显微镜观察（Gasol et al.，1995；Lovejoy et al.，1996）或流式细胞术检测（del Giorgio et al.，1997；Sieracki et al.，1999）。

CTC 活性检测方法需将 CTC 加入含细菌的自然海水中进行培养，使细菌吸收 CTC 生成 CTF，因此，CTC 的浓度和培养时间是该方法的关键条件。加入海水中的 CTC 的最佳终浓度为 5mmol/L，低于这个浓度得出的 CTC+比例将偏低，如若使用 0.75mmol/L 的 CTC 终浓度，则 CTC+细菌的比例为 1%～2%（Lovejoy et al.，1996），这个比例被认为过低。CTC 培养所需水样体积很小，只需用 0.5～1.5ml 水样室温培养即可，红色荧光在加入 CTC 后立刻产生，10min 后就可以进行检测，随着 CTF 结晶体积的增大，荧光强度也相应增强，50min 到 1h 后，红色结晶的数目不再增长，此时即可获得较准确的 CTC+细菌比例（Gasol and Arístegui，2007）。此后，虽然 CTC+细菌的数目不再增加，但是红色荧光的强度还会一直增加，甚至 10h 后还在增加（Gasol and Arístegui，2007）。

CTC+检测方法也有一定的局限性，并非所有细菌都能够利用 CTC，虽然 Sherr 等（1999）在海洋中分离的菌株都能降解 CTC，但是其他生境，如人体内（Smith and McFeters，1997）、厌氧环境（Bhupathiraju et al.，1999）、地下水（Hatzinger et al.，2003）、河流（Yamaguchi and Nasu，1997）的研究结果表明，有些细菌不能降解 CTC，所以海洋中也很可能存在不能降解 CTC 的细菌。CTC+细菌与其他细菌状态的指标也有对比实验，如 Karner 和 Fuhrman（1997）发现 CTC+细菌的数目低于显微放射自显影术得出的活跃细菌数目。

CTF 晶体在细菌细胞内累积可能会导致细菌破裂，使 CTF 释放到水体中。Gasol 等（1995）在用落射荧光显微镜观察时注意到，细菌细胞外也有 CTF 颗粒，Gasol 和 Arístegui（2007）用 SYTO 13 对 CTC 培养的细菌染色，用流式细胞术分别计数红色和绿色（SYTO 13 阳性）荧光颗粒，发现细菌破裂后释放到海水里的 CTF 结晶用流式细胞仪仍能检测到，会被计为一个细菌信号。

多项研究表明，CTC 或 CTF 对细菌具有毒性，如 CTC 染色后细菌放射性标记的生长率和呼吸率分别降低了 1%～14%和 4%～44%（Ullrich et al.，1996），用荧光细菌进行的毒性测试表明，浓度为 0.1～5μmol/L 的 CTC 在 15min 后，荧光强度减少 50%～100%（Ullrich et al.，1996）；加入 5mmol/L 的 CTC 培养 30min，细菌的丰度平均减少 22%（Gasol and Arístegui，2007）；加入 CTC 后，细菌的运动停止（Grossart et al.，2001）。也有研究认为，CTC 作为外来物质对细菌肯定产生影响，但其毒性可能很低，因为加入 CTC 数小时后 CTC+细菌的数目不再增加，但是单个细菌中的 CTF 荧光强度还在增加，说明仍有细菌降解 CTC，细菌并没有被立即杀死（del Giorgio et al.，1997；Gasol and Arístegui，2007）。由于毒性要在细菌吸收和降解 CTC 后才能发生，所以 CTC 的毒性对 CTC+细菌的检测数据没有影响，CTC+测试是有效的。

对于 CTC+细菌对细菌群落代谢的贡献有两种截然不同的观点：第一种观点认为，CTC+细菌是细菌群落代谢的主要贡献者，如虽然切萨皮克湾 CTC+细菌比例平均只有 14%，但是 CTC+细菌丰度和占总细菌数的比例都与细菌群落呼吸有很好的相关性（Smith，1998），而 Sherr 等（1999）发现，CTC+细菌是放射性标记物的主要吸收者；第二种观点认为，CTC+细菌的贡献并不那么大，Servais 等（2001）用放射性底物培养来自法国地中海沿岸海水的细菌，然后进行 CTC 染色并用流式细胞仪分选出 CTC+细菌，测定其放射性标记的量，结果发现与总细菌的放射性标记相比，CTC+细菌贡献了总细菌生产的近 60%，Longnecker 等（2005）使用同样的方法得出在美国俄勒冈外海 CTC+细菌贡献了总细菌生产的 7%~14%。然而，Gasol 和 Arístegui（2007）研究认为，第二种观点把释放在水体中的 CTF 颗粒也作为细菌进行了分选，从而导致 CTC+细菌的放射性标记培养结果偏低。

二、浮游细菌的呼吸活性

现场调查表明，海洋环境 CTC+细菌比例一般低于 10%，内陆架海区比外陆架海区高（Sherr et al.，2002），极地 CTC+细菌比例较高，平均为 32%~38%（Martin et al.，2008；Mosharova et al.，2017）。在北冰洋喀拉海河口，CTC+细菌比例随盐度升高（4.4~28.5）而增加（Mosharova et al.，2016），但在西非塞内加尔河口，CTC+细菌比例随盐度升高（0~20）而降低，但当盐度继续升高时 CTC+细菌比例增加（Bettarel et al.，2011）。CTC+细菌比例垂直分布基本为表层高、深层低，在加那利群岛附近海域表层 CTC+细菌比例为 5%~10%，1000m 处则小于 5%（Gasol et al.，2009），在比斯开湾夏季表层 CTC+细菌比例高，冬季则表层和深层无显著差异（Franco-Vidal and Morán，2011），在北卡罗来纳州外海陆架区（Sherr et al.，2002）和俄勒冈州外海（Longnecker et al.，2005），200m 以浅水柱内 CTC+细菌比例的垂直分布没有明显规律。关于 CTC+细菌比例季节和周年变化的研究很少，在比斯开湾 CTC+细菌占总细菌比例周年平均为 3.6%，最高值出现在 4 月，达 12%（Morán and Calvo-Díaz，2009），而在比斯开湾另一个海区，最高值（12%）出现在夏季（Franco-Vidal and Morán，2011）。

CTC+细菌比例可能受其摄食者鞭毛虫丰度的影响，CTC+细菌比例与异养鞭毛虫占鞭毛虫总丰度的比例相关，异养鞭毛虫在鞭毛虫总丰度中的比例高时，CTC+细菌比例降低，说明异养鞭毛虫选择摄食活跃的细菌（Lovejoy et al.，1996；Lovejoy et al.，2000），del Giorgio 等（1996）的培养实验也支持这一观点。

第四节 浮游细菌的核酸含量

一、核酸含量测定的原理

细菌经核酸染料染色后的荧光强度，在一定程度上代表了该细菌核酸含量的高低。早期研究中，可使用高分辨率荧光显微镜分辨荧光的强弱（Sieracki and Viles，1992）。随着流式细胞术在海洋生态学中的应用，有学者发现根据侧向散射光（side scatter，SSC，与细胞的密度相关）信号与绿色荧光（与细胞的核酸含量有关）信号的相对强度，可将浮游细菌在双参数流式图上划分为核酸含量不同的群（Li et al.，1995），这种分群现象在淡水和海水、寡营养和富营养水体中都有，是浮游细菌群落的普遍特征（Bouvier et al.，2007）。

各个研究对这些不同核酸含量的细菌分群有不同的名称，有研究称其为组（group）（Li et al.，1995），有的称其为亚群（subpopulation）（Gasol et al.，1999），也有的称为亚组（subgroup）（Gasol and del Giorgio，2000）。在自然海水样品中，浮游细菌经核酸染色大多可分为两群，通常称为高核酸含量群（high nucleic acid，HNA）和低核酸含量群（low nucleic acid，LNA），有的研究在 HNA 和 LNA 外还发现一个极高核酸含量群（very high nucleic acid，VHNA）。

用流式细胞仪分选 HNA 和 LNA 细菌进行 16S rDNA 测序以研究其物种组成，不同的学者得出了相互冲突的结论：有研究发现 HNA 和 LNA 在系统发生上类群组成没有区别（Bernard et al.，2000；Servais et al.，2003；Longnecker et al.，2005），而有的研究认为 HNA 和 LNA 在系统发生上的类群组成是不同的，或有的分类类群在一个核酸类群中占主导，如 LNA 主要与 SAR11 类群和 SAR86 类群有关，HNA 则主要与 γ-变形菌、红细菌目、SAR116 类群和拟杆菌门有关（Eilers et al.，2000；Fuchs et al.，2000；Zubkov et al.，2001；Zubkov et al.，2002；Fuchs et al.，2005；Mary et al.，2006；Schattenhofer et al.，2011；Vila-Costa et al.，2012）。

二、浮游细菌的核酸含量

自然海区浮游细菌中，HNA 的核酸含量大大高于 LNA，为后者的 4~6 倍（Li et al.，1995；Jellett et al.，1996；Sherr et al.，2006；Bouvier et al.，2007），HNA 和 LNA 两个群的平均荧光强度值成正比关系，即当 LNA 的荧光值增加时，HNA 的荧光值也增加（Bouvier et al.，2007）。自然海区中 HNA 丰度占总细菌丰度的百分比（HNA%）多在 30%~90%，表层的 HNA%随海区营养程度的升高而升高，如滨海湿地高于海洋（Bouvier et al.，2007），近岸海区高于陆坡和大洋区（Sherr et al.，2006），在澳大利亚南部大陆架上升流区 HNA%（84%～93%）高于其他区域（36%～43%）（Paterson et al.，2012）。

不同的海洋生境中 HNA%与叶绿素 a 浓度呈正相关，说明 HNA%随着有机物供给的增多而上升（Li et al.，1995）。HNA%与叶绿素 a 浓度有关，意味着 HNA%也很可能与海区生产力有关，但是叶绿素 a 浓度很高和很低的海区也会出现相近的 HNA%，表明除生产力之外还有其他影响因素（Bouvier et al.，2007）。

垂直方向上，在 200m 以浅的近岸海区，营养程度高的表层水体 HNA%低于营养程度低的底层水体（Jochem，2001；Calvo-Díaz and Morán，2006；Belzile et al.，2008），这与 Li 等（1995）的观点不同，原因尚不清楚。在大洋中，HNA%在 200m 以浅水柱中主要是随深度增大而稍有提高，但没有明显规律（Gasol et al.，2009；van Wambeke et al.，2011；Girault et al.，2015）。在地中海自 200m 以深 HNA%随深度增加而升高，从 0～250m 的 30%～50%升高至 3300m 的 50%～70%（van Wambeke et al.，2011）。在加那利群岛附近，0～1000m 水深 HNA%没有发生明显变化（Gasol et al.，2009）。在红海中部 200m 以浅为 LNA 占主导，自 200m 以深是 HNA 占主导（Calleja et al.，2018）。在东北大西洋深水中，从 70°N 到 30°N，HNA%降低，再到 10°S，HNA%稍微升高（Reinthaler et al.，2013）。近岸海区 HNA%随盐度变化的研究较少，在法国地中海沿岸罗讷河口，盐度变化范围为 15～36，HNA%在 40%～80%，但未发现随盐度变化 HNA%有明显的变化规律（Joux et al.，2005）。

表层 HNA%的季节变化一般为春季和冬季比例大，夏季比例小，如西班牙比斯开湾（Calvo-Díaz and Morán，2006）和亚得里亚海（Šantić et al.，2014），在北冰洋波弗特陆架区 HNA%在春季比冬季高（Belzile et al.，2008），而深水 HNA%没有明显的季节变化（Calvo-Díaz and Morán，2006；Belzile et al.，2008）。对影响 HNA%的因素目前所知甚少，培养实验中升温培养可以使细菌核酸含量降低（Huete-Stauffer et al.，2015），此外，不同浮游植物释放的 DOC 会影响 HNA%（Tada and Suzuki，2016），HNA 对浮游植物水华的反应比 LNA 明显（Gomes et al.，2015）。

LNA 和 HNA 一经发现，其存活状态和代谢活性上是否存在差异立即引起了学者的兴趣。Li 等（1995）发现，HNA 的分布和 DNA 含量与水体中叶绿素 a 浓度具有相关性，所以推测 HNA 是活跃细菌，而 LNA 是不活跃细菌。Jellett 等（1996）提出了活跃细胞指数（active cell index，ACI）概念，即 HNA 占总细菌丰度的比例，与 HNA%概念相同，他们发现 ACI 与细菌生产力的变化趋势一致，认为 ACI 可以表征细菌群落的活跃程度。

但是 HNA 和 LNA 是否活性不同仍需要直接的证据。第一项证据来自生长率培养实验，Gasol 等（1999）发现，稀释培养实验中 HNA 有较高的生长率，而

LNA 生长率较低，所以认为 HNA 是活跃细菌，LNA 是不活跃或死细菌，Vaqué 等（2001）同样支持这个观点。但也有培养实验发现 LNA 有较高的生长率，因而并非不活跃的细菌（Jochem，2001；Zubkov et al.，2004；Sherr et al.，2006）。第二项直接证据来自两类细菌对放射性标记底物的吸收能力的比较，有研究发现，HNA 能够活跃吸收底物（Lebaron et al.，2001；Lebaron et al.，2002；Servais et al.，2003），因此认为 HNA 是活跃细菌；但是也有一些研究发现 LNA 的活性和 HNA 相当，在单位细菌体积比较时甚至超过 HNA（Zubkov et al.，2001；Longnecker et al.，2005；Mary et al.，2006；Scharek and Latasa，2007；Wang et al.，2009；Longnecker et al.，2010；Talarmin et al.，2011；Huete-Stauffer and Morán，2012），HNA 中的细菌的活性也不尽相同（Morán et al.，2007）。由于生长率和对底物的吸收能力的实验都得出了矛盾的结果，目前普遍认为，LNA 和 HNA 的划分不能作为细菌是否活跃的指示，HNA 比 LNA 活跃一些，但是 LNA 中也有一些细菌是活跃的，能吸收有机底物并表现出一定的生长。

第五节 浮游细菌生存状态研究指标间的关系

已有相当多的研究分别使用不同的指标报道了浮游细菌的生存状态异质性，主要包括 NADS 法研究细菌存活状态、CTC 法研究细菌的代谢状态，以及 HNA/LNA 法研究细菌的化学组成，但对不同指标之间的关系研究相对较少。

采用 NADS 法观测到的活细菌比例通常大于该样品的 HNA%，这意味着 LNA 不全部是死细菌，如在南极半岛海区，61%的细菌是活的，这其中约 45%的细菌属于 HNA，也就是说有 16%的活细菌属于 LNA（Ortega-Retuerta et al.，2008）。空间分布方面，在加那利群岛附近海域这 3 个指标在垂直方向变化趋势并不一致（Gasol et al.，2009）。在同一海区的周年变化中，这些指标的变化也不同步，如在地中海西北部海区，HNA%在 3 月最高，但是活细菌比例在 9 月最高（Gomes et al.，2015）；在比斯开湾，HNA%和 CTC+细菌比例周年变化较为一致，但是活细菌比例与前两项指标不一致（Morán and Calvo-Díaz，2009）；在地中海西北部，CTC+细菌比例与 HNA%的日变化趋势一致（Lefort and Gasol，2014）；在围隔实验中，NADS 和 HNA%的变化趋势相似，而 CTC+细菌比例变化与前两者不一致（Baltar et al.，2016）。

参 考 文 献

Alonso-Sáez L, Gasol J M, Lefort T, et al. 2006. Effect of natural sunlight on bacterial activity and differential sensitivity of natural bacterioplankton groups in northwestern Mediterranean coastal waters. Applied and Environmental Microbiology, 72(9): 5806-5813.

Baltar F, Arístegui J, Gasol J M, et al. 2010. Mesoscale eddies: hotspots of prokaryotic activity and differential community structure in the ocean. The ISME Journal, 4: 975.

Baltar F, Palovaara J, Unrein F, et al. 2016. Marine bacterial community structure resilience to changes in protist predation under phytoplankton bloom conditions. The ISME Journal, 10: 568-581.

Belzile C, Brugel S, Nozais C, et al. 2008. Variations of the abundance and nucleic acid content of heterotrophic bacteria in Beaufort Shelf waters during winter and spring. Journal of Marine Systems, 74(3-4): 946-956.

Bernard L, Courties C, Servais P, et al. 2000. Relationships between bacterial cell size, productivity and genetic diversity in aquatic environments using cell sorting and flow cytometry. Microbial Ecology, 40(2): 148-158.

Bettarel Y, Bouvier T, Bouvier C, et al. 2011. Ecological traits of planktonic viruses and prokaryotes along a full-salinity gradient. FEMS Microbiology Ecology, 76(2): 360-372.

Bhupathiraju V K, Hernandez M, Landfear D, et al. 1999. Application of a tetrazolium dye as an indicator of viability in anaerobic bacteria. Journal of Microbiological Methods, 37(3): 231-243.

Boulos L, Prevost M, Barbeau B, et al. 1999. LIVE/DEAD® BacLight™: application of a new rapid staining method for direct enumeration of viable and total bacteria in drinking water. Journal of Microbiological Methods, 37(1): 77-86.

Bouvier T, del Giorgio P A, Gasol J M. 2007. A comparative study of the cytometric characteristics of high and low nucleic-acid bacterioplankton cells from different aquatic ecosystems. Environmental Microbiology, 9(8): 2050-2066.

Calleja M L, Ansari M I, Røstad A, et al. 2018. The mesopelagic scattering layer: a hotspot for heterotrophic prokaryotes in the red sea twilight zone. Frontiers in Marine Science, 5: 259.

Calvo-Díaz A, Morán X A G. 2006. Seasonal dynamics of picoplankton in shelf waters of the southern Bay of Biscay. Aquatic Microbial Ecology, 42(2): 159-174.

del Giorgio P A, Gasol J M, Vaqué D, et al. 1996. Bacterioplankton community structure: protists control net production and the proportion of active bacteria in a coastal marine community. Limnology and Oceanography, 41(6): 1169-1179.

del Giorgio P A, Prairie Y T, Bird D F. 1997. Coupling between rates of bacterial production and the number of metabolically active cells in lake bacterioplankton, measured using CTC reduction and flow cytometry. Microbial Ecology, 34(2): 144-154.

Eilers H, Pernthaler J, Amann R. 2000. Succession of pelagic marine bacteria during enrichment: a close look at cultivation-induced shifts. Applied and Environmental Microbiology, 66(11): 4634-4640.

Falcioni T, Papa S, Gasol J M. 2008. Evaluating the flow-cytometric nucleic acid double-staining protocol in realistic situations of planktonic bacterial death. Applied and Environmental Microbiology, 74(6): 1767-1779.

Franco-Vidal L, Morán X A G. 2011. Relationships between coastal bacterioplankton growth rates and biomass production: comparison of leucine and thymidine uptake with single-cell physiological characteristics. Microbial Ecology, 61(2): 328-341.

Fuchs B M, Woebken D, Zubkov M V, et al. 2005. Molecular identification of picoplankton populations in contrasting waters of the Arabian Sea. Aquatic Microbial Ecology, 39(2): 145-157.

Fuchs B M, Zubkov M V, Sahm K, et al. 2000. Changes in community composition during dilution cultures of marine bacterioplankton as assessed by flow cytometric and molecular biological techniques. Environmental Microbiology, 2(2): 191-201.

Gasol J M, Alonso-Sáez L, Vaqué D, et al. 2009. Mesopelagic prokaryotic bulk and single-cell heterotrophic activity and community composition in the NW Africa-Canary Islands coastal-transition zone. Progress in Oceanography, 83(1-4): 189-196.

Gasol J M, Arístegui J. 2007. Cytometric evidence reconciling the toxicity and usefulness of CTC as a marker of bacterial activity. Aquatic Microbial Ecology, 46(1): 71-83.

Gasol J M, del Giorgio P A, Massana R, et al. 1995. Active versus inactive bacteria: size-dependence in a coastal marine plankton community. Marine Ecology Progress Series, 128: 91-97.

Gasol J M, del Giorgio P A. 2000. Using flow cytometry for counting natural planktonic bacteria and understanding the structure of planktonic bacterial communities. Scientia Marina, 64(2): 197-224.

Gasol J M, Zweifel U L, Peters F, et al. 1999. Significance of size and nucleic acid content heterogeneity as measured by flow cytometry in natural planktonic bacteria. Applied and Environmental Microbiology, 65(10): 4475-4483.

Girault M, Arakawa H, Barani A, et al. 2015. Heterotrophic prokaryote distribution along a 2300 km transect in the North Pacific subtropical gyre during a strong La Niña conditions: relationship between distribution and hydrological conditions. Biogeosciences, 12(11): 3607-3621.

Gomes A, Gasol J M, Estrada M, et al. 2015. Heterotrophic bacterial responses to the winter-spring phytoplankton bloom in open waters of the NW Mediterranean. Deep Sea Research Part I: Oceanographic Research Papers, 96: 59-68.

Grégori G, Citterio S, Ghiani A, et al. 2001. Resolution of viable and membrane-compromised bacteria in freshwater and marine waters based on analytical flow cytometry and nucleic acid double staining. Applied and Environmental Microbiology, 67(10): 4662-4670.

Grossart H P, Riemann L, Azam F. 2001. Bacterial motility in the sea and its ecological implications. Aquatic Microbial Ecology, 25(3): 247-258.

Hatzinger P B, Palmer P, Smith R L, et al. 2003. Applicability of tetrazolium salts for the measurement of respiratory activity and viability of groundwater bacteria. Journal of Microbiological Methods, 52(1): 47-58.

Heissenberger A, Leppard G G, Herndl G J. 1996. Relationship between the intracellular integrity and the morphology of the capsular envelope in attached and free-living marine bacteria. Applied and Environmental Microbiology, 62(12): 4521-4528.

Hobbie J E, Daley R J, Jasper S. 1977. Use of Nuclepore filters for counting bacteria by fluorescence microscopy. Applied and Environmental Microbiology, 33(5): 1225-1228.

Huete-Stauffer T M, Arandia-Gorostidi N, Díaz-Pérez L, et al. 2015. Temperature dependences of growth rates and carrying capacities of marine bacteria depart from metabolic theoretical predictions. FEMS Microbiology Ecology, 91(10): fiv111.

Huete-Stauffer T M, Morán X A G. 2012. Dynamics of heterotrophic bacteria in temperate coastal waters: similar net growth but different controls in low and high nucleic acid cells. Aquatic Microbial Ecology, 67(3): 211-223.

Jellett J F, Li W K W, Dickie P M, et al. 1996. Metabolic activity of bacterioplankton communities assessed by flow cytometry and single carbon substrate utilization. Marine Ecology Progress Series, 136(1-3): 213-225.

Jochem F. 2001. Morphology and DNA content of bacterioplankton in the northern Gulf of Mexico: analysis by epifluorescence microscopy and flow cytometry. Aquatic Microbial Ecology, 25(2):

179-194.

Joux F, Servais P, Naudin J J, et al. 2005. Distribution of picophytoplankton and bacterioplankton along a river plume gradient in the Mediterranean Sea. Vie Et Milieu-Life and Environment, 55(3-4): 197-208.

Karner M, Fuhrman J A. 1997. Determination of active marine bacterioplankton: a comparison of universal 16S rRNA probes, autoradiography, and nucleoid staining. Applied and Environmental Microbiology, 63(4): 1208-1213.

Lasternas S, Agustí S, Duarte C M. 2010. Phyto-and bacterioplankton abundance and viability and their relationship with phosphorus across the Mediterranean Sea. Aquatic Microbial Ecology, 60(2): 175-191.

Lasternas S, Agustí S. 2014. The percentage of living bacterial cells related to organic carbon release from senescent oceanic phytoplankton. Biogeosciences, 11(22): 6377-6387.

Lebaron P, Servais P, Agogué H, et al. 2001. Does the high nucleic acid content of individual bacterial cells allow us to discriminate between active cells and inactive cells in aquatic systems? Applied and Environmental Microbiology, 67(4): 1775-1782.

Lebaron P, Servais P, Baudoux A C, et al. 2002. Variations of bacterial-specific activity with cell size and nucleic acid content assessed by flow cytometry. Aquatic Microbial Ecology, 28(2): 131-140.

Lefort T, Gasol J M. 2014. Short-time scale coupling of picoplankton community structure and single-cell heterotrophic activity in winter in coastal NW Mediterranean Sea waters. Journal of Plankton Research, 36(1): 243-258.

Li W K W, Jellett J F, Dickie P M. 1995. DNA distributions in planktonic bacteria stained with TOTO or TO-PRO. Limnology and Oceanography, 40(8): 1485-1495.

Longnecker K, Sherr B F, Sherr E B. 2005. Activity and phylogenetic diversity of bacterial cells with high and low nucleic acid content and electron transport system activity in an upwelling ecosystem. Applied and Environmental Microbiology, 71(12): 7737-7749.

Longnecker K, Wilson M J, Sherr E B, et al. 2010. Effect of top-down control on cell-specific activity and diversity of active marine bacterioplankton. Aquatic Microbial Ecology, 58(2): 153-165.

Lovejoy C, Legendre L, Klein B, et al. 1996. Bacterial activity during early winter mixing (Gulf of St. Lawrence, Canada). Aquatic Microbial Ecology, 10(1): 1-13.

Lovejoy C, Legendre L, Therriault J C, et al. 2000. Growth and distribution of marine bacteria in relation to nanoplankton community structure. Deep Sea Research Part II: Topical Studies in Oceanography, 47(3): 461-487.

Martin A, Hall J A, Toole R, et al. 2008. High single-cell metabolic activity in Antarctic sea ice bacteria. Aquatic Microbial Ecology, 52(1): 25-31.

Mary I, Heywood J L, Fuchs B M, et al. 2006. SAR11 dominance among metabolically active low nucleic acid bacterioplankton in surface waters along an Atlantic meridional transect. Aquatic Microbial Ecology, 45(2): 107-113.

Morán X A G, Bode A, Suárez L Á, et al. 2007. Assessing the relevance of nucleic acid content as an indicator of marine bacterial activity. Aquatic Microbial Ecology, 46(2): 141-152.

Morán X A G, Calvo-Díaz A. 2009. Single-cell vs. bulk activity properties of coastal bacterioplankton over an annual cycle in a temperate ecosystem. FEMS Microbiology Ecology, 67(1): 43-56.

Mosharova I V, Il'inskii V V, Mosharov S A. 2016. State of heterotrophic bacterioplankton of Yenisei estuary and the zone of Ob-Yenisei discharge in autumn in relation with environmental factors.

Water Resources, 43(2): 341-352.

Mosharova I V, Mosharov S A, Ilinskiy V V. 2017. Distribution of bacterioplankton with active metabolism in waters of the St. Anna Trough, Kara Sea, in autumn 2011. Oceanology, 57(1): 114-121.

Nescerecka A, Hammes F, Juhna T. 2016. A pipeline for developing and testing staining protocols for flow cytometry, demonstrated with SYBR Green I and propidium iodide viability staining. Journal of Microbiological Methods, 131: 172-180.

Ortega-Retuerta E, Reche I, Pulido-Villena E, et al. 2008. Exploring the relationship between active bacterioplankton and phytoplankton in the Southern Ocean. Aquatic Microbial Ecology, 52(1): 99-106.

Paterson J S, Nayar S, Mitchell J G, et al. 2012. A local upwelling controls viral and microbial community structure in South Australian continental shelf waters. Estuarine, Coastal and Shelf Science, 96: 197-208.

Reinthaler T, Álvarez S X A, Álvarez M, et al. 2013. Impact of water mass mixing on the biogeochemistry and microbiology of the Northeast Atlantic Deep Water. Global Biogeochemical Cycles, 27(4): 1151-1162.

Rodriguez G G, Phipps D, Ishiguro K, et al. 1992. Use of a fluorescent redox probe for direct visualization of actively respiring bacteria. Applied and Environmental Microbiology, 58(6): 1801-1808.

Roszak D, Grimes D, Colwell R. 1984. Viable but nonrecoverable stage of *Salmonella enteritidis* in aquatic systems. Canadian Journal of Microbiology, 30(3): 334-338.

Ruiz-Gonzalez C, Lefort T, Galí M, et al. 2012. Seasonal patterns in the sunlight sensitivity of bacterioplankton from Mediterranean surface coastal waters. FEMS Microbiology Ecology, 79(3): 661-674.

Šantić D, Šestanović S, Šolić M, et al. 2014. Dynamics of picoplankton community from coastal waters to the open sea in the Central Adriatic. Mediterranean Marine Science, 15(1): 179-188.

Scharek R, Latasa M. 2007. Growth, grazing and carbon flux of high and low nucleic acid bacteria differ in surface and deep chlorophyll maximum layers in the NW Mediterranean Sea. Aquatic Microbial Ecology, 46(2): 153-161.

Schattenhofer M, Wulf J, Kostadinov I, et al. 2011. Phylogenetic characterisation of picoplanktonic populations with high and low nucleic acid content in the North Atlantic Ocean. Systematic and Applied Microbiology, 34(6): 470-475.

Schaule G, Flemming H, Ridgway H. 1993. Use of 5-cyano-2, 3-ditolyl tetrazolium chloride for quantifying planktonic and sessile respiring bacteria in drinking water. Applied and Environmental Microbiology, 59(11): 3850-3857.

Servais P, Agogué H, Courties C, et al. 2001. Are the actively respiring cells (CTC+) those responsible for bacterial production in aquatic environments? FEMS Microbiology Ecology, 35(2): 171-179.

Servais P, Casamayor E O, Courties C, et al. 2003. Activity and diversity of bacterial cells with high and low nucleic acid content. Aquatic Microbial Ecology, 33(1): 41-51.

Sherr B F, del Giorgio P A, Sherr E B. 1999. Estimating abundance and single-cell characteristics of respiring bacteria via the redox dye CTC. Aquatic Microbial Ecology, 18(2): 117-131.

Sherr E B, Sherr B F, Longnecker K. 2006. Distribution of bacterial abundance and cell-specific nucleic acid content in the Northeast Pacific Ocean. Deep Sea Research Part I: Oceanographic

Research Papers, 53(4): 713-725.

Sherr E B, Sherr B F, Verity P G. 2002. Distribution and relation of total bacteria, active bacteria, bacterivory, and volume of organic detritus in Atlantic continental shelf waters off Cape Hatteras NC, USA. Deep Sea Research Part II: Topical Studies in Oceanography, 49(20): 4571-4585.

Sieracki M E, Cucci T L, Nicinski J. 1999. Flow cytometric analysis of 5-cyano-2, 3-ditolyl tetrazolium chloride activity of marine bacterioplankton in dilution cultures. Applied and Environmental Microbiology, 65(6): 2409-2417.

Sieracki M E, Viles C L. 1992. Distributions and fluorochrome-staining properties of submicrometer particles and bacteria in the North Atlantic. Deep Sea Research Part A. Oceanographic Research Papers, 39(11): 1919-1929.

Smith E M. 1998. Coherence of microbial respiration rate and cell-specific bacterial activity in a coastal planktonic community. Aquatic Microbial Ecology, 16(1): 27-35.

Smith J J, McFeters G A. 1997. Mechanisms of INT (2-(4-iodophenyl)-3-(4-nitrophenyl)-5-phenyl tetrazolium chloride), and CTC (5-cyano-2, 3-ditolyl tetrazolium chloride) reduction in *Escherichia coli* K-12. Journal of Microbiological Methods, 29(3): 161-175.

Tada Y, Suzuki K. 2016. Changes in the community structure of free-living heterotrophic bacteria in the open tropical Pacific Ocean in response to microalgal lysate-derived dissolved organic matter. FEMS Microbiology Ecology, 92(7): fiw099.

Talarmin A, van Wambeke F, Catala P, et al. 2011. Flow cytometric assessment of specific leucine incorporation in the open Mediterranean. Biogeosciences, 8(2): 253-265.

Ullrich S, Karrasch B, Hoppe H, et al. 1996. Toxic effects on bacterial metabolism of the redox dye 5-cyano-2, 3-ditolyl tetrazolium chloride. Applied and Environmental Microbiology, 62(12): 4587-4593.

van Wambeke F, Catala P, Pujo-Pay M, et al. 2011. Vertical and longitudinal gradients in HNA-LNA cell abundances and cytometric characteristics in the Mediterranean Sea. Biogeosciences, 8(7): 1853-1863.

Vaqué D, Casamayor E O, Gasol J M. 2001. Dynamics of whole community bacterial production and grazing losses in seawater incubations as related to the changes in the proportions of bacteria with different DNA content. Aquatic Microbial Ecology, 25(2): 163-177.

Vila-Costa M, Gasol J M, Sharma S, et al. 2012. Community analysis of high- and low-nucleic acid-containing bacteria in NW Mediterranean coastal waters using 16S rDNA pyrosequencing. Environmental Microbiology, 14(6): 1390-1402.

Wang Y, Hammes F, Boon N, et al. 2009. Isolation and characterization of low nucleic acid (LNA)-content bacteria. The ISME Journal, 3: 889-902.

Xu H S, Roberts N, Singleton F L, et al. 1982. Survival and viability of nonculturable *Escherichia coli* and *Vibrio cholerae* in the estuarine and marine environment. Microbial Ecology, 8(4): 313-323.

Yamaguchi N, Nasu M. 1997. Flow cytometric analysis of bacterial respiratory and enzymatic activity in the natural aquatic environment. Journal of Applied Microbiology, 83(1): 43-52.

Zubkov M V, Allen J I, Fuchs B M. 2004. Coexistence of dominant groups in marine bacterioplankton community-a combination of experimental and modelling approaches. Journal of the Marine Biological Association of the United Kingdom, 84(3): 519-529.

Zubkov M V, Fuchs B M, Burkill P H, et al. 2001. Comparison of cellular and biomass specific activities of dominant bacterioplankton groups in stratified waters of the Celtic Sea. Applied and

Environmental Microbiology, 67(11): 5210-5218.

Zubkov M V, Fuchs B M, Tarran G A, et al. 2002. Mesoscale distribution of dominant bacterioplankton groups in the northern North Sea in early summer. Aquatic Microbial Ecology, 29(2): 135-144.

Zweifel U L, Hagström Å. 1995. Total counts of marine bacteria include a large fraction of non-nucleoid-containing bacteria (ghosts). Applied and Environmental Microbiology, 61(6): 2180-2185.

第四章 微食物网生物的运动和趋化

第一节 运动和趋化的意义

异养细菌利用海洋中的溶解有机物（DOM）进行"二次生产"。海水中的 DOM 主要来源于浮游植物渗出和颗粒有机物（POM）分解。尽管海水中 DOM 的浓度总体来看很低，但在水体中并非平均分布，而是在浮游植物和 POM 周围形成浓度梯度。海水中 DOM 浓度最高的地方被称为热点（hot spot）（Mitchell et al.，1985），其周围的高浓度区域称为微域（microzone），浮游植物的微域又称为藻际微环境（phycosphere）（Bell and Mitchell，1972；Seymour et al.，2017）。浮游植物和其他 POM 会持续向外释放 DOM，因此这类 DOM 热点存留的时间较长。海水中还有两种稍纵即逝的热点：第一种是浮游植物细胞裂解，向水体释放 DOM；第二种是当浮游植物和 POM 下沉或随水流移动的时候，在其运动轨迹上产生 DOM 残留，称为尾羽（plume，图 4-1）（Kiørboe and Jackson，2001）。随着时间的流逝，所有的热点最终会消失，因此细菌需要进行趋向运动，寻找并利用这些存留时间很短的 DOM。

图 4-1 半径为 0.5cm 的颗粒物以 1mm/s 的速度下沉时的尾羽示意图[改绘自 Kiørboe and Jackson（2001）]

尾羽以对应的氨基酸浓度高于背景 30nmol/L 为边界，尾羽长度以颗粒物半径（r）为单位在海水中浮游生物构成极其稀薄的悬浮体,按体积计算通常小于 10μl/L(Wolfe,2000),因此对于以这些浮游生物为食的浮游动物来说,海洋中的食物处于稀缺的状态。以细菌、鞭毛虫和纤毛虫为例,假设它们是边长分别为 0.5μm、3μm 和 10μm 的立方体,且丰度分别为 10^6cell/ml、10^3ind./ml 和 1ind./ml,则 1nl(100μm 的立方)水体中有 1 个细菌,1μl(1000μm 的立方)水体中有 1 个鞭毛虫。若这些浮游生物在水体中均匀分布,那么它们之间的空间关系为:每 2 个细菌之间的距离为 100μm,其间分布有 2~3 个病毒;每 2 个鞭毛虫之间的距离为 1mm,其间分布有 10 个细菌;每 2 个纤毛虫之间的距离为 1cm,其间分布有 10 个鞭毛虫、100 个细菌。按体积计算,细菌、鞭毛虫和纤毛虫在水中的浓度极低,仅分别为 0.125μl/L、0.027μl/L 和 0.001μl/L。因此,细菌的摄食者需要运动,从而主动接近细菌进行摄食。

由于 DOM 热点的存在,浮游细菌都偏向于集中到这些热点区域,而细菌的摄食者也尾随而至。寻找 DOM 热点和饵料颗粒满足自己的食物需求是微食物网生物面临的主要生存挑战。

即使没有趋化行为,单是运动本身就可以增加生物探索空间的大小。假设一个直径 0.4μm 的细菌不运动,布朗运动将使它在 10min 内探索 35μm^3(即每天 430μm^3)的空间;若这个细菌以 50μm/s 的速度运动,则 10min 内可探索 0.8mm^3 的空间,即每天 1cm^3(Stocker,2012)。

运动和趋化增加了颗粒之间的会遇率(encounter rate,E)。摄食者和饵料颗粒的会遇率可以表示为

$$E = \frac{\pi}{3} \times \left(r_g + r_P\right)^2 \times N_P \times \left(3v_g^2 + v_P^2\right) / v_g \tag{4-1}$$

式中,r_g 和 r_P 分别表示摄食者和饵料的半径,v_g 和 v_P 分别表示摄食者和饵料的速度,N_P 表示饵料的丰度。

假设存在这样一个体系,含有丰度为 10^6cell/ml、直径 1μm 的细菌,以及丰度 10^3ind./ml、直径 15μm 的浮游植物。如果细菌不运动,一个细菌遇到浮游植物的概率为 0.0035/d,一个浮游植物遇到细菌的概率为 3.5/d。如果其中有 10%的细菌是运动的,那么一个运动细菌遇到浮游植物的概率为 9/d,一个浮游植物遇到细菌的概率为 900/d(Seymour et al.,2017)。显然,运动和趋化行为同时增加了生物与其饵料、配偶和摄食者的会遇率。会遇率与运动速度和运动模式(motility pattern)有关,因此生物对运动模式的选择体现了对饵料和摄食两种压力的平衡(Wolfe,2000)。

运动给生物带来的好处远不止有助于摄食,还能够通过运动更新自己的环境,远离饵料减少而排泄物增加的原环境。对于群体来说,运动可以使得种群扩散到更大的生存空间,有利于生殖类群中不同个体相遇进行有性生殖。

第二节 运动原理和运动器官

浮游生物生活在水中,其运动符合流体力学原理。液体和空气等流体运动由纳维-斯托克斯方程(Navier-Stokes equation)描述,生物在流体中的运动受到惯性力和黏滞力的影响。该方程中惯性力和黏滞力的比值定义了雷诺数(Reynolds number,Re),Re 是流体力学中表征流体流动情况的无量纲数:

$$\mathrm{Re} = \frac{uL}{v} = \frac{uL\rho}{\eta} \tag{4-2}$$

式中,u 表示物体与流体的相对速度(m/s),L 表示物体长度(m),v 表示流体的运动黏度(kinematic viscosity,m²/s),ρ 表示流体的密度(kg/m³),η 表示流体的动力黏度[dynamic viscosity,Pa·s 或 kg/(m·s)]。

Re 值较大时,流体运动中的惯性力占主导,Re 值较小时则黏滞力占主导。式(4-2)中最重要的是物体长度和速度的乘积,表明二者是协同工作的,而不是相互抵消。对于生物来说,个体小几乎总意味着运动慢,而个体大几乎总意味着运动快。这就是为什么生物的雷诺数变化范围(约 16 个数量级)远远超过了生物个体尺寸变化(7~8 个数量级)(表 4-1)。人在游泳时 Re 约为 10^4,金鱼约为 10^2(Purcell,1977),细菌约为 10^{-5}(Vogel,1994),其他浮游生物在 10^{-8}~10^1(Kiørboe,1993)。

表 4-1 不同生物在海水或空气中运动时的雷诺数

生物	体长/粒径(m)	速度(m/s)	雷诺数	参考文献
大型鲸	20~30	10	3×10^8	
金枪鱼		10	3×10^7	
鸭子		20	3×10^5	
蜻蜓		7	3×10^4	
桡足类		0.2	3×10^2	Vogel (1994)
小型飞行昆虫			3×10^1	
无脊椎动物幼体	3×10^{-4}	1×10^{-3}	3×10^{-1}	
海胆精子	2×10^{-4}		3×10^{-2}	
细菌		1×10^{-5}	1×10^{-5}	
浮游生物	1×10^{-6}		2.73×10^{-8}	Kiørboe (1993)
	1×10^{-5}		2.73×10^{-5}	
	1×10^{-4}		2.73×10^{-2}	
	1×10^{-3}		2.73×10^{1}	

诺贝尔物理学奖得主 Edward M. Purcell 以扇贝为例描述了宏观和微观尺度下生物在流体中的运动(图 4-2)。宏观尺度下,生物在流体中的运动很大程度上依

赖于周围流体的惯性。Purcell 认为扇贝在水中张开壳又迅速合拢，这一动作往复进行，水从外套膜腔挤出去产生的冲击力使扇贝在水中游动（图 4-2a）（Purcell，1977）。

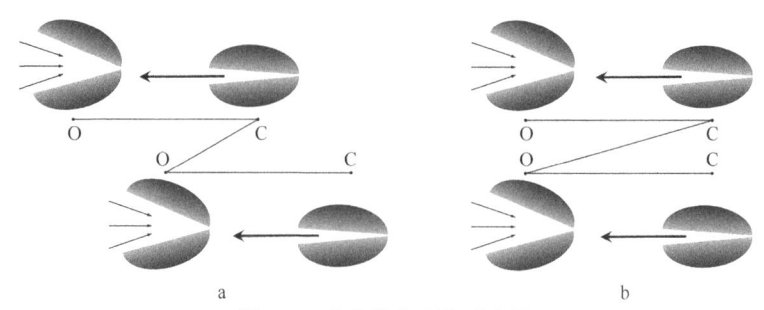

图 4-2　"扇贝定理"示意图

a. Re 值≥1 时，重复贝壳张开（O）和闭拢（C）的两步过程会导致生物在宏观尺度上的净运动；b. 在低 Re 值（≪1）情况下，开闭贝壳的往复式运动具有时间可逆性，造成生物在原地往复运动

然而这种运动策略在微观尺度上注定是失败的。假设扇贝被缩小到只有几微米，此时扇贝处于低 Re 值（≪1）的环境，惯性力在黏滞力面前几乎可以忽略不计。开闭贝壳的往复式运动（reciprocal motion）具有时间可逆性，贝壳张开和闭拢的形变过程在时间上是对称的，贝壳张开时产生位移，但在合拢时会产生相同距离的反方向位移，最终造成的结果是生物在原地往复运动（图 4-2b），并不产生位移。低 Re 值情况下，生物依靠刚性运动器官的往复式运动无法在流体中产生位移，这一现象被称为"扇贝定理"（scallop theorem）（Purcell，1977）。类似地，鱼鳍周期性摆动、鱿鱼或水母的喷射式前进均无法在低 Re 值情形下驱动。

要想在低 Re 值的微观情形中游动，微型生物需要打破运动器官形变在时间上的对称性。微型生物的应对策略是采用柔性运动器官，包括鞭毛（flagellum，复数 flagella）和纤毛（cilium，复数 cilia）来进行时间不可逆性运动。

鞭毛是细菌和鞭毛虫的运动器官。浮游细菌大多具有一根鞭毛，而鞭毛虫通常有两根鞭毛，大多数时候还不等长。细菌的鞭毛直径有 10～20nm，在立体空间内做螺旋运动，这种运动可产生一个向尾部传播的形变波，类似葡萄酒瓶塞的螺旋开瓶器（图 4-3a）。鞭毛虫的鞭毛直径约 200nm，通过从身体传向鞭毛尾端的行波状变形，产生与波传播方向相反的推进力，这种变形与蛇的运动方式类似（图 4-3b）。

鞭毛虫鞭毛的摆动方式使得在任一时刻，鞭毛的某些部分（转角）不能产生推动力，有的部分甚至产生后拉力，降低了效率。细菌鞭毛的螺旋式运动使得鞭毛的每一个部分都产生推动力，从这一角度看，细菌鞭毛比鞭毛虫的鞭毛使用效率高，但是像直升机需要尾桨抵消主桨的角动量一样，细胞本身也会产生旋转，从而耗费一些能量。

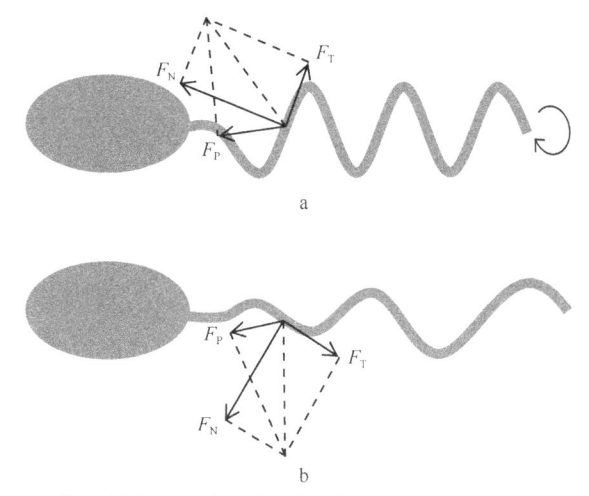

图 4-3　鞭毛的摆动方式和推进机制 [改绘自 Ghanbari（2020）]

a. 细菌通过鞭毛螺旋转动推进；b. 鞭毛虫通过鞭毛的行波状变形推进。F_T 和 F_N 分别表示切向和法向阻力，F_P 为鞭毛摆动产生的使生物运动的推进力，环形箭头表示细菌鞭毛螺旋转动的方向

纤毛虫的运动器官是纤毛。纤毛往复运动拨动水体，在拨水时，纤毛完全伸开，能带动大体积的水体，纤毛回复原位时，紧贴细胞，拨动的水体较少（图 4-4）。两次拨动的综合效果是产生反作用力，推动纤毛虫运动，其原理类似人在蛙泳时双臂的动作。单根纤毛的拨动只会使细胞旋转，纤毛虫体表着生多根纤毛，甚至排列成纤毛列，通过调整纤毛的拨水时间来协同调整虫体的运动姿态。

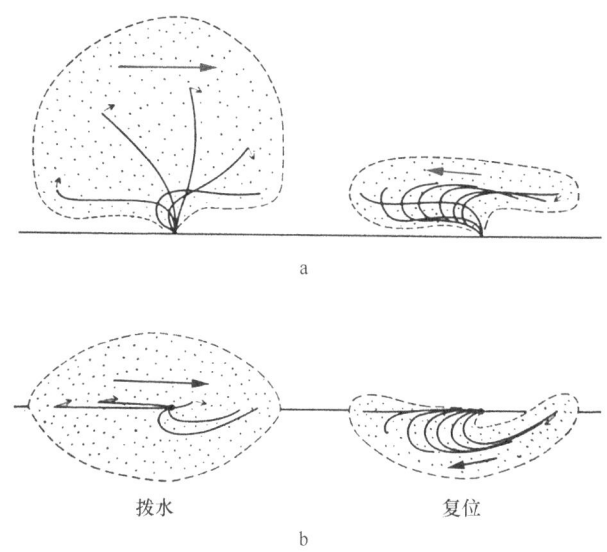

图 4-4　纤毛摆动示意图[改绘自 Cosson 和 Prokopchuk（2014）]

a. 侧视图；b. 俯视图

不具有鞭毛和纤毛的微型生物也可能有运动能力。Waterbury 等在 20 世纪 80 年代中期从寡营养海区中分离获得了几株有游泳能力的聚球藻属蓝细菌

（Waterbury et al.，1985），它们的细胞为棒状，体长 1~2.5μm，直径 0.7~0.9μm，游泳时绕长轴旋转，不轻易转向，没有鞭毛等明显的运动细胞器。目前从海水中分离的聚球藻中约 1/3 有运动能力，它们以 5~25μm/s 的速度在海水中游动，同时以大约 1Hz 的频率围绕其长轴旋转。聚球藻的游泳以钠动力为直接能量来源，但这种神秘运动形式的具体机制尚不清晰。通过基因组分析已经确定聚球藻具有可参与运动的细胞表面蛋白（Brahamsha，1996；McCarren and Brahamsha，2007），在一些菌株中也观察到从细胞延伸出来的细针状（spicule）突起，可能是运动细胞器（Samuel et al.，2001）。Ehlers 和 Oster（2012）提出了解释聚球藻运动的模型，认为聚球藻细胞表面针状突起进行桨状运动在细胞表面产生行波，进而推进细胞移动（图 4-5）。

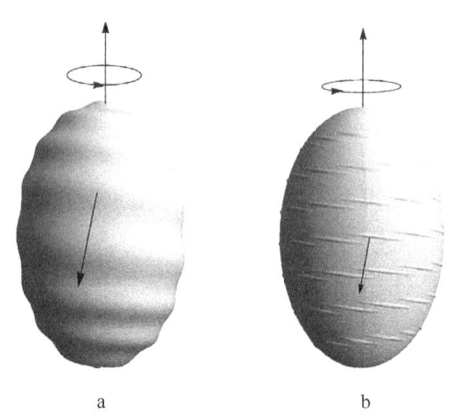

图 4-5　聚球藻细胞表面产生行波以推进细胞移动的两种模型[改绘自 Ehlers 和 Oster（2012）]
a. 固定马达模型；b. 行进马达模型

海洋生物通常可分为 9 个粒级，从 10^{-2}μm 到 10^{7}μm。其中，浮游生物的粒级范围为 10^{-2}~10^{5}μm，占据了 7 个量级，分别称为极微型（femto）、微微型（pico）、微型（nano）、小型（micro）、中型（meso）、大型（macro）和巨型（mega）浮游生物。微食物网生物粒径小于 2×10^{2}μm，占据 4 个量级（图 4-6）。

尽管微食物网的大部分生物类群都能运动，但是被认为属于浮游动物的只有鞭毛虫和纤毛虫。甲藻常被视作浮游植物来研究，在本书中涉及较少。虽然鞭毛虫和纤毛虫的粒级包括了 nano 和 micro 两个粒级，但它们常被统称为 microzooplankton（Beers and Stewart，1967），多被译为微型浮游动物，也有的译为小型浮游动物，本书采用微型浮游动物这一译名。

随着体长和运动速度的增加，各海洋生物类群运动时面对的雷诺数逐渐增加。如前所述，Re 值较小时则黏滞力占主导，Re 值较大时惯性力占主导，因此可依据 Re 值将海洋生物分为 3 个世界（图 4-6）：①Re＜1 时，海水的黏度对生物运动起决定作用，生物处于浮游状态，称为黏性浮游世界，黏性浮游世界的生物体长

小于 $2\times10^2\mu m$，主要为微食物网生物；②1＜Re＜2000 时，惯性起决定作用，生物仍然处于浮游状态，称为惯性浮游世界，惯性浮游世界生物的最大体长为 5~6cm，优势浮游动物为桡足类和磷虾类，使用 5 对联动的腹足游动；③Re＞2000 时，生物运动受到湍流的影响，5 对腹足的游动方式效率不高，生物主要是靠尾鳍推动的游泳生物，称为惯性游泳世界。

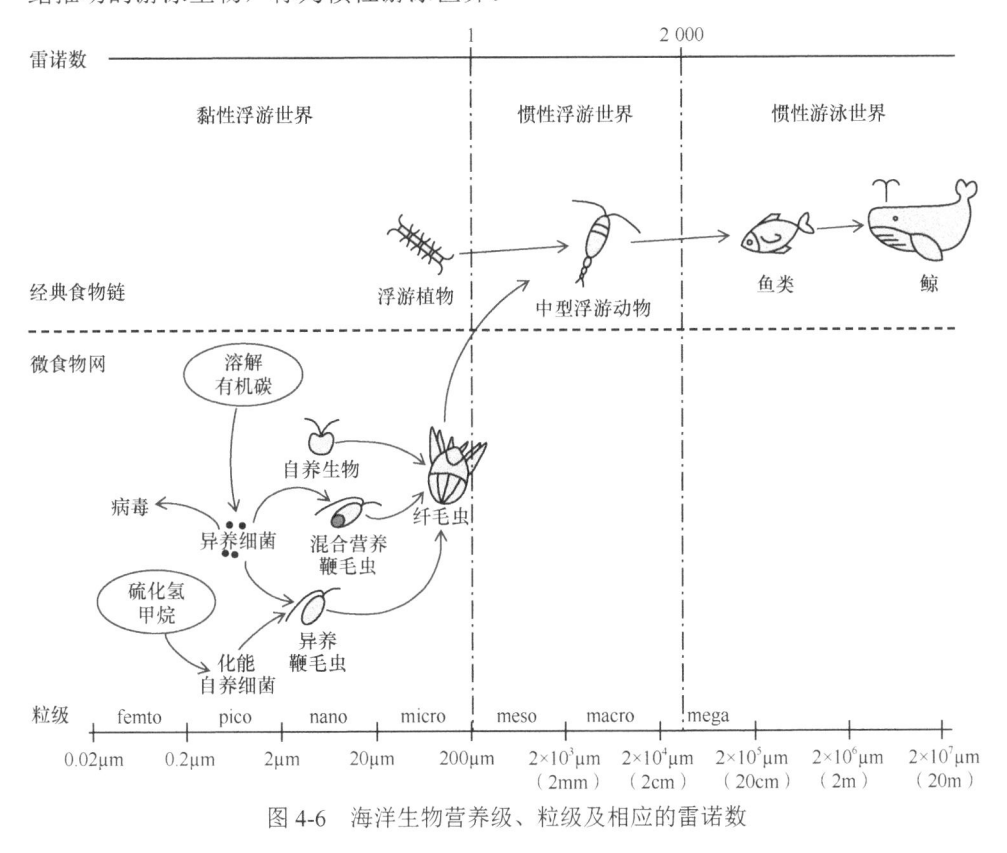

图 4-6 海洋生物营养级、粒级及相应的雷诺数

第三节 微食物网生物的趋化性和其他趋性

海洋微食物网生物的趋化性研究最初借鉴的是毛细管法（capillary assay），即在一端封口的毛细管内装入受试物质，放入含受试生物的培养液一段时间后，比较进入实验毛细管（装有受试物质）和对照毛细管（没有受试物质，其他条件相同）中受试生物的数量，以检验生物被受试物质吸引的情况（图 4-7）。在测试时间内，管内的液体和受试生物培养液之间的扩散混合是不可避免的，但是毛细管法的一端封闭和较细的管径（200μm）减少了由于对流造成的混合（Adler，1973）。毛细管法操作简单，结果具有显著的重复性，并且毛细管中物质扩散产生的梯度可用数学表征。然而毛细管中的浓度梯度随着时间的推移会变小，直至最终消失，而且暴露在可检测浓度梯度下的细胞数量相对较低，这

些缺点限制了毛细管法的敏感性。

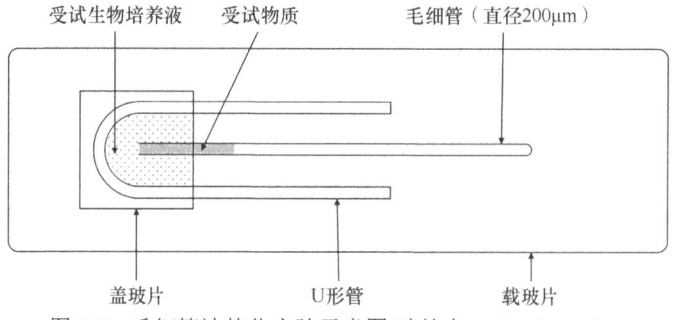

图 4-7 毛细管法趋化实验示意图[改绘自 Adler（1973）]

近年来，微流控法（microfluidic assay）发展迅速，微流控装置的透明性、生物相容性和简单易用性使其非常适合于微型生物生态学研究。该方法使用微型注射器向微通道中注射化学诱导剂，从而产生一个 100μm 宽的趋化物质条带，受试生物由微型注射泵沿着趋化物质带的两侧平流注入（图 4-8）。注射泵分离时微通道中的流动停止，趋化物质带以分子扩散速率横向扩展。通过外接显微镜来观察受试生物在通道上的位置，拍照记录每个时间点受试生物在通道内（X 轴方向）的分布情况，即可观测受试生物的趋化反应（Seymour et al.，2008）。与传统的毛细管法相比，微流控法可在群体和单细胞水平上实时评估和量化受试生物对潜在吸引物的趋化性，从而能够快速筛选多种化学物质。此外，通过跟踪单个生物体的运动轨迹，可以获得关于趋化行为的详细信息。

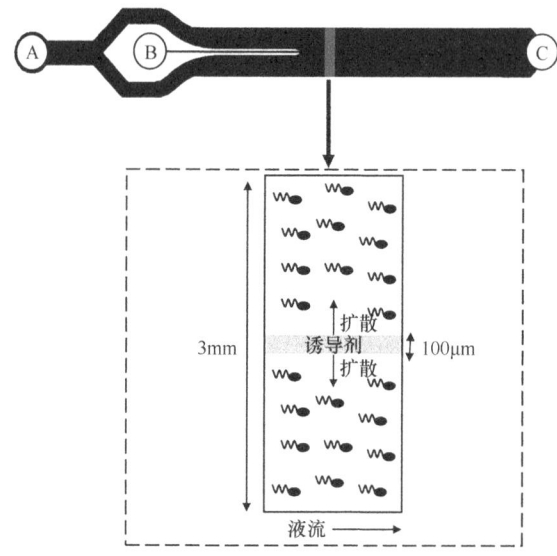

图 4-8 微流控法趋化实验示意图 [改绘自 Seymour 等（2010）]
上图：微流控装置，受试生物（细菌）和化学诱导剂分别通过入口 A 和 B 注入微通道，C 为微通道出口，灰色框为微注射器针尖下游 3mm 处的数据采集区。下图：上图灰色框所示区域放大示意图

一、浮游细菌趋化

Bell 和 Mitchell（1972）首先用毛细管法研究了海洋浮游细菌对浮游植物胞外渗出物的趋化行为。他们发现不同种类的浮游细菌在毛细管内的分布不同，有的细菌很快就运动到毛细管的顶端，有的细菌则只分布在毛细管的开口处。不同受试物质对细菌有不同的吸引力，细菌对氨基酸有最强的趋向性，对葡萄糖的趋向性不强，大多数细菌对蔗糖有趋向反应，对甘油和甘露醇没有趋向反应。

据估算，细菌对氨基酸和糖类的趋化浓度阈值为 $10^{-6}\sim 10^{-5}$mol/L。而氨基酸在浮游植物细胞周围的浓度为 $10^{-8}\sim 10^{-6}$mol/L，浓度高值已达到细菌趋化阈值，因此海洋浮游细菌对浮游植物培养液中的胞外渗出物可做出趋向反应，尤其是培养时间长的培养液有更强的吸引力，可能是由于培养后期的藻液中吸引细菌产生趋化运动的物质的浓度逐渐升高（Bell and Mitchell，1972）。

Willey 和 Waterbury（1989）研究了聚球藻的趋化性，该研究使用的 Blind-well 趋化实验装置由体积为 200μl 和 800μl 的两个培养皿状的容器组成，试验时在大容器内装入受试物质，在小容器内装入聚球藻，在两个容器之间置入孔径为 3μm 的滤膜从而将两个容器连接起来，聚球藻可以通过滤膜进入到大的容器中。培养一段时间后，比较对照组和实验组大容器中聚球藻的数量以估计趋化性。该研究共实验了 22 种物质，包括糖、氨基酸、氮化合物等，浓度范围为 $10^{-10}\sim 10^{-5}$mol/L。该研究揭示，聚球藻 Synechococcus sp. WH8113 对氨、硝酸盐、丙氨酸、甘氨酸和尿素都有趋向运动，趋化反应的浓度阈值在 $10^{-10}\sim 10^{-9}$mol/L。

Seymour 等（2010）通过微流控趋化试验发现，游海假交替单胞菌 Pseudoalteromonas haloplanktis、溶藻弧菌 Vibrio alginolyticus 和矽杆菌 Silicibacter sp. TM1040 对聚球藻 Synechococcus elongatus 的胞外产物具有趋化反应。其中，游海假交替单胞菌的趋化反应最为强烈和快速，在 S. elongatus 胞外产物带内聚集的浓度高出背景水平 9 倍。游海假交替单胞菌和矽杆菌对原绿球藻 Prochlorococcus marinus MED 4Ax 的胞外产物具有显著趋化反应，但溶藻弧菌对 P. marinus 的趋化较弱。

二、鞭毛虫趋化

原生动物的趋化行为研究开始得较早，在 19 世纪 80 年代即有研究发现波豆虫（Bodo sp.）有化感能力。在海洋微食物网研究早期，Goldman（1984）提出假说认为海洋鞭毛虫能通过化感和趋化作用定位饵料或饵料浓度大的微环境。Sibbald 等（1987）利用 T 形迷宫装置（T-maze apparatus）实验首次证实海洋鞭毛虫有趋化行为，发现鞭毛虫 Pseudobodo tremulans 对一株海洋细菌的渗出物质有趋向性。Sibbald 等（1987）还在 T 形迷宫装置中单独添加了不同化学物质，发现铵、组氨酸、甘氨酸和苏氨酸对 *P. tremulans* 有吸引作用，硝酸盐和亚硝酸盐对 *P. tremulans* 有排斥作用，天门冬氨酸、脯氨酸、苯丙氨酸、丝氨酸、葡萄糖、乙酸、尿素和叶酸等对 *P. tremulans* 没有诱导作用。

毛细管法也可以用于研究鞭毛虫对细菌的趋化性。Mohapatra 和 Fukami（2007）使用毛细管法发现鞭毛虫 *Jakoba libera* 对假单胞菌 *Pseudomonas* sp.的化感吸引力最高，其次是黄杆菌 *Flavobacterium* sp.和气单胞菌 *Aeromonas* sp.。进一步的实验表明，*J. libera* 对假单胞菌细胞壁的某种（某些）化合物具有很强的趋化性。这些细菌表面化合物(bacterial surface compound)的分子质量较高（>10kDa），这意味着与 *P. tremulans* 不同，*J. libera* 的趋化物质并非氨基酸和短肽等小分子量化合物（Mohapatra and Fukami，2007）。

鞭毛虫在摄食过程中，可以与细菌相互作用产生水体中原来没有的氨基酸，从而吸引更多鞭毛虫前来摄食。例如，食细菌鞭毛虫 *Rhynchomonas nasuta* 与细菌的共同培养液中积累了甘氨酸、L-丙氨酸、D-丙氨酸、丝氨酸和 L-谷氨酸，且这些氨基酸的积累量与 *R. nasuta* 的丰度呈正相关，表明 *R. nasuta* 参与了它们的生成。在这 4 种氨基酸中，*R. nasuta* 对甘氨酸的趋化反应最强；而 L-丙氨酸和 D-丙氨酸的影响与鞭毛虫生长阶段有关，仅在幼龄鞭毛虫中引起显著趋化反应（Ayo et al.，2010）。

浮游植物也会引起鞭毛虫的趋化运动。Menden-Deuer 和 Grünbaum（2006）在异养鞭毛虫 *Oxyrrhis marina* 培养液中放置了一个厚度为 5mm 的薄层装置，并在薄层中引入微藻 *Isochrysis galbana* 细胞或其胞外渗出物。实验表明，*O. marina* 能被 *I. galbana* 细胞或其胞外渗出物吸引，转向速率显著增加。*O. marina* 的游泳速度在对完整的猎物细胞作出趋化反应时显著增加，但对 *I. galbana* 的胞外渗出物没有显著趋化性。4h 后，薄层内的捕食者丰度比引入猎物前高出 20 倍。

三、纤毛虫趋化

许多海洋浮游纤毛虫都具有化学感应能力，使它们能够通过积极的趋化运动在微藻和细菌群落中识别最适宜的饵料。砂壳纤毛虫 *Tintinnopsis dadayi* 和无壳纤毛虫 *Strombidium* sp.能够找到并识别它们喜欢的猎物，且饥饿的个体比饱食的个体有更好的趋向性。*T. dadayi* 和 *Strombidium* sp.对硅藻、定鞭藻和甲藻有趋向性，但是对绿藻和蓝细菌趋向性不强。*T. dadayi* 会躲避赤潮藻 *Olisthodiscus*。对于同一种藻，*T. dadayi* 喜欢指数生长期的个体，而躲避平台期的个体（Verity，1988，1991）。

当在砂壳纤毛虫 *Favella* sp.培养液中加入鞭毛藻 *Heterocapsa triquetra* 或其渗

出物时，*Favella* sp.表现出迅速的行为变化，包括游动速度下降、转向率增加。*Favella* sp.对鞭毛藻 *H. niei*、*H. pygmaea* 也会表现出游动速度降低，但转向率无显著变化。对其他鞭毛藻，如 *Gonyaulax polyedra*、*Tetraselmis chuii*、*Isochrysis galbena*，*Favella* sp.无论是游动速度或转向率都无变化。这些趋化行为可促使 *Favella* sp.趋向聚集在适宜饵料周边进行摄食（Buskey and Stoecker，1989）。

海洋纤毛虫 *Strombidium sulcatum* 能够感应到培养基中高密度的细菌斑块，通过降低游动速度并将游动模式从螺旋路径转变为封闭的圆圈，使 *S. sulcatum* 在几分钟内即可聚集到饵料源周围（Fenchel and Jonsson，1988）。

四、趋化作用的空间尺度

因为细菌对化学物质的感应阈值浓度高于该物质在海水中的浓度，因此细菌的趋化能力不能在大尺度内发生作用，而是使细菌在撞见这个化学物质斑块后，能向其源头运动（Bell and Mitchell，1972）。类似地，纤毛虫的趋化作用在大尺度的化学梯度内可能也不重要，纤毛虫不能去寻找大的饵料斑块，而只能在几毫米的小尺度内接近饵料时发生作用（Jonsson，1989）。

物质从细胞内向外扩散的能力取决于该物质在胞外胞内的浓度差以及扩散速度。若这两个因素一致，则来自大细胞的物质扩散距离远，而来自小细胞的物质扩散距离近。当细胞小到只有 2~2.5μm 时，细菌的化学探测能力就探测不到了。因此对于鞭毛虫和纤毛虫而言，趋化作用能使它们找到饵料形成的斑块，但不能帮助它们定位单个饵料（Jackson，1987）。

五、其他趋性

淡水纤毛虫 *Loxodes* spp.和底栖纤毛虫 *Remanella* spp.可通过平衡囊（statocyst）感知重力方向，因此具有真趋地性（true geotaxis）（Fenchel and Finlay，1986）。海洋浮游纤毛虫的趋地性研究极少。自然环境中，许多海洋浮游纤毛虫聚集在海面或温盐跃层附近。将纤毛虫自然种群或人工养殖种群置于实验培养水柱中，类似的垂直分布模式也会重复出现。但在海洋浮游纤毛虫中尚未发现平衡囊类的重力感受器，因此这类趋性行为被称为被动趋地性（passive geotaxis）（Jonsson，1989）。由于纤毛虫虫体的重心偏下，在停止游动悬浮状态下虫体会自动转向，使虫体前部指向上方，这样虫体的游动就指向水体上层，导致海洋浮游纤毛虫的被动趋地行为。

迄今尚未确定微食物网生物是否有真趋光性（true phototaxis），但光照强度对它们的行为有影响。纤毛虫 *Strombidium reticulatum* 在光线强度突然降低后，运动速度也下降，直线运动距离减少；1h 后，*S. reticulatum* 的运动速度才慢慢恢复，但仍低于强光下的运动速度（Jonsson，1989）。从强光照转入弱光 10min 后，聚球藻的运动速度减半，恢复光照 2min 后，运动速度复原（Willey and Waterbury，1989）。聚球藻没有真趋光性的原因可能是其运动速度低，以聚球藻运动速度 25μm/s 来估计，一天可运动 2m，在自然海区这段距离内没有光线强弱的本质区别。这一解释可能也适用于细菌、鞭毛虫和纤毛虫等其他微食物网生物。

第四节 微食物网生物的运动及其与细胞的营养吸收

一、运动模式

微食物网生物的运动是为了找寻周围的饵料。细菌、鞭毛虫和纤毛虫具有感知海水中化学梯度的机制：它们对自己所处环境的化学浓度有短暂的记忆，通过移动一段距离并比较两处的差异，从而判断是否存在化学梯度。当没有饵料生物和敌害生物的线索时，生物处于搜索模式，此时运动的目的是避免重复搜索同一个空间。

浮游细菌和部分鞭毛虫类群在搜索过程中使用随机游动（random walk）模式，运动轨迹由直行+转向+直行……组成，每次直行的距离和转向的角度都是随机的，不形成趋化运动。浮游纤毛虫和另一部分鞭毛虫（如甲藻）则采用螺旋式游动（helical swimming）模式（图 4-9），这种运动方式在空间的三个维度都有速度。随机游动和螺旋式游动的共同特点是使生物避免重复进入同一个空间。

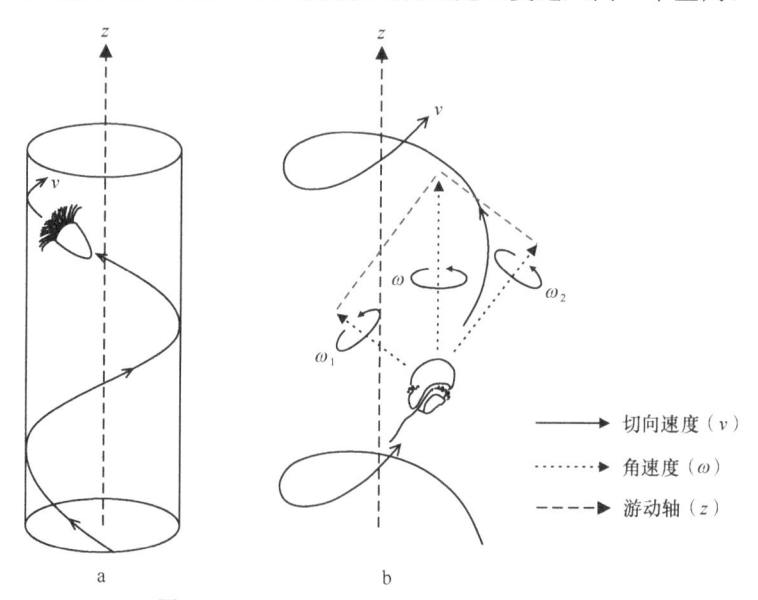

图 4-9 纤毛虫和鞭毛虫的螺旋式游动示意图

a. 纤毛虫的螺旋式游动，改绘自 Fenchel 和 Blackburn（1999）；b. 鞭毛虫的螺旋式游动，改绘自 Esteban 和 Fenchel（2020）

当在直行过程中探测到信号物质浓度梯度时，微食物网生物运动模式会发生变化。若生物原本即朝向自己喜爱的热点信号源，此时生物需要进行的转向较少，直行距离较长；若生物背向这个信号源，需要执行的转向较多且偏向源头，直行距离较短，才能朝向热点运动。这种改变后的随机游动被称为偏向性随机游动（biased random walk）（图 4-10a）。偏向性随机游动可能达到的实际速度约是其直行速度的 10%（Mitchell et al., 1985）。当接近热点信号源时，直行的距离变短，速度逐渐变为 0，生物就聚集在热点信号源的周围。

图 4-10 细菌和纤毛虫在发现热点信号源后向源接近的运动轨迹[改绘自 Christensen-Dalsgaard （2006）]
a. 细菌；b. 纤毛虫

90%以上的海洋浮游细菌具单鞭毛，鞭毛可以通过正转和反转调节进退和转向，造成的细菌转向游动模式被称为 flick（图 4-11）。当前进中的细菌检测到化学梯度后，先停止鞭毛正转从而停止运动，然后鞭毛旋转方向改变为反转使得细菌后退，后退停止后鞭毛和细菌细胞主轴方向发生改变，鞭毛恢复正转促成细菌转向。细菌的游泳速度越快，转向越频繁，获得的实际速度越大，越早接近化学物质的源头（Son et al., 2016）。

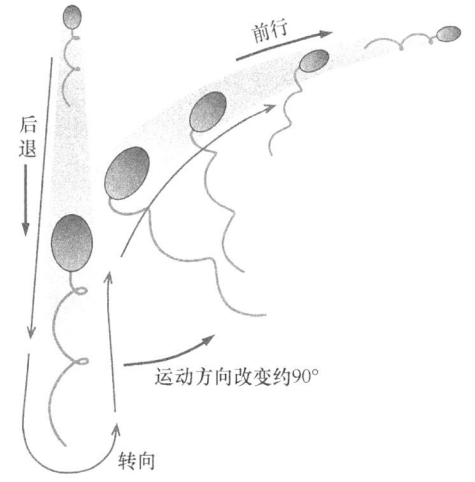

图 4-11 细菌的 flick 转向游动[改绘自 Son 等（2015）]

部分鞭毛虫和纤毛虫采用的螺旋式前进形成像弹簧的运动轨迹一样，弹簧的轴向是生物运动的方向（图 4-10b）。生物可以改变弹簧的轴向改变前进方向直接转向信息源头，也可以通过改变弹簧各圈层之间的距离改变前进速度，距离增大时，生物以接近直线的运动轨迹前进，当距离减小时，生物运动减慢，直至停止。这种运动方式又称为螺旋斜趋性（helical klinotaxis）(Fenchel and Blackburn，1999)。

二、运动速度

速度是一段时间内起始点与终点的距离和时间的比值，此处所指的运动速度是直线运动速度。在随机游动模式下，一段时间内运动开始和结束两点间的直线距离（l）和运动轨迹（L）之间的比值被称为净位移与总位移比（net to gross displacement ratio, NGDR）（图 4-12）(Buskey and Stoecker, 1988)。NGDR 是生物运动轨迹曲折程度的指标，介于 0 和 1 之间：接近 1 的 NGDR 值对应直线运动，根据速度公式得出的是生物真正的运动速度；而当 NGDR 值越接近 0，生物运动的路径就越曲折和复杂，计算获得的速度大大降低，无法体现生物真正的运动速度。

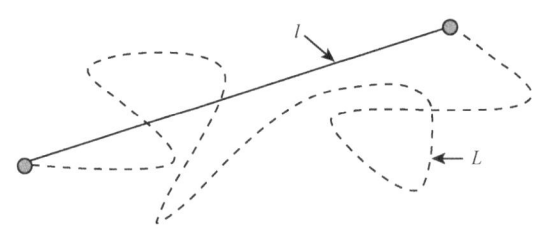

图 4-12　随机游动模式下生物运动的直线距离（l）和运动轨迹（L）[改绘自 Visser 和 Kiørboe（2006）]

海水中的大多数细菌直径通常小于 0.6μm，因此需要较高的游泳速度（约 100μm/s）以对抗由于布朗运动而引起的重新定位。自由游动的细菌在寡营养海洋环境中比例可高达总细菌数量的 70%，它们并不是一直在游动，也不是以恒定的速度游动。Mitchell 等（1995b）第一个测量了自然海水中细菌的运动速度，发现最高可达 230μm/s。后续研究观测到海洋细菌的运动速度有很大的范围，速度最低小于 10μm/s，最高超过 500μm/s（Grossart et al., 2001）。

实验室培养细菌的最大运动速度通常低于自然海区样品。Kiørboe 等（2002）测定了 8 株实验室内培养的细菌，发现它们的运动速度为 20~60μm/s，转向频率每秒 0.076~2.06 次。Johansen 等（2002）测定了 84 株培养细菌，发现它们的运动速度为 11~38μm/s，最大可达 75μm/s，每次直行的时间为 0.105~0.323s，70% 的运动转向角度超过 150°，基本上表明游泳方向在停止和转向后发生了逆转，细菌的平均运动速度和生长率呈显著正相关（Johansen et al.，2002）。对于同一株细菌，在饥饿状态下的运动速度会降低（Yam and Tang，2007），这与自然海水添加有机质后运动细菌比例增加（Grossart et al.，2001）可能是一个道理。

不同种类鞭毛虫的游泳速度差异较大。以随机游动模式缓慢游泳的 *Bodo designis* 运动速度为 35μm/s，而采用螺旋式前进模式的 *Spumella* sp.的直行速度为 107μm/s，其螺旋运动轨迹的振幅和频率有很大的变化，并且相互成反比，分别为 20～120μm 及 0.1～1.3 次/s（Kiørboe et al.，2004）。异养鞭毛虫 *Oblea rotunda* 的运动速度约为 360μm/s，相当于每秒运动 16 个体长的距离（Strom and Buskey，1993）。

不同生命阶段鞭毛虫的游泳行为可能不同。有毒甲藻 *Alexandrium fundyense* 的营养细胞笔直游动，并绕着顶端轴旋转，直到撞到其他物体后转向并继续笔直游动，游动速度约为(239±49)μm/s，并具有较高的 NGDR(0.87±0.12)。*A. fundyense* 的配子运动速度较慢，游动速度约为（61±40）μm/s，并且以频繁转向和环形运动为特征，其 NGDR 值较低（0.37±0.24）（Persson et al.，2013）。

20 世纪 60～70 年代即有对海洋砂壳纤毛虫运动速度的研究，Vitiello（1964）测定的砂壳纤毛虫的游泳速度约为 5m/h，Zaika 和 Ostrovskaya（1972）的测定结果为约 4.5m/h。在已有的文献报道中，纤毛虫的游泳速度在微食物网生物中最高，如 *Strombidium* sp.的平均游动速度在 315～800μm/s，但其最高运动速度可高出平均速度 2 个数量级，在 32 370～51 200μm/s（Vandromme et al.，2010）。*Favella* sp. 的平均游动速度在 740～1060μm/s，并具有较高的 NGDR(0.70～0.80)（Buskey and Stoecker，1988）。

浮游生活的海洋微食物网生物比其他环境中同类生物的运动速度高。由于大多数热点是短暂的，趋化是一场与时间的赛跑。海洋细菌的平均游泳速度往往超过 60～80μm/s，而微生物运动研究模式生物大肠杆菌的运动速度为 15～30μm/s。传统微生物学观点认为，对于细菌来说运动是低成本高收益的，这是针对在营养丰富的环境（如肠道）中缓慢游动的细菌（如大肠杆菌）提出的，并不适用于海洋环境。在寡营养的海水中，营养物质的浓度要比营养丰富的环境低几个数量级，而所需的推进力（与速度平方成正比）要大 10 倍以上。快速运动是海洋微食物网生物在需消耗的能量成本和可获取的能量之间权衡的结果，造就了海水中丰富的寡营养生物（Stocker，2012）。

整体来看，微食物网生物的运动速度与其粒径大小成正比，随着细胞增大，其绝对运动速度增加。以甲藻和纤毛虫为例，体长从 10μm 到 150μm，游泳速度从 0.2mm/s 升高到 1.0mm/s。但是由于运动耗能增加，微食物网生物以体长倍数表达的相对速度变小（Fenchel，1987），鞭毛虫的相对速度相当于每秒 20～80 倍体长，而纤毛虫的相对速度则仅有每秒 0.2～0.5 倍体长。

三、运动与细胞的营养吸收

海水具有一定的黏性,因此在微食物网生物的细胞外存在一个扩散边界层(diffusion boundary layer),其大小一般为细胞直径的数倍(Stocker,2012)。扩散边界层是细胞和周围海水物质交换(吸收营养、排出废物)的中间层,营养物质的浓度自边界层的外边界向内边界逐渐降低。当营养盐的扩散限制了细胞吸收营养时,减小边界层的厚度可以增加细胞的营养供应,提高细胞的吸收能力。当细胞在海水中运动时,边界层可产生形变,运动方向前端的边界层被压缩从而增加细胞的吸收能力,运动方向后端的边界层则被拉伸从而降低吸收能力。因为细胞对营养的吸收本已经处于受限状态,所以边界层拉伸导致的吸收能力降低效果大于边界层压缩导致的吸收能力增加,因此总的效果是减少了营养的吸收。

在营养物质均匀分布的水体中,若直径 1μm、运动速度 30μm/s 的细菌想通过提高运动速度实现营养吸收增加一倍,需要其运动速度达到 3000μm/s(Berg and Purcell,1977)。这一结果表明,运动不能明显增加小细胞(<10μm)对物质的获取。但在营养物质非均匀分布的水体中,生物体可以利用趋化运动,寻找更有利的区域。

第五节 细菌向热点的聚集

当细菌的粒径低于 0.6μm 时,其运动主要受分子布朗运动的影响,此时对外界热点信号无法产生有效响应。这些小的不运动的细菌的表面积与体积的比值更大,可以依靠较高的单位生物量的营养物质吸收率在低背景浓度溶解有机碳(DOC)中存活,它们的染色体中缺少运动和趋化的功能基因和调节基因。

大的可运动的细菌的生存策略则是游向热点,利用转瞬即逝的高浓度的 DOC。由于 DOC 是由热点释放出来的,DOC 的浓度高说明热点的数量多,因此当 DOC 浓度高时,运动细菌占的比例增加。海水中运动细菌的比例一般低于 10%,但是增加有机质后,经过 7~12h 的时滞,运动细菌的比例增加,并在 15~30h 达到最大值(Mitchell et al.,1995a)。Grossart 等(2001)在美国加州海岸的 10 个月的监测表明,秋冬季运动细菌的比例为 5%~25%,而春夏季为 40%~70%。在中尺度围隔培养中,水华消退时,运动细菌所占比例急剧增加(Smriga et al.,2016)。

一、细菌向浮游植物热点聚集

浮游植物细胞本身就是一个可以长时间存在的热点,不断向周围水体释放有机质。若以海水中有机物质的浓度为本底,将比本底高 10%的位置定义为藻际微域的外缘（内缘为细胞壁），那么浮游植物的微域的半径大约为 1mm。由于理论上细菌有能力锁定直径≥2.5μm 的热点（Jackson，1987），这意味着浮游植物可能会吸引细菌积聚到其周围。

有部分研究认为，细菌不会向浮游植物热点周围聚集。例如，Vaqué 等（1989）将浮游植物和细菌共培养液离心，获得上清液中的自由生活的细菌，然后将离心沉降的浮游植物用无菌过滤海水再悬浮并过滤到孔径为 2μm 的滤膜上，用落射荧光显微镜计数浮游植物以及浮游植物附生细菌，通过研究浮游植物、自由生活细菌和浮游植物附生细菌的丰度关系，发现浮游植物附生细菌的丰度是由浮游植物丰度和自由细菌相遇概率决定的，没有明显的细菌向浮游植物聚集效应。Müller-Niklas 等（1996）在自然海水中多次取样，发现存在浮游植物的样品和不存在浮游植物的样品中细菌数量没有显著差异。

上述实验条件未考虑现实海洋中的扰动和采样造成的扰动，而这种扰动可使浮游植物细胞周围的高浓度 DOC 与周围水体混合，导致细菌不能积聚，或在采样中将原本积聚的细菌混合到水体中。因此在自然海水中直接取样观察细菌积聚的难度很大，需要其他的方法对此进行验证（Fenchel，2002）。例如，使用延时摄影的方法，在自然海水样品中浮游植物、死亡的桡足类和桡足类粪球周围都可观察到细菌积聚，积聚过程和消失过程都很迅速，短则几分钟，长则一小时（Smriga et al.，2016）。

细菌可通过两种运动策略提高对营养的摄取效率：第一是运动到高营养浓度的热点并停留在此处；第二是高速游动以增加营养物质的扩散通量。理论上，跟踪浮游植物热点的细菌可以同时使用这两种方式，故而存留在热点周围但不固着在热点上是细菌的最佳选择，此时在热点周围将出现细菌丰度的最大条带或环带（Barbara and Mitchell，2003b）。然而，Smriga 等（2016）的观察结果是细菌聚集在硅藻的周围但未形成环带，这意味着细菌仅采取了第一种方式，即向浮游植物热点聚集以提高对营养的摄取。

在现实当中，海水里的颗粒始终处于运动状态，要么被水流带动，要么在静止的水体中下沉。有实验观察到培养体系中细菌会追逐移动的硅藻（Barbara and Mitchell，2003a）。但该实验中观察到的细菌转向也可能是一种被动的水动力效应而非主动追逐（Locsei and Pedley，2009），因此对于海洋当中细菌是否能跟上运动中的浮游生物还没有明确的观察结论。

二、细菌对稍纵即逝的热点的反应

由于没有持续的营养物质补充，稍纵即逝的热点斑块很快就会被物理作用稀释，细菌若想利用这类热点，它们到达这些斑块的时间需要小于斑块稀释殆尽的时间。

来自细胞裂解或原生动物排粪的点状源营养物质释放后，逐渐扩散到周围水体中，点状源周边的细菌先是变多，从背景丰度 10^6cell/ml 增加到 10^7cell/ml，在 100~300s 达到丰度高峰，积聚带的直径大于 1mm。随着营养源浓度逐渐降低，细菌丰度降低，10min 后积聚消失（Blackburn et al.，1998）。

细菌也会被沉降颗粒（如海雪和浮游动物粪便）形成的尾羽中的营养物质吸引（图 4-13）。当营养颗粒以 66μm/s 的速度沉降时，其周边有直径 0.6μm、丰度为 10^5cell/ml 的细菌积聚带，据估计颗粒的前 20mm 羽流内 37%的 DOM 将被细菌吸收同化，若颗粒移动速度提高至 220μm/s 或 660μm/s，这一比例分别下降到 7%和 2%（Stocker et al.，2008）。

图 4-13　尾羽中营养物质对细菌的吸引作用[改绘自 Stocker 等（2008）]
营养物质颗粒以 110μm/s 的速度自上至下运动，深色背景表示较高的营养物浓度，每个黑点代表一个细菌 *Pseudoalteromonas haloplanktis*

参 考 文 献

Adler J. 1973. A method for measuring chemotaxis and use of the method to determine optimum conditions for chemotaxis by *Escherichia coli*. Journal of General Microbiology, 74(1): 77-91.

Ayo B, Txakartegi A, Baña Z, et al. 2010. Chemosensory response of marine flagellate towards L- and D- dissolved free amino acids generated during heavy grazing on bacteria. International Microbiology, 13(3): 151-158.

Barbara G M, Mitchell J G. 2003a. Bacterial tracking of motile algae. FEMS Microbiology Ecology, 44(1): 79-87.

Barbara G M, Mitchell J G. 2003b. Marine bacterial organisation around point-like sources of amino acids. FEMS Microbiology Ecology, 43(1): 99-109.

Beers J R, Stewart G L. 1967. Micro-zooplankton in the euphotic zone at five locations across the California current. Journal of the Fisheries Research Board of Canada, 24(10): 2053-2068.

Bell W, Mitchell R. 1972. Chemotactic and growth responses of marine bacteria to algal extracellular products. The Biological Bulletin, 143(2): 265-277.

Berg H C, Purcell E M. 1977. Physics of chemoreception. Biophysical Journal, 20(2): 193-219.

Blackburn N, Fenchel T, Mitchell J G. 1998. Microscale nutrient patches in planktonic habitats shown by chemotactic bacteria. Science, 282(5397): 2254-2256.

Brahamsha B. 1996. An abundant cell-surface polypeptide is required for swimming by the nonflagellated marine cyanobacterium *Synechococcus*. Proceedings of the National Academy of Sciences of the United States of America, 93(13): 6504-6509.

Buskey E J, Stoecker D K. 1988. Locomotory patterns of the planktonic ciliate *Favella* sp.: adaptations for remaining within food patches. Bulletin of Marine Science, 43(3): 783-786.

Buskey E J, Stoecker D K. 1989. Behavioral responses of the marine tintinnid *Favella* sp. to phytoplankton: influence of chemical, mechanical and photic stimuli. Journal of Experimental Marine Biology and Ecology, 132(1): 1-16.

Christensen-Dalsgaard K K. 2006. Implications of microbial motility on water column ecosystems//Herrel A, Speck T, Rowe N P. Ecology and Biomechanics: A Mechanical Approach to the Ecology of Animals and Plants. Boca Raton: CRC Press: 271-299.

Cosson J, Prokopchuk G. 2014. Wave propagation in flagella//Rocha L G M. Wave Propagation. Cheyenne: Academy Publish: 541-583.

Ehlers K, Oster G. 2012. On the mysterious propulsion of *Synechococcus*. PLoS One, 7(5): e36081.

Esteban G F, Fenchel T M. 2020. Orientation in the environment//Esteban G F, Fenchel T M. Ecology of Protozoa: The Biology of Free-living Phagotrophic Protists. Cham: Springer International Publishing: 23-31.

Fenchel T. 2002. Microbial behavior in a heterogeneous world. Science, 296(5570): 1068-1071.

Fenchel T, Blackburn N. 1999. Motile chemosensory behaviour of phagotrophic protists: mechanisms for and efficiency in congregating at food patches. Protist, 150(3): 325-336.

Fenchel T, Finlay B J. 1986. Photobehavior of the ciliated protozoon *Loxodes*: taxic, transient, and kinetic responses in the presence and absence of oxygen. The Journal of Protozoology, 33(2): 139-145.

Fenchel T, Jonsson P R. 1988. The functional biology of *Strombidium sulcatum*, a marine oligotrich ciliate(Ciliophora, Oligotrichina). Marine Ecology Progress Series, 48: 1-15.

Fenchel T M. 1987. Ecology of Protozoa. The Biology of Free-Living Phagotrophic Protists. Brock Springer Series in Contemporary BioScience. Berlin: Springer-Verlag Berlin Heidelberg: 197.

Ghanbari A. 2020. Bioinspired reorientation strategies for application in micro/nanorobotic control. Journal of Micro-Bio Robotics, 16(2): 173-197.

Goldman J C. 1984. Conceptual role for microaggregates in pelagic waters. Bulletin of Marine Science, 35(3): 462-476.

Grossart H P, Riemann L, Azam F. 2001. Bacterial motility in the sea and its ecological implications. Aquatic Microbial Ecology, 25(3): 247-258.

Jackson G A. 1987. Simulating chemosensory responses of marine microorganisms. Limnology and Oceanography, 32(6): 1253-1266.

Johansen J E, Pinhassi J, Blackburn N, et al. 2002. Variability in motility characteristics among marine bacteria. Aquatic Microbial Ecology, 28(3): 229-237.

Jonsson P R. 1989. Vertical distribution of planktonic ciliates - an experimental analysis of swimming behavior. Marine Ecology Progress Series, 52: 39-53.

Kiørboe T. 1993. Turbulence, phytoplankton cell size, and the structure of pelagic food webs. Advances in Marine Biology, 29: 1-72.

Kiørboe T, Grossart H P, Ploug H, et al. 2002. Mechanisms and rates of bacterial colonization of sinking aggregates. Applied and Environmental Microbiology, 68(8): 3996-4006.

Kiørboe T, Grossart H P, Ploug H, et al. 2004. Particle-associated flagellates: swimming patterns, colonization rates, and grazing on attached bacteria. Aquatic Microbial Ecology, 35(2): 141-152.

Kiørboe T, Jackson G A. 2001. Marine snow, organic solute plumes, and optimal chemosensory behavior of bacteria. Limnology and Oceanography, 46(6): 1309-1318.

Locsei J T, Pedley T J. 2009. Bacterial tracking of motile algae assisted by algal cell's vorticity field. Microbial Ecology, 58(1): 63-74.

McCarren J, Brahamsha B. 2007. SwmB, a 1.12-megadalton protein that is required for nonflagellar swimming motility in *Synechococcus*. Journal of Bacteriology, 189(3): 1158-1162.

Menden-Deuer S, Grünbaum D. 2006. Individual foraging behaviors and population distributions of a planktonic predator aggregating to phytoplankton thin layers. Limnology and Oceanography, 51(1): 109-116.

Mitchell J G, Okubo A, Fuhrman J A. 1985. Microzones surrounding phytoplankton form the basis for a stratified marine microbial ecosystem. Nature, 316(6023): 58-59.

Mitchell J G, Pearson L, Bonazinga A, et al. 1995a. Long lag times and high velocities in the motility of natural assemblages of marine bacteria. Applied and Environmental Microbiology, 61(3): 877-882.

Mitchell J G, Pearson L, Dillon S, et al. 1995b. Natural assemblages of marine bacteria exhibiting high-speed motility and large accelerations. Applied and Environmental Microbiology, 61(12): 4436-4440.

Mohapatra B R, Fukami K. 2007. Chemical detection of prey bacteria by the marine heterotrophic nanoflagellate *Jakoba libera*. Basic and Applied Ecology, 8(5): 475-481.

Müller-Niklas G, Agis M, Herndl G J. 1996. Microscale distribution of bacterioplankton in relation to phytoplankton: results from 100-nl samples. Limnology and Oceanography, 41(7): 1577-1582.

Persson A, Smith B C, Wikfors G H, et al. 2013. Differences in swimming pattern between life cycle stages of the toxic dinoflagellate *Alexandrium fundyense*. Harmful Algae, 21-22: 36-43.

Purcell E M. 1977. Life at low Reynolds number. American Journal of Physics, 45(1): 3-11.

Samuel A D T, Petersen J D, Reese T S. 2001. Envelope structure of *Synechococcus* sp. WH8113, a nonflagellated swimming cyanobacterium. BMC Microbiology, 1(1): 4.

Seymour J R, Ahmed T, Durham W M, et al. 2010. Chemotactic response of marine bacteria to the extracellular products of *Synechococcus* and *Prochlorococcus*. Aquatic Microbial Ecology, 59(2): 161-168.

Seymour J R, Ahmed T, Marcos, et al. 2008. A microfluidic chemotaxis assay to study microbial behavior in diffusing nutrient patches. Limnology and Oceanography: Methods, 6(9): 477-488.

Seymour J R, Amin S A, Raina J B, et al. 2017. Zooming in on the phycosphere: the ecological interface for phytoplankton-bacteria relationships. Nature Microbiology, 2: 17065.

Sibbald M J, Albright L J, Sibbald P R. 1987. Chemosensory responses of a heterotrophic microflagellate to bacteria and several nitrogen compounds. Marine Ecology Progress Series, 36: 201-204.

Smriga S, Fernandez V, Mitchell J G, et al. 2016. Chemotaxis toward phytoplankton drives organic matter partitioning among marine bacteria. Proceedings of the National Academy of Sciences of the United States of America, 113(6): 1576-1581.

Son K, Brumley D R, Stocker R. 2015. Live from under the lens: exploring microbial motility with dynamic imaging and microfluidics. Nature Reviews Microbiology, 13(12): 761-775.

Son K, Menolascina F, Stocker R. 2016. Speed-dependent chemotactic precision in marine bacteria. Proceedings of the National Academy of Sciences of the United States of America, 113(31): 8624-8629.

Stocker R. 2012. Marine microbes see a sea of gradients. Science, 338(6107): 628-633.

Stocker R, Seymour J R, Samadani A, et al. 2008. Rapid chemotactic response enables marine bacteria to exploit ephemeral microscale nutrient patches. Proceedings of the National Academy of Sciences of the United States of America, 105(11): 4209-4214.

Strom S L, Buskey E J. 1993. Feeding, growth, and behavior of the thecate heterotrophic dinoflagellate *Oblea rotunda*. Limnology and Oceanography, 38(5): 965-977.

Vandromme P, Schmitt F G, Souissi S, et al. 2010. Symbolic analysis of plankton swimming trajectories: case study of *Strobilidium* sp.(Protista) helical walking under various food conditions. Zoological Studies, 49(3): 289-303.

Vaqué D, Duarte C M, Marrase C. 1989. Phytoplankton colonization by bacteria: encounter probability as a limiting factor. Marine Ecology Progress Series, 54(1): 137-140.

Verity P G. 1988. Chemosensory behavior in marine planktonic ciliates. Bulletin of Marine Science, 43(3): 772-782.

Verity P G. 1991. Aggregation patterns of ciliates from natural assemblages in response to different prey. Marine Microbial Food Webs, 5(1): 115-128.

Visser A W, Kiørboe T. 2006. Plankton motility patterns and encounter rates. Oecologia, 148(3): 538-546.

Vitiello P. 1964. Contribution a l'etude des tintinnides de la baie d'Alger. Pelagos, 2(2): 5-41.

Vogel S. 1994. Life in Moving Fluid: The Physical Biology of Flow. Second Edition, Revised and Expanded. Princeton: Princeton University Press: 467.

Waterbury J B, Willey J M, Franks D G, et al. 1985. A cyanobacterium capable of swimming motility. Science, 230(4721): 74-76.

Willey J M, Waterbury J B. 1989. Chemotaxis toward nitrogenous compounds by swimming strains of marine *Synechococcus* spp. Applied and Environmental Microbiology, 55(8): 1888-1894.

Wolfe G V. 2000. The chemical defense ecology of marine unicellular plankton: constraints, mechanisms, and impacts. The Biological Bulletin, 198(2): 225-244.

Yam E M, Tang K W. 2007. Effects of starvation on aggregate colonization and motility of marine bacteria. Aquatic Microbial Ecology, 48(3): 207-215.

Zaika V Y, Ostrovskaya N A. 1972. Pattern of diurnal variations in microzooplankton abundance in the surface layer of the Mediterranean Sea. Oceanology, 12: 725-729.

第二部分

第五章　微食物网生物的异养营养

第一节　异养细菌利用溶解有机碳

20 世纪 30 年代，人们就已经意识到海洋中的大部分有机物质既不是在生物的细胞之中，也不是在它们形成的碎屑之中，而是处于溶解状态，即溶解有机碳（dissolved organic carbon，DOC）（Krogh and Keys，1934；Keys et al.，1935）。全球海洋的 DOC 含量约为 662Pg C（1Pg = 10^{15}g），超过海水中所有生命体生物量之和的 200 倍（Hansell et al.，2009）。海洋表层 DOC 浓度范围为 40～80μmol/kg，在热带和亚热带海区最高，在南大洋最低，这可能是南大洋缺乏 DOC 的深层水上涌造成的（Roshan and DeVries，2017）。

在 20 世纪初，人们曾经认为 DOC 能被所有的生命体利用。到了 20 世纪 30 年代，有学者发现大部分生物都不能以 DOC 为生，但是异养细菌可以依靠海洋中的 DOC 存活（Keys et al.，1935）。异养细菌是海水中 DOC 的主要利用者。根据对 DOC 浓度的适应程度，细菌可以分为富营养型（copiotroph）和寡营养型（oligotroph）（Cavicchioli et al.，2003）。

并不是所有的 DOC 都能被海洋异养细菌利用，其中可以被细菌利用的部分称为活性 DOC（liable DOC，LDOC）。尽管 LDOC 的通量很大，但其周转速度很快，因此仅占海洋总 DOC 含量的不到 1%（Hansell et al.，2009）。不能被细菌利用的 DOC 被称为惰性 DOC（recalcitrant DOC，RDOC）。对于 RDOC 的来源和去向，"惰性假说"认为微食物网生物是 RDOC 的主要贡献者，其通过 3 个基本途径把容易被利用的 LDOC 或可被缓慢降解的半活性 DOC（semi-liable DOC）转化为难以被降解或利用的 RDOC：①微食物网生物通过自身机制直接代谢分泌 RDOC；②病毒裂解作用释放微食物网生物细胞内的大分子物质；③细菌降解有机底物后留下无法被利用的残留化合物。这 3 个途径共同导致海洋中 RDOC 的浓度增加，而 RDOC 可以长期储碳（焦念志，2012；张传伦，2018）。微型生物碳泵（microbial carbon pump）（Jiao et al.，2010）的概念和理论框架便源于此。

Arrieta 等（2015）从另一个角度解释了 RDOC 不能被异养细菌利用的原因，即稀释假说（dilution hypothesis）：RDOC 由极为复杂的多种组分组成，与此相对应，海洋中存在具有吸收利用每个组分的相应的细菌类群。尽管 RDOC 的总浓度很高，看上去应该可以被细菌利用，但是其中各个组分的浓度都低于其对应的细菌可以利用的浓度，细菌针对其中的某个组分产生的酶的收益都低于付出，即浓度过于稀释，造成各个组分都没有被利用。稀释假说不支持微型生物碳泵的概念。

目前，稀释假说和惰性假说的争论还没有定论（Jiao et al.，2015）。但这两个假说并不互相排斥，可能在自然界同时存在，即 RDOC 中有一些组分是不能被细菌降解的，但同时也有一些组分是因为浓度过低逃过了细菌的利用。

除异养细菌外，生命体能否利用 DOC 的观点争论一直持续到 20 世纪 80 年代。Sherr（1988）首次证明异养鞭毛虫可以利用高分子量的多糖，但是不能作为主要营养方式。有些类群的纤毛虫，如四膜虫（*Tetrahymena*）能在实验室培养条件下利用 DOC 繁殖种群，但在自然界中绝大多数纤毛虫无法依靠利用 DOC 生存。实验室内添加荧光标记右旋糖酐作为辅助饵料的底栖纤毛虫尾丝虫（*Uronema marinum*）生长率并没有提高，其利用右旋糖酐的方式不是细胞膜的渗透营养而是形成食物泡，即通过摄食的方式（First and Hollibaugh，2009）。

近年来有研究发现了一类单细胞真核异养生物 *Picomonas judraskeda* 能摄食<150nm 的有机颗粒，被称为皮胆虫（picozoa），对这一类生物的研究刚刚开始（Seenivasan et al.，2013；Moreira and López-García，2014）。

第二节　病毒对微食物网生物的下行控制

水体中的病毒要感染宿主，首先要与宿主相接触。病毒的颗粒较小，布朗运动对其影响较大，因此病毒和宿主的接触符合扩散规律。病毒与宿主颗粒的接触概率（contact rate，CR）可以根据式（5-1）估算。

$$CR = (Sh \times 2\pi \times d \times D_v) \times V \times H \tag{5-1}$$

式中，Sh 表示舍伍德数（Sherwood number），与宿主（细菌）的运动速度有关；d 表示宿主直径；D_v 表示病毒的扩散系数；V 表示病毒丰度；H 表示宿主丰度（Murray and Jackson，1992）。

由式（5-1）可见，个体大、运动快、丰度高的宿主与病毒的接触机会较大。宿主群落中丰度高的宿主（即所谓"赢家"）与病毒的接触概率大，理论上讲病毒杀死这类宿主的概率也较高，这就是"杀死赢家"（kill the winner，KtW）假说（Thingstad，2000）。根据 KtW 假说，海水中病毒和宿主（细菌）的总丰度波动都不大，但其中不同病毒和细菌类群的丰度有较大的变化。每当一个细菌类群的丰度升高，杀死该类群的病毒丰度随之变大，将该类细菌的丰度降低，该类病毒的丰度也随之降低（图 5-1）。浮游病毒通过 KtW 机制调节宿主种群的丰度，进而调控宿主群落多样性。在没有病毒调控的情况下，群落中的优势种类（winner）迅速生长，占用有限的资源，使劣势种类无法生长，导致整个群落呈现单一物种状态，不利于群落的生存和生态系统的稳定。在 KtW 机制下，优势种类被病毒裂解致死（kill），从而允许劣势种类获得足够资源进行生长存活，以保持群落结构的多样性和生态系统的稳定性。

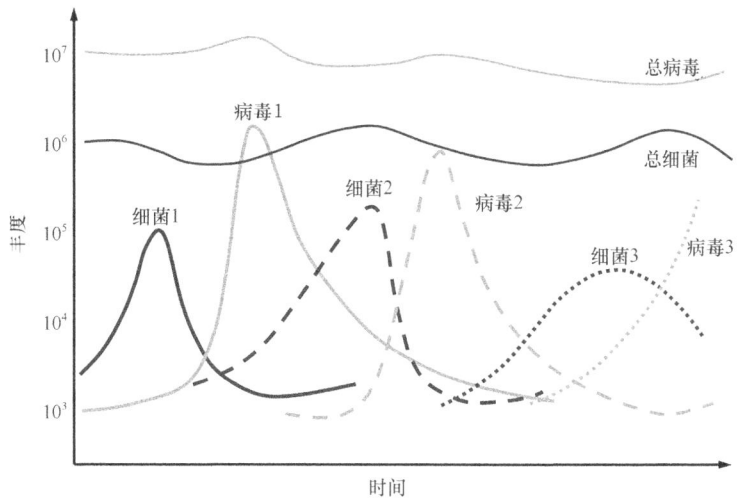

图 5-1 "杀死赢家"假说示意图[改绘自 Breitbart 等（2008）]
丰度单位：细菌，cell/ml；病毒，particle/ml

接触概率不是病毒杀死宿主的唯一影响因子，病毒与宿主接触后的感染率也起显著作用。在墨西哥湾的近岸环境中，每天 80%的聚球藻与有感染性的噬藻体（cyanophage，即蓝细菌病毒）接触，但每天只有 1%的接触会导致感染，但在远岸环境中聚球藻与噬藻体的接触基本都导致了感染。这一差异的可能原因是在远岸环境中病毒与宿主丰度都较低，高侵染率有利于保持病毒种群丰度（Suttle and Chan，1994）。

病毒侵染宿主后可通过两个途径来维持种群丰度，第一是在宿主体内大量复制、装配后，裂解宿主并释放子代病毒。第二个途径中，病毒可能不立即复制产生子代病毒，而是将其核酸整合到宿主基因组中潜伏起来，称为原病毒，并随着宿主的分裂从母代宿主传递到子代宿主，称为病毒的溶原性（lysogeny）。原病毒可自发或受环境因素（如温度、盐度、营养条件等）诱导，脱离宿主基因组进入裂解周期，产生并释放子代病毒。

近年来有学者提出了另一个海洋病毒-宿主动力学关系模型，称为"搭载赢家"（piggyback the winner，PtW）假说（Knowles et al.，2016）。PtW 假说考虑了溶原性病毒和裂解性病毒对宿主的不同影响，认为溶原性病毒与宿主间的关系并非敌对，而更类似伙伴之间的共生关系。PtW 假说提出在高宿主生长率和密度条件下，溶原性病毒比裂解性病毒丰度更高，宿主可能会主动招募溶原性病毒，通过重复感染排除（superinfection exclusion）过程保护宿主免受其他病毒侵害。大量宿主被"搭载"（piggyback）而非被杀死，"赢家"的高密度得以维持。

传统观点（即 KtW 假说）认为在寡营养水体中，如深海和大洋表层，因为缺乏营养物质使宿主丰度较低，不足以支撑病毒快速侵染裂解的生活方式，病毒的溶原状态占主导；随着营养水平提高，宿主丰度增加，病毒的裂解性感染增加，导致浮游生态系统中的病毒丰度升高（图 5-2a）。但近年来有研究显示，在部分海洋环境，如寡营养的表层海水（Lara et al.，2017）、珊瑚礁（Knowles et al.，2016）、深渊（Jian et al.，2021）中，在宿主丰度较高的情况下，病毒的溶原性生产占主导地位，这些现象支持了 PtW 假说（图 5-2b）。KtW 假说和 PtW 假说看似相互矛盾，但也可在培养系统内切换和共存（Chen et al.，2019），表明 KtW 机制和 PtW 机制共同发生，而不是相互排斥，这两个病毒控制机制共同塑造了宿主群落的演替和动态。

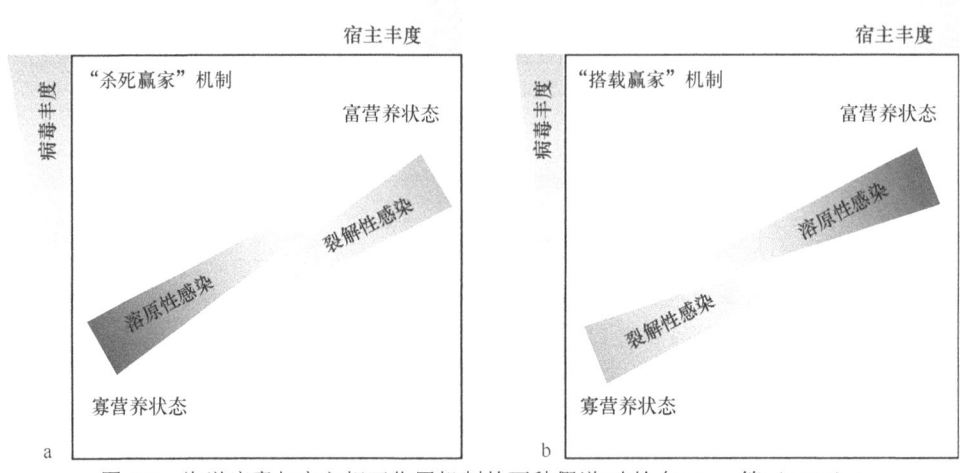

图 5-2 海洋病毒与宿主相互作用机制的两种假说[改绘自 Chen 等（2021）]

在病毒侵染细菌的后期，成熟的病毒会在细菌体内排列，这类含有大量病毒颗粒的细菌可通过透射电镜观察分辨，称为可见的被感染细菌（visibly infected bacteria，VIB）。通过测定可见感染时长在细菌感染周期中所占的比例，可以估计样品中被感染细菌的数目和比例。通过可见的被感染细菌占总细菌丰度比例，可以估计细菌被病毒裂解导致的死亡率。

已有的研究表明，噬菌体侵染裂解是导致海洋细菌死亡的主要因素之一，海洋环境中可以有高达 84% 的细菌被噬菌体侵染（Weinbauer et al.，2003），海洋中每天平均有 20%～30% 的细菌被浮游病毒裂解死亡（Bratbak and Heldal，2000）。由噬藻体介导的蓝细菌死亡率比细菌略低，为 5%～15%（Suttle，1994）。浮游病毒对真核藻类的种群调控研究多集中在水华过程，如在 *Emiliania huxleyi* 水华消退期，最高可有 50% 的藻细胞被病毒侵染，能导致 12%～100% 的藻细胞死亡（Jacquet et al.，2002）。

病毒裂解会使宿主细胞内的物质释放到环境中，这些溶解有机物和颗粒有机物可供细菌使用重新进入海洋生物地球化学循环，从而形成微食物网的"病毒回路"（viral shunt），海洋中 25%的有机碳会流经病毒回路（Breitbart et al.，2008）。在南极近岸海域，夏季会出现细菌和异养鞭毛虫丰度降低、而病毒丰度升高的现象，同时异养鞭毛虫对细菌的摄食减少，而病毒的生产增加，这时病毒回路的作用增加（Evans et al.，2021）。

病毒也可以侵染并杀死异养鞭毛虫（Garza and Suttle，1995），但是目前这类研究很少，仅有的研究表明,病毒对异养鞭毛虫的致死影响较小(Tsai et al., 2016)。迄今尚无病毒导致浮游纤毛虫死亡的报道。

第三节　微型浮游动物的摄食行为

一、鞭毛虫的摄食行为

鞭毛虫的摄食行为主要分为 4 个阶段，分别为接触期、处理期、吞入期和不应期（图 5-3）。

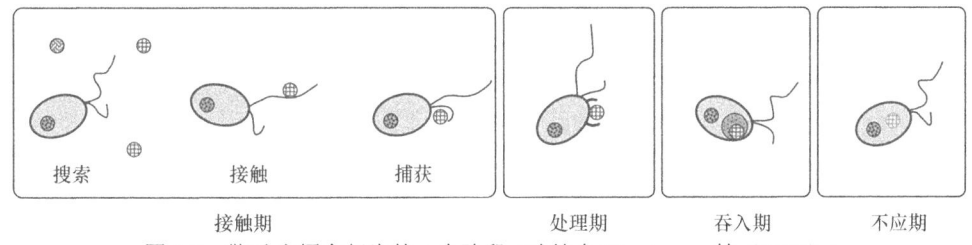

图 5-3　鞭毛虫摄食行为的 4 个阶段 [改绘自 Montagnes 等（2008）]

趋化作用能引导鞭毛虫寻找到饵料浓度高的斑块，但是到达斑块之后，趋化作用的精准度不高，不足以帮助鞭毛虫锁定单个猎物（Jackson，1987）。这意味着鞭毛虫并无法精准地接触目标饵料，与饵料接触是鞭毛虫不能确定的概率事件。浮游鞭毛虫接触饵料的方式主要分为 3 种（图 5-4）：①过滤摄食（filter feeding），通过挥舞鞭毛拨动水流，使水中的饵料颗粒过滤到套笼等过滤器官上；②抓取摄食（interception feeding），使用鞭毛接触并抓取颗粒；③捕食（raptorial feeding），鞭毛仅用来游泳，用胞咽（pharynx）或伪足（pseudopod）等专用摄食器官接触并捕获食物。其中,抓取摄食是浮游鞭毛虫摄食的主要方式(Zhukov,1993；Boenigk and Arndt，2002）。

此外，还有很多鞭毛虫没有明显的摄食器官，通过细胞表面摄食饵料。但并不是鞭毛虫细胞表面的任何一个地方都可以摄食，而是存在特殊的区域，称为敏感区。饵料与鞭毛虫接触后，鞭毛虫将饵料运移到敏感区的时间称为处理期（processing phase）。

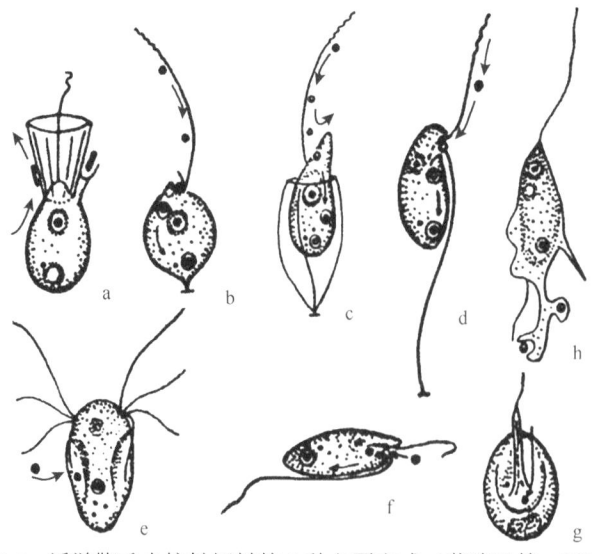

图 5-4　浮游鞭毛虫接触饵料的 3 种主要方式（黄凌风等，2013）
a. 过滤摄食，领鞭毛虫（*Monosiga* sp.）；b~e：抓取摄食，b. 金滴虫（*Spumella* sp.），c. 杯鞭虫（*Bicosoeca* sp.），d. 舞行波豆虫（*Bodo saltans*），e. 双滴虫类；f~h：捕食，f. 波豆虫类，g. 有孔虫类，h. 尾滴虫（*Cercomonas* sp.）

饵料颗粒到敏感区后，鞭毛虫形成食物泡的过程称为吞入期（ingestion phase）。在形成食物泡并吞入饵料后，鞭毛虫有一段时间对接触到的新饵料没有反应，称为不应期（refractory phase）。

不同种类鞭毛虫各个摄食阶段的时间长度可能有较大的差异，吻滴虫（*Rhynchomonas*）、金滴虫（*Spumella*）和棕鞭藻（*Ochromonas*）可以在 5s 内完成整个摄食过程，而部分鞭毛虫，如领鞭毛虫（*Monosiga*）和餐厅虫（*Cafeteria*）的摄食过程可超过 1min（图 5-5）。可以消化的颗粒形成食物泡后到排出需要大约 30min 的时间，不能消化的颗粒在 2~3min 就会被排出体外（Boenigk et al., 2001；Boenigk and Arndt, 2002）。

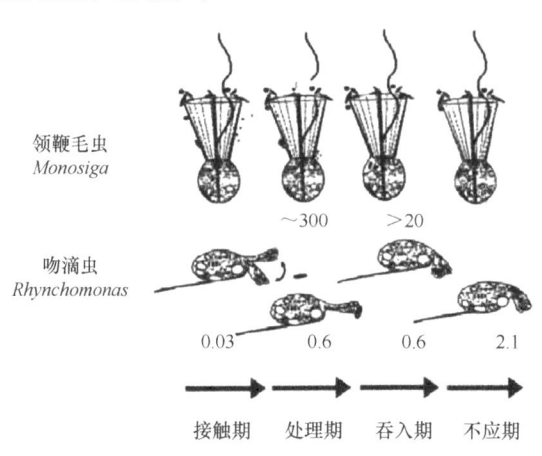

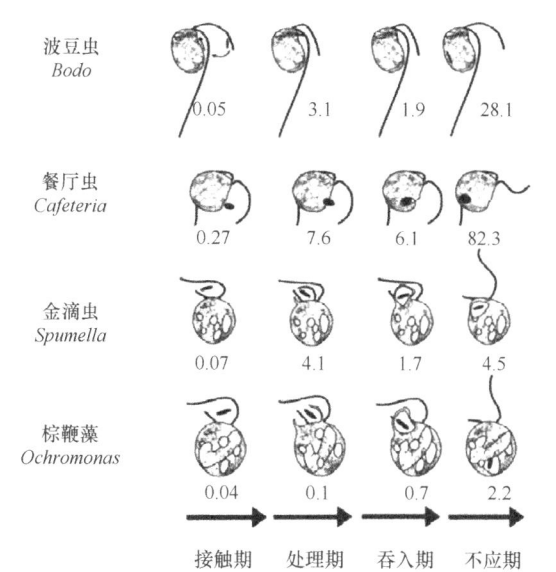

图 5-5　不同鞭毛虫各个摄食阶段的时间长度（单位：s）[改绘自 Boenigk 和 Arndt（2002）]

二、鞭毛虫对饵料的选择

（一）鞭毛虫摄食饵料的最低浓度

鞭毛虫每小时清滤的体积为 5～15nl（1nl = $10^6 \mu m^3$ = 10^{-9}L），相当于其身体体积的 $10^5 \sim 10^6$ 倍（Fenchel，1987；Suzuki-Tellier et al.，2022），是所有海洋浮游动物中最高的（Hansen et al.，1997）。以海洋浮游细菌的平均丰度 10^6cell/ml（即 1cell/nl）来估算，鞭毛虫每小时可摄食 5～15 个细菌。若细菌的丰度为 $2 \times 10^5 \sim 3 \times 10^5$cell/ml（即 0.2～0.3cell/nl），鞭毛虫在一个小时内可摄食的细菌就很少。因此，通常认为细菌丰度 10^6cell/ml 是维持鞭毛虫生长的最低丰度，低于这个数值鞭毛虫可能会处于饥饿状态（Jürgens and DeMott，1995；Boenigk and Arndt，2002）。

（二）鞭毛虫对饵料生理特征的选择性

鞭毛虫对不同饵料的选择性研究主要通过培养实验进行。部分研究使用具备某种特性的活体细菌，如处于饥饿和非饥饿状态的菌株（Anderson et al.，2011），作为饵料与鞭毛虫共培养，测试鞭毛虫的摄食和生长情况。更多的研究将不同特性（大小、死活、是否运动）的细菌制成荧光标记细菌（fluorescence labeled bacteria，FLB），通过观察鞭毛虫体内 FLB 的有无和数量评估鞭毛虫对饵料的选择性（Sherr et al.，1987）。

培养实验发现鞭毛虫偏向于摄食颗粒大、运动且代谢活跃的细菌（González et al.，1990；Landry et al.，1991；González et al.，1993b）。将同一种类但分别处于指数生长期和饥饿期的细菌热杀死制作成 FLB 与鞭毛虫共培养时，鞭毛虫对指数生长期的细菌的清滤率高于饥饿期细菌，消化饥饿期细菌的时间长于指数生长期细菌，但是摄食饥饿期细菌的鞭毛虫的生长效率较高（González et al., 1993b）。另一个研究发现，使用饥饿期细菌为饵料时，有的鞭毛虫类群生长没有受到影响，但部分鞭毛虫类群则不生长甚至丰度降低（Anderson et al., 2011）。

鞭毛虫可能对不同营养价值的细菌具有选择性。低 C：N 的细菌具有较高的营养价值，鞭毛虫更倾向于摄食该类细菌（Shannon et al., 2007; Ng and Liu, 2016）。DNA 含量高的细菌也容易被摄食（Vaqué et al., 2001）。

培养实验表明，鞭毛虫对细菌表面的疏水性和电荷并没有明显的选择性，但是对疏水性人工颗粒的捕获效率和摄食效率都显著降低，表明对饵料其他特征的选择性要大于对细菌表面的疏水性和电荷的选择性（Matz and Jürgens, 2001; Matz et al., 2002）。

对特定细菌类群的选择性摄食研究不多，目前已经知道一些鞭毛虫会选择性摄食古菌（Ballen-Segura et al., 2017）。

上述研究评估的都是鞭毛虫对细菌某一特征的选择性摄食，但实际环境中细菌的不同特征往往相互关联。细菌生活的环境变化多样，当细菌处于饥饿状态时，其体积缩小数倍，代谢降低，运动变慢甚至停止，体内 DNA 含量和元素比例会变化，单位体积内的蛋白质和碳含量升高，细胞表面也可能发生变化。所以，对这些研究的结论要持保守态度，鞭毛虫选择不同饵料的原因很复杂，在大多数情况下，很难确定是针对哪一特征的选择。

（三）鞭毛虫摄食行为对饵料的选择性

1. 鞭毛虫与饵料接触概率的影响

选择性摄食是通过对比饵料的丰度关系和被摄食的数量关系来进行评估的。丰度的大小不是决定接触概率的唯一因素，当饵料里大细菌和小细菌的丰度接近时，鞭毛虫与大细菌的接触概率大，意味着鞭毛虫可能会选择性地摄食更多大细菌，而水体中会保留较多小细菌。运动的细菌与鞭毛虫接触概率大，所以运动细菌被选择性摄食的概率也更高。

2. 饵料适口性的影响

在与饵料接触后，鞭毛虫可能会选择不处理该饵料或处理后不摄食。这一过程可能涉及受体介导的饵料颗粒识别（Wootton et al., 2007），是选择性摄食过程中最活跃的步骤之一。Matz 等（2002）将金滴虫（*Spumella* sp.）固定在培养容器底部，用显微摄影技术研究金滴虫对所遇的每一个饵料的捕获、处理情况，发现金滴虫对遇到的不同菌株饵料具有不同的捕获概率，对捕获的不同饵料有不同的进食概率。这表明鞭毛虫在捕获和处理两个环节都有选择性，但细菌表面和内部有哪些化学成分导致适口性选择摄食还不明确。

对饵料适口性的选择与鞭毛虫的生理状态有关。用粒径相同的乳胶微球（latex beads）和 FLB 作为饵料投喂给舞行波豆虫（*Bodo saltans*）和金滴虫，当鞭毛虫处于饥饿状态时对 FLB 和乳胶微球均有较高的清除率。此时若在培养体系中添加 FLB 饵料，两种鞭毛虫对乳胶微球的清除率迅速降低，主要选择摄食 FLB。当培养 24h 体系中的 FLB 浓度降低后，两种鞭毛虫对乳胶微球的清除率又逐渐回升（Jürgens and DeMott，1995）。

3. 饵料的动量和拉力的影响

饵料的动量和拉力可能是摄食行为产生的刺激因子之一。当颗粒的动量和拉力处于一定的阈值范围内，才会导致鞭毛虫将颗粒包裹进自己的细胞。这意味着对于特定鞭毛虫种类，存在一个最适饵料粒径范围，很少摄食大于和小于这个粒径的饵料。细菌的形状对鞭毛虫的摄食也可能有影响，最初普遍认为丝状细菌（filamentous bacteria）不是鞭毛虫的合适饵料，但后来研究发现鞭毛虫可以摄食丝状细菌（Wu et al.，2004）。

4. 鞭毛虫摄食能力的影响

鞭毛虫细胞表面形成食物泡的区域大小决定了鞭毛虫可以吞入的颗粒的大小。当鞭毛虫处于饥饿状态时，细胞的体积变小，敏感区的面积相应缩小，这意味着鞭毛虫的适口粒径降低，此时鞭毛虫倾向于摄食较小的饵料（Pfandl et al.，2004）。

上述摄食行为对饵料的选择性可以分为两类：对细菌粒级（含形状）的选择和对适口性（化学信号）的选择。在粒级选择性方面，接触机会导致的对大颗粒的选择被称为被动选择，而其他的选择都是鞭毛虫具有主动性，称为主动选择。

三、纤毛虫的摄食行为

与鞭毛虫摄食相比，纤毛虫的摄食行为研究较少。浮游纤毛虫的摄食和运动是关联在一起的，大多数浮游纤毛虫种类通过围口小膜（membranelle）推动水流，使得虫体前行或偶尔后退。每片小膜都由多根纤毛融合形成，水流在通过小膜间的空隙时，其中的饵料颗粒可被小膜捕获并被送入口区（图 5-6），然后被摄食消化。

纤毛虫的摄食过程可以分为 3 个阶段（图 5-7）：首先是接触期，纤毛虫使用围口小膜搜索并与饵料相接触；第二阶段为捕获期，饵料被小膜捕获，通过纤毛摆动将饵料运送到口区；最后是处理期，通过胞咽将颗粒包裹成食物泡，吞入体内进行消化。在处理食物的过程中，纤毛虫虽然保持运动状态，但是对新触碰到的饵料不再进行捕获。

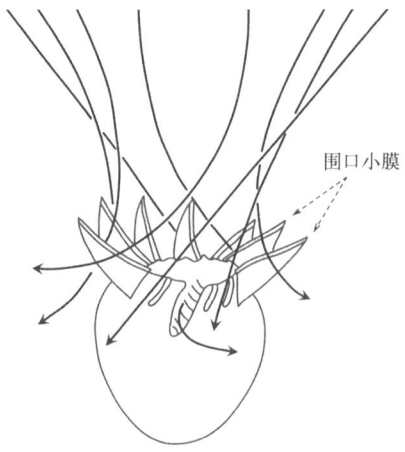

图 5-6 纤毛虫滤食产生的水流（图中实线箭头）示意[改绘自 Fenchel（1987）]

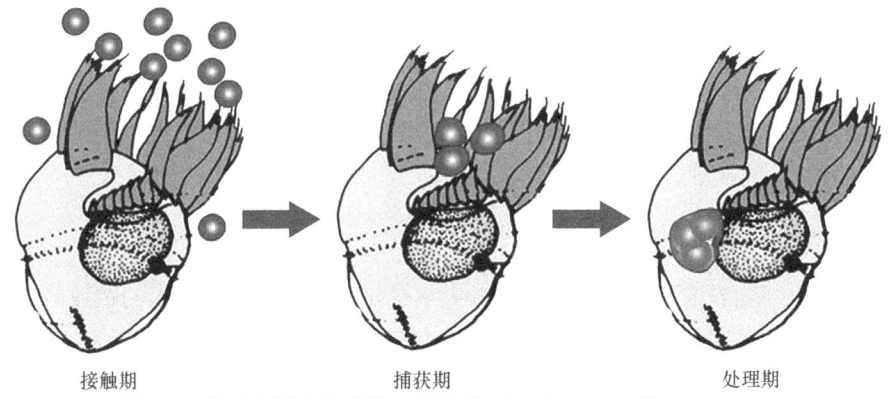

图 5-7 纤毛虫摄食过程的 3 个阶段[改绘自 Weisse 等（2016）]

多种因素可以在不同的摄食阶段影响纤毛虫对饵料的选择和摄食效率，如摄食和游动行为的流体动力学、围口小膜摆动的模式，以及饵料自身的特性。高速摄影研究表明，砂壳纤毛虫 *Codonellopsis orthoceras* 和 *Climacocylis scalaria* 的围口小膜摆动频率为 10~50Hz，摆动的频率越高，产生的水流速度越大，最高为 2mm/s。正对砂壳纤毛虫口区的区域流速最大，是水流中心（图 5-8）。在水流中心的饵料颗粒被捕获的概率约为 75%，而离开水流中心的 1.5 倍口区直径之外，捕获概率骤降为 5%（Wandel and Holzman，2022）。这表明砂壳纤毛虫与饵料的有效相遇率由小膜摆动频率，以及饵料相对于中心流线的位置决定。

总体而言，与纤毛虫相遇的颗粒被成功捕获并移入口区的概率约为 14%，在捕获的颗粒中 50%的颗粒在初次捕获后被主动放弃，所以总的摄食成功率为所有接触颗粒的 7%左右。砂壳纤毛虫在捕获和处理阶段对饵料都表现出选择性。在这两个阶段，砂壳纤毛虫都喜好较大的饵料颗粒，对饵料的游泳速度没有明显偏好，对于大小接近的自然饵料和人工合成颗粒也没有选择性。在捕获阶段砂壳纤毛虫

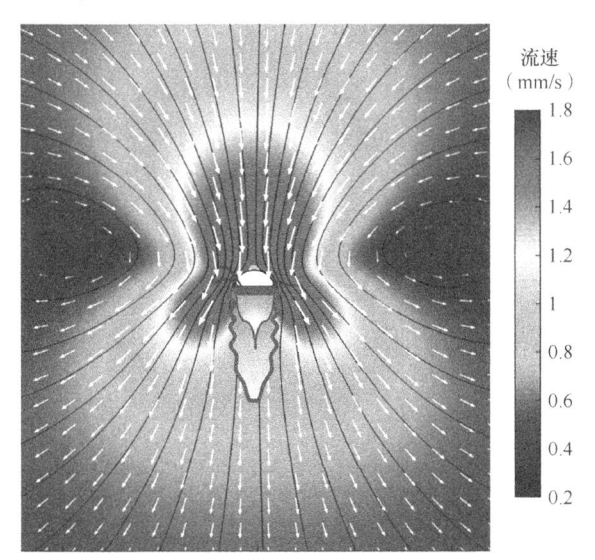

图 5-8　砂壳纤毛虫身体周围的流场速度和方向[改绘自 Wandel 和 Holzman（2022）]

对饵料的形状没有选择性；在处理阶段，砂壳纤毛虫则更偏好圆形颗粒，长宽比大于 1.5 的藻类的摄取率较低。在一段相对较长时间的正常游泳后，纤毛虫会产生反向水流，持续时间较短。当产生反向水流时，摄食偏好也发生改变，从偏好大颗粒改为偏好小颗粒（Wandel and Holzman，2022）。

尽管纤毛虫对饵料粒径的形状有一定选择，但有时长型饵料仍会被摄食，甚至充斥纤毛虫的整个身体。图 5-9 展示的是纤毛虫吞食一个链状硅藻，硅藻具壳故形态不可改变，纤毛虫原来的身体形态被硅藻撑得发生了改变。

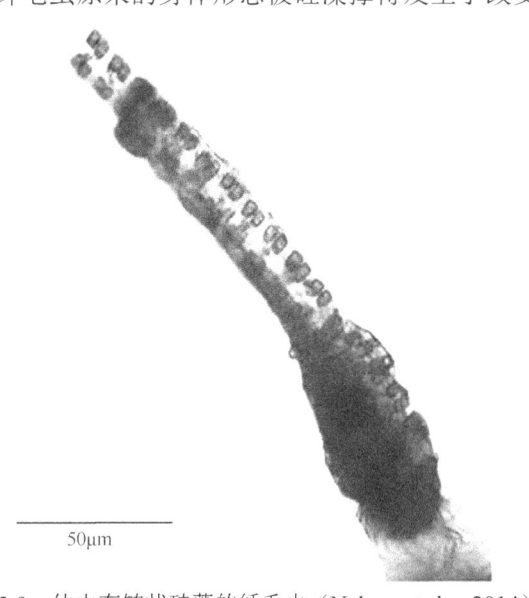

图 5-9　体内有链状硅藻的纤毛虫（Nelson et al.，2014）

鞭毛虫体内的饵料体积占比不大，而与鞭毛虫不同，纤毛虫是"贪得无厌"的摄食者，如图 5-9 所示，硅藻占据了纤毛虫身体的很大体积。类似地在东海的原甲藻赤潮水体中，可观察到纤毛虫体内有很多的原甲藻（图 5-10），几乎充满了虫体。

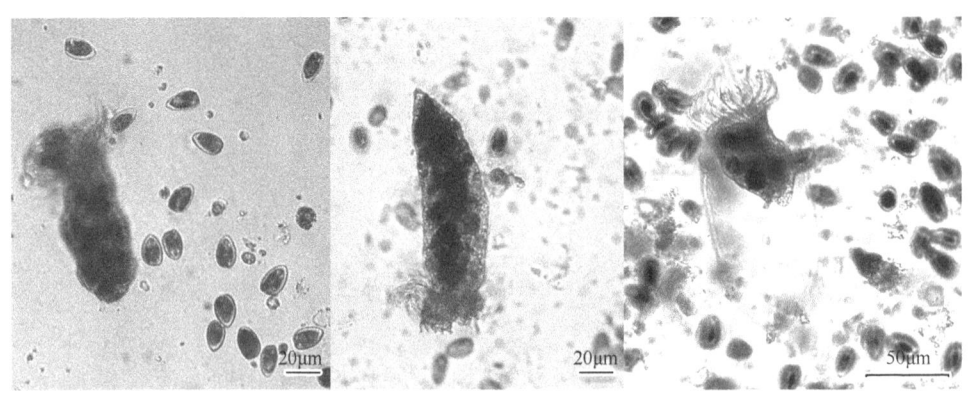

图 5-10　体内含有原甲藻的纤毛虫（张武昌和于莹 摄）

关于纤毛虫对饵料的选择性和摄食强度，将在本书第九章详细介绍。

第四节　微型浮游动物的摄食速率

一、摄食速率的估算方法

摄食速率[ingestion rate，I，prey/（grazer·h）]指的是单个摄食者（grazer）在单位时间内摄入的饵料（prey）数量（或质量，通常以碳生物量表示）。清滤速率[clearance rate，F，nl/（grazer·h）]则是单个摄食者在单位时间内摄入饵料所需滤除水的体积，可通过摄食速率除以培养期间饵料的平均丰度来计算。摄食压力（grazing pressure，P，%）指单位时间内摄食者的摄食量占饵料丰度（或生物量、生产力）的比例，通常以百分比形式体现。

研究鞭毛虫和纤毛虫的摄食速率，主要使用体内颗粒增加法或水体颗粒减少法。

（一）体内颗粒增加法

想确认饵料是否被鞭毛虫和纤毛虫摄食，最直观的方法是检查摄食者体内是否含有饵料颗粒。但在普通光学显微镜下观察时，通常难以将被摄入的饵料与摄食者的体内成分区别开，因此需要对饵料进行标记。对饵料进行标记的方法主要有两种：放射性标记法和荧光标记法。

放射性标记法通常是在细菌培养基中添加同位素标记的底物，如 ^{3}H-胸腺嘧啶和 ^{14}C-亮氨酸等，细菌吸收同化这些底物后，即可获得放射性标记细菌

（radioactively labelled bacteria，RLB）。将 RLB 投喂给摄食者，每隔一定时间监测摄食者体内的放射性的增加量，即可估算摄食速率（Zubkov et al.，1998）。但这一方法在研究中使用相对较少。

目前研究中使用较多的是荧光标记法，即使用 FLB 投喂鞭毛虫和纤毛虫，在荧光显微镜下观察计数摄食者体内的 FLB 有无及个数。FLB 通常是通过对加热灭活的细菌进行荧光染色来制备的（Sherr et al.，1987），但有的摄食者会倾向于摄食活的饵料而忽视死饵料，因此 FLB 方法并不完美（Landry et al.，1991）。考虑到鞭毛虫和纤毛虫对与自然饵料大小接近的人工合成颗粒通常选择性不强，部分实验也采用带有荧光的人工合成颗粒，如聚苯乙烯荧光微球替代 FLB。

除 FLB、荧光微球外，还有其他几种颗粒可用于摄食速率研究，如由碳水化合物和蛋白质制成的人工颗粒（Hammer et al.，2001），免疫荧光标记细菌（fluorescent antibody labelled bacteria，FALB）（Christoffersen et al.，1997），基因改造的含绿色荧光蛋白的细菌（Parry et al.，2001）等。原生生物摄食细菌后体内的溶菌酶水平和体内的 FLB 数目有很好的关联，因此也可以用溶菌酶活性估算摄食量（González et al.，1993a）。但是上述技术的样品制备过程复杂、成本较高，没有被广泛使用。

自然海区鞭毛虫对细菌摄食速率的研究主要使用 FLB 或荧光微球等荧光颗粒示踪。将荧光颗粒按该海区细菌丰度的一定比例（通常为 10%）加入海水中，培养大约几十分钟到几小时，培养期间每隔一段时间采样计数鞭毛虫体内的荧光颗粒的数量。培养开始后，鞭毛虫体内的荧光颗粒的数量逐渐增加，到一定时间后鞭毛虫摄食和排出荧光颗粒的速度一致，此时鞭毛虫体内的荧光颗粒数目不再增加。可根据鞭毛虫体内颗粒数目线性增长的情况获得鞭毛虫摄食荧光颗粒的速度，再根据水体中荧光颗粒和细菌丰度的比例估算该海区鞭毛虫对细菌的摄食速率。

（二）水体颗粒减少法

通过监测培养体系内饵料的减少，也可以估算鞭毛虫和纤毛虫的摄食速率。大肠杆菌的变异菌株能进行异常分裂，形成直径为 $0.2\sim0.9\mu m$ 的小细胞（minicell）。用放射性标记或荧光染料染色小细胞后，与现场细菌按一定丰度比例投入到自然海水中，小细胞被微型浮游动物摄食后消失。根据小细胞丰度减少或放射性的消失情况，可估计微型浮游动物摄食情况，称为小细胞示踪法（Wikner et al.，1986；Bochdansky and Clouse，2015）。

在细菌培养基中添加 ^{3}H-胸腺嘧啶，可达到对细菌 DNA 进行放射性标记的目的。DNA 在细菌体内降解的速度很慢，实验过程中每个细菌中的 DNA 中放射性不变，因此可使用 DNA 放射性强度作为评估细菌丰度的指标。将标记后的细菌与摄食者共培养，培养体系中 DNA 放射性的减少是由细菌被摄食和被病毒杀死造成的。用孔径 2μm 的滤膜过滤去除摄食者后重复该实验,此时 DNA 放射性的减少主要由病毒杀死细菌造成。两个培养实验之差即为鞭毛虫和纤毛虫对细菌的摄食致死情况(Servais et al., 1985; Servais et al., 1989)。

二、鞭毛虫对细菌的摄食强度和压力

在海上现场评估鞭毛虫对细菌的摄食速率主要使用 FLB 法。大多数研究发现,每个鞭毛虫每小时摄食 0~76 个细菌(表 5-1),摄食速率通常与细菌的丰度成正比。

表 5-1 鞭毛虫对异养细菌的现场摄食速率(FLB 法)

摄食者	海区	季节	个体摄食速率 [prey/(grazer·h)]	清滤速率 [nl/(grazer·h)]	群体摄食速率	参考文献
异养鞭毛虫	北大西洋,萨佩洛岛近岸河口	夏季	—	3.2	$1.0×10^6$~$5.8×10^6$ prey/(ml·d)	Sherr 等(1987); Sherr 等(1989)
	北大西洋,哈利法克斯港	夏季	13~76	—	—	Novitsky (1990)
	北大西洋,Vineyard 海湾	冬季(12月)至夏季(7月)	—	—	$4×10^4$ prey/(ml·d)(12月)~$2×10^6$ prey/(ml·d)(7月)	Marrasé 等(1992)
	北大西洋,比斯开省近岸	冬季(1月)至夏季(8月)	0.9(4月)~37.2(8月),平均14	—	$6.7×10^5$ prey/(L·h)(1月)~$1.572×10^7$ prey/(L·h)(8月),平均 $7.35×10^6$ prey/(L·h)	Barcina 等(1992)
	大西洋	春季	1.3~10.4	—	$1.3×10^3$~$1.15×10^5$ prey/(ml·d)	Rocke 等(2015)
	亚热带东北大西洋	春季和秋季	—	—	$2×10^3$~$8.6×10^5$ prey/(ml·d)	Vaqué 等(2013)
	西太平洋,韩国近海	春季和夏季	—	0.7~10.4	—	Choi 等(2003)
	西太平洋,韩国近海	春季、夏季、秋季	1.0~5.6	2.3~11.7	$1×10^5$~$2.33×10^7$ prey/(L·h)	Cho 等(2000)
	西太平洋,黄海	秋季	3.92~15.57,平均 8.02			Lin 等(2014)
	南太平洋,提克豪环礁	秋季		0.8~4.5	$3.86×10^6$ prey/(L·h)	González 等(1998)
	北冰洋中部	夏季	—	2.5~5.6	$1.2×10^3$~$4.6×10^4$ prey/(ml·d)	Sherr 等(1997)
	北冰洋,雷索卢特海峡	春季	3~64,平均13	3~86,平均12		Laurion 等(1995)
	南大洋,普里兹湾	夏季	0.06~8.28	0.06~9.32		Leakey 等(1996)
	南大洋,大西洋扇区 6°W 断面	春季	粒径<5μm 的 HNF:最大速率 1.1;粒径>5μm 的 HNF:最大速率 7.3			Becquevort (1997)

续表

摄食者	海区	季节	个体摄食速率 [prey/(grazer·h)]	清滤速率 [nl/(grazer·h)]	群体摄食速率	参考文献
异养鞭毛虫	南极半岛莱德湾	夏季	—	—	$1.1×10^5$~$8.3×10^5$ prey/(ml·d)	Evans 等 (2021)
混合营养鞭毛虫和异养鞭毛虫	北大西洋,波士顿港	秋季	HNF: 0.6, MNF: 0.7~1.8	HNF: 5.3, MNF: 6.2~14.6	—	Epstein 和 Shiaris (1992)
	丹麦近海 Aarhus 湾	夏季	HNF: 2.9~4.1, MNF: 0.4~10	HNF: 2.3~4.2, MNF: 0.4~7.9	—	Havskum 和 Riemann (1996)
	西太平洋,中国台湾东北部近岸	春季、夏季、秋季	HNF: 0.2~2.0, MNF: 0.5~5.2	—	HNF: $0.28×10^4$~$2.62×10^4$ prey/(ml·d), MNF: $0.32×10^4$~$5.22×10^4$ prey/(ml·d)	Tsai 等 (2011)
	南太平洋,新西兰东部辐聚区	夏季	—	HNF: 3.99~6.03, MNF: 2.68~3.53	—	Safi 和 Hall (1999)
	南太平洋,智利近岸上升流区	春/夏和冬季	HNF: 2~7.5, MNF: 7.3~30.7	—	HNF: $0.5×10^5$~$9.4×10^5$ prey/(ml·d), MNF: $0.04×10^5$~$1×10^5$ prey/(ml·d)	Vargas 等 (2012)
	地中海东部	秋季	HNF: 0.63~5.78, MNF: 0.17	HNF: 9.4~18.3, MNF: 0.3	HNF: $2.74×10^2$~$1.39×10^5$ prey/(ml·d), MNF: $2.53×10^3$ prey/(ml·d)	Pachiadaki 等 (2016)
	地中海西北部	周年	HNF: 4.0~15.4, MNF: 1.1~1.3	HNF: 2.2~6.1, MNF: 1.2~1.5	—	Unrein 等 (2007)

注：HNF 表示异养鞭毛虫，MNF 表示混合营养鞭毛虫；—表示无数据

根据鞭毛虫的丰度和摄食速率可以估计鞭毛虫群体的摄食量及其摄食压力（即摄食量占现存量或生产量的比例）。在南大洋的深层海水中，鞭毛虫等对异养细菌的摄食压力范围在 3.79%~31.14%（Rocke et al., 2015）。在北冰洋，鞭毛虫每天摄食 1%~25.2% 的细菌生物量（Sanders and Gast, 2012）。在智利近海的上升流区，混合营养鞭毛虫和异养鞭毛虫对细菌生产力的摄食压力分别为每天 1%~51% 和 24%~100%（Vargas et al., 2012）。在黄海，异养鞭毛虫摄食占细菌生物量的 22.75%，细菌生产力的 6.65%（Lin et al., 2014）。

鞭毛虫的摄食可能存在昼夜变化。在南海的研究表明，异养鞭毛虫白天的摄食速率比晚上高（Ng and Liu, 2016）。在北太平洋亚热带环流的研究结果与此不同，混合营养鞭毛虫对细菌的摄食没有明显的昼夜变化，而异养鞭毛虫对

细菌的摄食在晚上比在白天高（Connell et al.，2020）。但在美国伍兹霍尔的Vineyard 海湾，有研究并未发现光照对鞭毛虫的摄食有影响（Marrasé et al.，1992）。

用 FLB 示踪法可以区别异养鞭毛虫和混合营养鞭毛虫的摄食。方法是在荧光显微镜检查鞭毛虫体内 FLB 的同时观察虫体内有无叶绿素的红色荧光，含有叶绿素的即可认为是混合营养鞭毛虫。在多数研究中，混合营养鞭毛虫的摄食量占总鞭毛虫摄食量的比例大于异养鞭毛虫。在丹麦的 Aarhus 湾，混合营养鞭毛虫对细菌的摄食量占所有鞭毛虫摄食量的 86%（Havskum and Riemann，1996）。在地中海的近岸寡营养海区，混合营养鞭毛虫的摄食量占鞭毛虫总摄食量的约 50%（Unrein et al.，2007）。除 FLB 法，也有其他研究方法发现混合营养鞭毛虫是异养细菌的主要摄食者，其摄食量占细菌被摄食量的 50%以上（Zubkov and Tarran，2008；Ptacnik et al.，2016）。在部分海区，虽然单个异养鞭毛虫的摄食速率高于混合营养鞭毛虫，但由于混合营养鞭毛虫的丰度高于异养鞭毛虫，所以对细菌的群体摄食量仍是混合营养鞭毛虫占优势（Unrein et al.，2007；Zubkov and Tarran，2008）。

光照可影响混合营养鞭毛虫对细菌的摄食速率。当光照强度适合混合营养鞭毛虫进行光合自养生长时，其对异养细菌的摄食速率适中；当光照强度过低时，其对细菌的摄食速率最高；而当光照强度过高时，其对细菌的摄食速率降至最低（Anderson and Hansen，2020）。

混合营养鞭毛虫摄食细菌的速度与环境中营养盐的浓度成反比，当水体营养盐不足时，混合营养鞭毛虫倾向于大量摄食细菌（Zubkov and Tarran，2008；Tsai et al.，2011；Unrein et al.，2014）。这些细菌所含的营养盐直接进入混合营养鞭毛虫体内参与光合作用，而不是以溶解态营养盐形式释放回水体再被浮游植物利用，这被称为微食物网中的营养盐捷径（Ptacnik et al.，2016）。

不同粒级鞭毛虫的摄食能力有差异，通常大个体鞭毛虫摄食细菌的速度较快。在北大西洋，粒径 5μm 的混合营养鞭毛虫摄食速率高于粒径 2μm 的混合营养鞭毛虫（Zubkov and Tarran，2008）。在地中海西北部，粒径 3～5μm 和 5～20μm 的混合营养鞭毛虫摄食细菌的平均速度分别为 1.1prey/（grazer·h）和 1.3prey/（grazer·h），粒径<5μm 和 5～20μm 的异养鞭毛虫摄食速率分别为 4.0prey/（grazer·h）和 15.4prey/（grazer·h）（Unrein et al.，2007）。

在自然海区，鞭毛虫对细菌粒径的选择性摄食导致细菌的粒级变小。当把摄食者去除后，可以观察到培养体系中大个体细菌数量出现增长，再向海水中加入鞭毛虫后，细菌的粒级又将变小（Suzuki，1999；Strom，2000）。增加摄食者还可以改变细菌的群落组成结构（Jürgens et al.，1999；Suzuki，1999；Hahn and Höfle，2001；Gasol et al.，2002；Jürgens and Matz，2002）。

鞭毛虫和纤毛虫都是微食物网中的摄食者，异养细菌是它们的共同饵料。但是由于荧光颗粒法较少用于纤毛虫的摄食研究，因此比较鞭毛虫和纤毛虫对细菌摄食压力的研究不多。在日本近海，纤毛虫对细菌的摄食速率为 7～34prey/（grazer·h），大大高于鞭毛虫的 0.3～2.2prey/（grazer·h），但是由于纤毛虫丰度较低，鞭毛虫对细菌现存量的摄食压力（1%～15%/d）反而高于纤毛虫的摄食压力（0～0.63%/d）（Ichinotsuka et al.，2006）。在东北大西洋，纤毛虫摄食细菌生产力的摄食压力为<10%，而鞭毛虫的摄食压力可以高达 83%（Karayanni et al.，2008），所以该海区鞭毛虫是细菌的重要摄食者，纤毛虫的影响较小。在北大西洋波士顿港的半咸水中，鞭毛虫的摄食速率为 0.6～0.9prey/（grazer·h），清滤速率约为 5.3nl/（grazer·h）；而纤毛虫的摄食速率为 12～20prey/（grazer·h），清滤速率为 90nl/（grazer·h），纤毛虫是异养细菌的主要摄食者，对细菌的摄食压力为 15%～90%/d（Epstein and Shiaris，1992）。鞭毛虫和纤毛虫对细菌的摄食压力通常冬季较低，夏季较高（Marrasé et al.，1992）。随着海区生产力的提高，鞭毛虫和纤毛虫对细菌的摄食压力增加（Thelaus et al.，2008）。

参 考 文 献

黄凌风, 林施泉, 熊源, 等. 2013. 微型异养鞭毛虫摄食选择性的研究进展. 海洋学报, 35(2): 1-8.

焦念志. 2012. 海洋固碳与储碳——并论微型生物在其中的重要作用. 中国科学: 地球科学, 42: 1473.

张传伦. 2018. 微型生物碳泵——海洋生物地球化学研究的新模式. 中国科学: 地球科学, 48: 805-808.

Anderson R, Hansen P J. 2020. Meteorological conditions induce strong shifts in mixotrophic and heterotrophic flagellate bacterivory over small spatio-temporal scales. Limnology and Oceanography, 65(6): 1189-1199.

Anderson R, Kjelleberg S, McDougald D, et al. 2011. Species-specific patterns in the vulnerability of carbon-starved bacteria to protist grazing. Aquatic Microbial Ecology, 64(2): 105-116.

Arrieta J M, Mayol E, Hansman R L, et al. 2015. Dilution limits dissolved organic carbon utilization in the deep ocean. Science, 348(6232): 331-333.

Ballen-Segura M, Felip M, Catalan J. 2017. Some mixotrophic flagellate species selectively graze on archaea. Applied and Environmental Microbiology, 83(2): e02317-02316.

Barcina I, Ayo B, Unanue M, et al. 1992. Comparison of rates of flagellate bacterivory and bacterial production in a marine coastal system. Applied and Environmental Microbiology, 58(12): 3850-3856.

Becquevort S. 1997. Nanoprotozooplankton in the Atlantic sector of the Southern Ocean during early spring: biomass and feeding activities. Deep Sea Research Part II: Topical Studies in Oceanography, 44(1-2): 355-373.

Bochdansky A B, Clouse M A. 2015. New tracer to estimate community predation rates of phagotrophic protists. Marine Ecology Progress Series, 524: 55-69.

Boenigk J, Arndt H. 2002. Bacterivory by heterotrophic flagellates: community structure and feeding strategies. Antonie Van Leeuwenhoek, 81(1-4): 465-480.

Boenigk J, Matz C, Jürgens K, et al. 2001. The influence of preculture conditions and food quality on the ingestion and digestion process of three species of heterotrophic nanoflagellates. Microbial Ecology, 42(2): 168-176.

Bratbak G, Heldal M. 2000. Viruses rule the waves-the smallest and most abundant members of marine ecosystems. Microbiology Today, 27: 171-173.

Breitbart M, Middelboe M, Rohwer F. 2008. Marine viruses: community dynamics, diversity and impact on microbial processes//Kirchman D L. Microbial Ecology of the Oceans. Second Edition. Hoboken: John Wiley & Sons: 443-479.

Cavicchioli R, Ostrowski M, Fegatella F, et al. 2003. Life under nutrient limitation in oligotrophic marine environments: an eco/physiological perspective of *Sphingopyxis alaskensis*(formerly *Sphingomonas alaskensis*). Microbial Ecology, 45(3): 203-217.

Chen X, Ma R, Yang Y, et al. 2019. Viral regulation on bacterial community impacted by lysis-lysogeny switch: a microcosm experiment in eutrophic coastal waters. Frontiers in Microbiology, 10: 01763.

Chen X, Weinbauer M G, Jiao N, et al. 2021. Revisiting marine lytic and lysogenic virus-host interactions: Kill-the-Winner and Piggyback-the-Winner. Science Bulletin, 66(9): 871-874.

Cho B C, Na S C, Choi D H. 2000. Active ingestion of fluorescently labeled bacteria by mesopelagic heterotrophic nanoflagellates in the East Sea, Korea. Marine Ecology Progress Series, 206: 23-32.

Choi D H, Chung Y H, Byung C C. 2003. Comparison of virus- and bacterivory-induced bacterial mortality in the eutrophic Masan Bay, Korea. Aquatic Microbial Ecology, 30(2): 117-125.

Christoffersen K, Nybroe O, Jürgens K, et al. 1997. Measurement of bacterivory by heterotrophic nanoflagellates using immunofluorescence labelling of ingested cells. Aquatic Microbial Ecology, 13(1): 127-134.

Connell P E, Ribalet F, Armbrust E V, et al. 2020. Diel oscillations in the feeding activity of heterotrophic and mixotrophic nanoplankton in the North Pacific Subtropical Gyre. Aquatic Microbial Ecology, 85: 167-181.

Epstein S S, Shiaris M P. 1992. Size-selective grazing of coastal bacterioplankton by natural assemblages of pigmented flagellates, colorless flagellates, and ciliates. Microbial Ecology, 23(3): 211-225.

Evans C, Brandsma J, Meredith M P, et al. 2021. Shift from carbon flow through the microbial loop to the viral shunt in coastal Antarctic waters during austral summer. Microorganisms, 9(2): 460.

Fenchel T M. 1987. Ecology of Protozoa. The Biology of Free-Living Phagotrophic Protists. Brock Springer Series in Contemporary BioScience. Berlin: Springer-Verlag Berlin Heidelberg: 197.

First M R, Hollibaugh J T. 2009. The model high molecular weight DOC compound, dextran, is ingested by the benthic ciliate *Uronema marinum* but does not supplement ciliate growth. Aquatic Microbial Ecology, 57(1): 79-87.

Garza D R, Suttle C A. 1995. Large double-stranded DNA viruses which cause the lysis of a marine heterotrophic nanoflagellate(*Bodo* sp.) occur in natural marine viral communities. Aquatic Microbial Ecology, 9: 203-210.

Gasol J M, Pedrós-Alió C, Vaqué D. 2002. Regulation of bacterial assemblages in oligotrophic plankton systems: results from experimental and empirical approaches. Antonie Van Leeuwenhoek, 81(1-4):

435-452.

González J M, Sherr B F, Sherr E. 1993a. Digestive enzyme activity as a quantitative measure of protistan grazing: the acid lysozyme assay for bacterivory. Marine Ecology Progress Series, 100: 197-206.

González J M, Sherr E B, Sherr B F. 1990. Size-selective grazing on bacteria by natural assemblages of estuarine flagellates and ciliates. Applied and Environmental Microbiology, 56(3): 583-589.

González J M, Sherr E, Sherr B F. 1993b. Differential feeding by marine flagellates on growing versus starving, and on motile versus nonmotile, bacterial prey. Marine Ecology Progress Series, 102: 257-267.

González J M, Torréton J P, Dufour P, et al. 1998. Temporal and spatial dynamics of the pelagic microbial food web in an atoll lagoon. Aquatic Microbial Ecology, 16(1): 53-64.

Hahn M W, Höfle M G. 2001. Grazing of protozoa and its effect on populations of aquatic bacteria. FEMS Microbiology Ecology, 35(2): 113-121.

Hammer A, Grüttner C, Schumann R. 2001. New biocompatible tracer particles: use for estimation of microzooplankton grazing, digestion, and growth rates. Aquatic Microbial Ecology, 24(2): 153-161.

Hansell D A, Carlson C A, Repeta D J, et al. 2009. Dissolved organic matter in the ocean: a controversy stimulates new insights. Oceanography, 22(4): 202-211.

Hansen P J, Bjørnsen P K, Hansen B W. 1997. Zooplankton grazing and growth: scaling within the 2-2000-μm body size range. Limnology and Oceanography, 42(4): 687-704.

Havskum H, Riemann B. 1996. Ecological importance of bacterivorous, pigmented flagellates(mixotrophs) in the Bay of Aarhus, Denmark. Marine Ecology Progress Series, 137: 251-263.

Ichinotsuka D, Ueno H, Nakano S. 2006. Relative importance of nanoflagellates and ciliates as consumers of bacteria in a coastal sea area dominated by oligotrichous *Strombidium* and *Strobilidium*. Aquatic Microbial Ecology, 42(2): 139-147.

Jackson G A. 1987. Simulating chemosensory responses of marine microorganisms. Limnology and Oceanography, 32(6): 1253-1266.

Jacquet S, Heldal M, Iglesias-Rodriguez D, et al. 2002. Flow cytometric analysis of an *Emiliana huxleyi* bloom terminated by viral infection. Aquatic Microbial Ecology, 27: 111-124.

Jian H, Yi Y, Wang J, et al. 2021. Diversity and distribution of viruses inhabiting the deepest ocean on Earth. The ISME Journal, 15(10): 3094-3110.

Jiao N, Herndl G J, Hansell D A, et al. 2010. Microbial production of recalcitrant dissolved organic matter: long-term carbon storage in the global ocean. Nature Reviews Microbiology, 8(8): 593-599.

Jiao N, Legendre L, Robinson C, et al. 2015. Comment on "dilution limits dissolved organic carbon utilization in the deep ocean". Science, 350(6267): 1483.

Jürgens K, DeMott W R. 1995. Behavioral flexibility in prey selection by bacterivorous nanoflagellates. Limnology and Oceanography, 40(8): 1503-1507.

Jürgens K, Matz C. 2002. Predation as a shaping force for the phenotypic and genotypic composition of planktonic bacteria. Antonie Van Leeuwenhoek, 81(1-4): 413-434.

Jürgens K, Pernthaler J, Schalla S, et al. 1999. Morphological and compositional changes in a planktonic bacterial community in response to enhanced protozoan grazing. Applied and Environmental Microbiology, 65(3): 1241-1250.

Karayanni H, Christaki U, Thyssen M, et al. 2008. Heterotrophic nanoflagellate and ciliate bacterivorous activity and growth in the Northeast Atlantic Ocean: a seasonal mesoscale study. Aquatic Microbial Ecology, 51(2): 169-181.

Keys A, Christensen E H, Krogh A. 1935. The organic metabolism of sea-water with special reference to the ultimate food cycle in the sea. Journal of the Marine Biological Association of the United Kingdom, 20(2): 181-196.

Knowles B, Silveira C B, Bailey B A, et al. 2016. Lytic to temperate switching of viral communities. Nature, 531(7595): 466-470.

Krogh A, Keys A. 1934. Methods for the determination of dissolved organic carbon and nitrogen in sea water. The Biological Bulletin, 67(1): 132-144.

Landry M R, Lehner-Fournier J M, Sundstrom J A, et al. 1991. Discrimination between living and heat-killed prey by a marine zooflagellate, *Paraphysomonas vestita* (Stokes). Journal of Experimental Marine Biology and Ecology, 146(2): 139-151.

Lara E, Vaqué D, Sà E L, et al. 2017. Unveiling the role and life strategies of viruses from the surface to the dark ocean. Science Advances, 3(9): e1602565.

Laurion I, Demers S, Vézina A F. 1995. The microbial food web associated with the ice algal assemblage: biomass and bacterivory of nanoflagellate protozoans in Resolute Passage(High Canadian Arctic). Marine Ecology Progress Series, 120: 77-87.

Leakey R J G, Archer S D, Grey J. 1996. Microbial dynamics in coastal waters of East Antarctica: bacterial production and nanoflagellate bacterivory. Marine Ecology Progress Series, 142: 3-17.

Lin S, Huang L, Lu J. 2014. Weak coupling between heterotrophic nanoflagellates and bacteria in the Yellow Sea Cold Water Mass area. Acta Oceanologica Sinica, 33(9): 125-132.

Marrasé C, Lim E L, Caron D A. 1992. Seasonal and daily changes in bacterivory in a coastal plankton community. Marine Ecology Progress Series, 82(3): 281-289.

Matz C, Boenigk J, Arndt H, et al. 2002. Role of bacterial phenotypic traits in selective feeding of the heterotrophic nanoflagellate *Spumella* sp. Aquatic Microbial Ecology, 27(2): 137-148.

Matz C, Jürgens K. 2001. Effects of hydrophobic and electrostatic cell surface properties of bacteria on feeding rates of heterotrophic nanoflagellates. Applied and Environmental Microbiology, 67(2): 814-820.

Montagnes D J S, Barbosa A B, Boenigk J, et al. 2008. Selective feeding behaviour of key free-living protists: avenues for continued study. Aquatic Microbial Ecology, 53: 83-98.

Moreira D, López-García P. 2014. The rise and fall of picobiliphytes: how assumed autotrophs turned out to be heterotrophs. BioEssays: News and Reviews in Molecular, Cellular and Developmental Biology, 36(5): 468-474.

Murray A G, Jackson G A. 1992. Viral dynamics: a model of the effects of size, shape, motion and abundance of single-celled planktonic organisms and other particles. Marine Ecology Progress Series, 89(2-3): 103-116.

Nelson R J, Ashjian C J, Bluhm B A, et al. 2014. Biodiversity and biogeography of the lower trophic taxa of the Pacific Arctic region: sensitivities to climate change//Grebmeier J M, Maslowski W. The Pacific Arctic Region: Ecosystem Status and Trends in a Rapidly Changing Environment. Dordrecht: Springer Netherlands: 269-336.

Ng W H A, Liu H. 2016. Diel periodicity of grazing by heterotrophic nanoflagellates influenced by prey cell properties and intrinsic grazing rhythm. Journal of Plankton Research, 38(3): 636-651.

Novitsky J A. 1990. Protozoa abundance, growth, and bacterivory in the water column, on sedimenting particles, and in the sediment of Halifax Harbor. Canadian Journal of Microbiology, 36(12): 859-863.

Pachiadaki M G, Taylor C, Oikonomou A, et al. 2016. *In situ* grazing experiments apply new technology to gain insights into deep-sea microbial food webs. Deep Sea Research Part II: Topical Studies in Oceanography, 129: 223-231.

Parry J D, Heaton K, Drinkall J, et al. 2001. Feasibility of using GFP-expressing *Escherichia coli*, coupled with fluorimetry, to determine protozoan ingestion rates. FEMS Microbiology Ecology, 35(1): 11-17.

Pfandl K, Posch T, Boenigk J. 2004. Unexpected effects of prey dimensions and morphologies on the size selective feeding by two bacterivorous flagellates(*Ochromonas* sp. and *Spumella* sp.). Journal of Eukaryotic Microbiology, 51(6): 626-633.

Ptacnik R, Gomes A, Royer S J, et al. 2016. A light-induced shortcut in the planktonic microbial loop. Scientific Reports, 6: 29286.

Rocke E, Pachiadaki M G, Cobban A, et al. 2015. Protist community grazing on prokaryotic prey in deep ocean water masses. PLoS One, 10(4): e0124505.

Roshan S, DeVries T. 2017. Efficient dissolved organic carbon production and export in the oligotrophic ocean. Nature Communications, 8(1): 2036.

Safi K A, Hall J A. 1999. Mixotrophic and heterotrophic nanoflagellate grazing in the convergence zone east of New Zealand. Aquatic Microbial Ecology, 20(1): 83-93.

Sanders R W, Gast R J. 2012. Bacterivory by phototrophic picoplankton and nanoplankton in Arctic waters. FEMS Microbiology Ecology, 82(2): 242-253.

Seenivasan R, Sausen N, Medlin L, et al. 2013. *Picomonas judraskeda* Gen. Et Sp. Nov.: the first identified member of the Picozoa Phylum Nov., a widespread group of picoeukaryotes, formerly known as 'picobiliphytes'. PLoS One, 8(3): e59565.

Servais P, Billen G, Martinez J, et al. 1989. Estimating bacterial mortality by the disappearance of ^{3}H-labeled intracellular DNA. FEMS Microbiology Letters, 62(2): 119-125.

Servais P, Billen G, Rego J V. 1985. Rate of bacterial mortality in aquatic environments. Applied and Environmental Microbiology, 49(6): 1448-1454.

Shannon S P, Chrzanowski T H, Grover J P. 2007. Prey food quality affects flagellate ingestion rates. Microbial Ecology, 53(1): 66-73.

Sherr B F, Sherr E B, Fallon R D. 1987. Use of monodispersed, fluorescently labeled bacteria to estimate *in situ* protozoan bacterivory. Applied and Environmental Microbiology, 53(5): 958-965.

Sherr B, Sherr E B, Pedrós-Alió C. 1989. Simultaneous measurement of bacterioplankton production and protozoan bactivory in estuarine water. Marine Ecology Progress Series, 54(3): 209-219.

Sherr E B. 1988. Direct use of high molecular weight polysaccharide by heterotrophic flagellates. Nature, 335(6188): 348-351.

Sherr E B, Sherr B F, Fessenden L. 1997. Heterotrophic protists in the Central Arctic Ocean. Deep Sea Research Part II: Topical Studies in Oceanography, 44(8): 1665-1682.

Strom S L. 2000. Bacterivory: interactions between bacteria and their grazers//Kirchman D L. Microbial Ecology of the Oceans. New York: Wiley-Liss: 351-386.

Suttle C A. 1994. The significance of viruses to mortality in aquatic microbial communities. Microbial Ecology, 28(2): 237-243.

Suttle C A, Chan A M. 1994. Dynamics and distribution of cyanophages and their effect on marine

Synechococcus spp. Applied and Environmental Microbiology, 60(9): 3167-3174.

Suzuki M T. 1999. Effect of protistan bacterivory on coastal bacterioplankton diversity. Aquatic Microbial Ecology, 20(3): 261-272.

Suzuki-Tellier S, Andersen A, Kiørboe T. 2022. Mechanisms and fluid dynamics of foraging in heterotrophic nanoflagellates. Limnology and Oceanography, 67(6): 1287-1298.

Thelaus J, Haecky P, Forsman M, et al. 2008. Predation pressure on bacteria increases along aquatic productivity gradients. Aquatic Microbial Ecology, 52(1): 45-55.

Thingstad T F. 2000. Elements of a theory for the mechanisms controlling abundance, diversity, and biogeochemical role of lytic bacterial viruses in aquatic systems. Limnology and Oceanography, 45(6): 1320-1328.

Tsai A Y, Gong G C, Sanders R W, et al. 2011. Importance of bacterivory by pigmented and heterotrophic nanoflagellates during the warm season in a subtropical western Pacific coastal ecosystem. Aquatic Microbial Ecology, 63(1): 9-18.

Tsai A Y, Gong G C, Chang W L. 2016. Control of nanoflagellate abundance by microzooplankton and viruses in a coastal ecosystem of the subtropical western Pacific. Journal of Experimental Marine Biology and Ecology, 481: 57-62.

Unrein F, Gasol J M, Not F, et al. 2014. Mixotrophic haptophytes are key bacterial grazers in oligotrophic coastal waters. The ISME Journal, 8(1): 164-176.

Unrein F, Massana R, Alonso-Sáez L, et al. 2007. Significant year-round effect of small mixotrophic flagellates on bacterioplankton in an oligotrophic coastal system. Limnology and Oceanography, 52(1): 456-469.

Vaqué D, Alonso-Sáez L, Arístegui J, et al. 2013. Bacterial production and losses to predators along an open ocean productivity gradient in the subtropical North East Atlantic Ocean. Journal of Plankton Research, 36(1): 198-213.

Vaqué D, Casamayor E O, Gasol J M. 2001. Dynamics of whole community bacterial production and grazing losses in seawater incubations as related to the changes in the proportions of bacteria with different DNA content. Aquatic Microbial Ecology, 25(2): 163-177.

Vargas C A, Contreras P Y, Iriarte J L. 2012. Relative importance of phototrophic, heterotrophic, and mixotrophic nanoflagellates in the microbial food web of a river-influenced coastal upwelling area. Aquatic Microbial Ecology, 65(3): 233-248.

Wandel H, Holzman R. 2022. Modulation of cilia beat kinematics is a key determinant of encounter rate and selectivity in tintinnid ciliates. Frontiers in Marine Science, 9: 845903.

Weinbauer M G, Brettar I, Höfle M G. 2003. Lysogeny and virus-induced mortality of bacterioplankton in surface, deep, and anoxic marine waters. Limnology and Oceanography, 48(4): 1457-1465.

Weisse T, Anderson R, Arndt H, et al. 2016. Functional ecology of aquatic phagotrophic protists-concepts, limitations, and perspectives. European Journal of Protistology, 55: 50-74.

Wikner J, Andersson A, Normark S, et al. 1986. Use of genetically marked minicells as a probe in measurement of predation on bacteria in aquatic environments. Applied and Environmental Microbiology, 52(1): 4-8.

Wootton E C, Zubkov M V, Jones D H, et al. 2007. Biochemical prey recognition by planktonic protozoa. Environmental Microbiology, 9(1): 216-222.

Wu Q L, Boenigk J, Hahn M W. 2004. Successful predation of filamentous bacteria by a nanoflagellate challenges current models of flagellate bacterivory. Applied and Environmental

Microbiology, 70(1): 332-339.

Zhukov B F. 1993. Atlas Presnovodnykh Geterotrofnykh Zhgutikonostsev(Biologiya, Ekologiya i Sistematika) [Atlas of Freshwater Heterotrophic Flagellates(Biology, Ecology, and Taxonomy)]. Rybinsk: Russian Academy of Sciences: 160.

Zubkov M V, Sleigh M A, Burkill P H. 1998. Measurement of bacterivory by protists in open ocean waters. FEMS Microbiology Ecology, 27(1): 85-102.

Zubkov M V, Tarran G A. 2008. High bacterivory by the smallest phytoplankton in the North Atlantic Ocean. Nature, 455(7210): 224-226.

第六章　自然海区的混合营养浮游鞭毛虫和纤毛虫

混合营养（mixotrophy）是指一种生物既能进行光合自养，又能进行摄食异养。微食物网中各种生物的摄食营养关系纵横交织，微型浮游生物的混合营养现象使得营养关系更加错综复杂。混合营养生物既是生产者，又是消费者，因此对传统的食物链和生态模型研究是个挑战。

海洋浮游生物包括浮游植物和浮游动物。在经典的生物营养模式分类中，浮游植物是自养生物，利用阳光合成有机物、固定能量；浮游动物是异养生物，利用浮游植物合成的有机物进行生长。但是人们很早就意识到海洋中的原生生物（protist）在同一个体中同时存在自养和异养两种营养方式，这种模式被称为混合营养。异养的方式有两种，一种是渗透营养（osmotrophy），一种是吞噬营养（phagotrophy）。广义的混合营养所指的异养方式可以是其中的任何一种。根据这个定义，几乎所有的浮游植物都是混合营养者，因为它们都可以通过渗透营养吸收溶解有机物分子（Lewin，1953；Eiler，2006），如最小的自养生物原绿球藻也能吸收溶解氨基酸（Zubkov et al.，2004）。吞噬营养方式则意味着饵料生物的死亡，是生物之间营养关系的具体体现。Flynn 等（2013）建议用自养和吞噬营养共同存在于一种生物的营养方式来定义混合营养，本书即采取这个定义。

在自然海区浮游生物里的混合营养生物主要是鞭毛虫和纤毛虫。鞭毛虫最初被认为是自养生物，但是后来发现可以进行异养；而纤毛虫最初被认为是异养生物，后来发现部分种类可以进行光合自养。Stoecker（1998）将混合营养分成 3 种类型：①"理想的混合营养"，即自养和异养的贡献一样多，在自然界中这一类型非常罕见；②以自养为主的藻类；③以异养为主的原生动物。Caron（2000）则将混合营养仅分为后两类，并且认为藻类中的混合营养者是具叶绿体的原生生物，其在后来获得吞噬营养的能力，是像动物的植物，这些类群主要有混合营养的鞭毛虫和甲藻；而原生动物中的混合营养者是异养的原生动物摄食藻类细胞后，保留了其叶绿体，并使得叶绿体在一段时间内参与光合作用，因此是像植物的动物，这些类群主要有浮游纤毛虫。

甲藻是浮游植物的重要类群，其混合营养在浮游植物研究中有广泛的开展。混合营养鞭毛虫和纤毛虫的研究则落后很多，在自然海区的丰度和分布情况还没有充分的研究。混合营养的海洋浮游生物在自然海区的分布是理解它们在自然海区生态功能的基础，本章将介绍自然海区混合营养浮游生物的重要类群——鞭毛虫和纤毛虫，包括自然海区混合营养鞭毛虫、纤毛虫分别在总鞭毛虫和总纤毛虫丰度和生物量中的比例。

第一节 混合营养鞭毛虫

一、混合营养鞭毛虫的发现

在20世纪50年代，人们就意识到了藻类存在异养营养方式。一些海洋浮游植物在无菌培养时，仅靠无机营养盐不能培养成功（Caron，2000）。这些浮游植物能合成大部分有机物质，但是需要其他生物（通常是细菌）的存在才能生长。细菌为藻类提供了维生素、特殊脂肪酸等生长因子，而藻类可能为细菌制造有机质从而作为生长的基质。意识到这一现象后，科学家在藻类无菌培养液中加入有机质，从而成功实现很多藻类的无菌培养（Caron，2000）。但是这一时期，人们认为藻类摄入有机营养的方式为胞饮作用（pinocytosis）。现在已经知道，许多藻类也能进行吞噬摄食。

在海洋微食物网概念提出以前，就有含色素的鞭毛虫体内有外来颗粒的报道（Sanders and Porter，1988）。最早是 Kofoid 和 Swezy（1921）发现一些藻类中含有食物泡，后有其他学者在自养鞭毛虫体内发现有机颗粒。但这些结果也有可能是共生或寄生，并不是混合营养的直接证据（Porter，1988）。微食物网概念首次提出细菌可被异养鞭毛虫摄食，此后，学者也开始认真审视含色素鞭毛虫对细菌的摄食。

首先要确认含色素鞭毛虫能摄食细菌，进行异养营养。尽管尚无切实的证据，但是 Porter 等已将含色素鞭毛虫列入细菌的摄食者中，"含色素鞭毛虫和无色素鞭毛虫分类学上的亲缘关系很近，因此结构也很相似，故含色素鞭毛虫可能和它的异养亲缘类群一样通过利用颗粒状的饵料进行异养"（Porter et al.，1985）。Bird 和 Kalff（1986）在实验室内用荧光微球投喂含色素的金藻类（chrysophytes），并用荧光显微镜观察到藻体内具有荧光微球，这意味着这类金藻有摄食行为。Estep 等（1986）将从马尾藻海分离的14个含色素鞭毛虫 chrysomonad 进行克隆，再将其放入含细菌的培养液中，发现它们和在藻类培养液中生长得一样好，电镜观察表明 chrysomonad 克隆体内有被摄食的细菌，从而证实含色素鞭毛虫能够摄食细菌。

鉴定海洋中的浮游藻类哪些具有吞噬行为一直是浮游微食物网研究的重要内容。有研究发现，颗石藻 *Emiliania huxleyi*、*Calcidiscus leptoporus*、*Coccolithus braarudii* 和 *Calyptrosphaera sphaeroidea* 有摄食细菌的能力，但是它们可能不经常进行吞噬营养，只在少于1%的个体中观察到吞噬的荧光标记细菌（Avrahami and Frada，2020）。

二、混合营养鞭毛虫的丰度和生物量比例

最初计数鞭毛虫是使用光学显微镜和电子显微镜,因为这种方法对含色素鞭毛虫和异养鞭毛虫区分不明显,所以研究者大多将所有的鞭毛虫都归为自养。20世纪 80 年代,人们开始探寻更好的鉴别和计数含色素鞭毛虫的方法。Davis 和 Sieburth(1982)首先使用吖啶橙染色和落射荧光显微镜观察法,先在紫外光激发光下找到吖啶橙染色的鞭毛虫,再切换到蓝光激发光下观察,此时鞭毛虫体内的色素体呈红色,用这一方法可以区分含色素鞭毛虫和异养鞭毛虫,对自然海区含色素鞭毛虫的鉴别和计数工作从此开展起来。Davis 和 Sieburth(1984)通过将研究样品先用落射荧光显微镜检查鉴别自养和异养,再用电镜检查可自养种的形态,从而区分其中的纯自养鞭毛虫和含色素鞭毛虫。

含色素的鞭毛虫并不一定摄食饵料进行异养营养,因此含色素的鞭毛虫并不都是混合营养,只有那些摄食饵料的含色素鞭毛虫才是混合营养鞭毛虫。Arenovski 等(1995)进行了第一个估计混合营养鞭毛虫丰度的研究,先通过落射荧光显微镜检查含色素鞭毛虫和异养鞭毛虫的丰度,再通过添加荧光标记蓝细菌饵料进行培养实验,确定摄食饵料的含色素鞭毛虫在含色素鞭毛虫丰度中的比例,从而估计混合营养鞭毛虫在含色素鞭毛虫中的比例(Arenovski et al.,1995)。目前用这种方法进行的研究较少(表 6-1),据估计混合营养鞭毛虫丰度占含色素鞭毛虫丰度的比例最大为 50%。这个比例随水深增加而减少,如在马尾藻海,表层混合营养鞭毛虫丰度占含色素鞭毛虫丰度的比例为 50%,但在叶绿素最大层只有不到 0.5%(Sanders et al.,2000)。

表 6-1 不同海区混合营养鞭毛虫占含色素鞭毛虫的比例

海区	水层	季节	培养时间	示踪物	丰度比例(%)	生物量比例(%)	参考文献
马尾藻海	表层	4月、8月	24h	荧光标记蓝细菌	0~50		Arenovski 等(1995)
	深层叶绿素最大层	4月、8月	24h	荧光标记蓝细菌	<0.5		
亚热带西太平洋	表层	周年	1h	荧光标记蓝细菌	1.3~3.8		Chan 等(2009)
爱琴海	0~100m	春季、夏季	1h	荧光标记细菌	6.2~12		Christaki 等(1999)
智利南部 Aysen 湾	1~25m	春季	50min	荧光标记细菌	0		Czypionka 等(2011)
	1~25m	冬季	50min	荧光标记细菌	约 50		
丹麦 Aarhus 湾	表层	6月	20min	荧光标记细菌		49 或 9	Havskum 和 Riemann(1996)

续表

海区	水层	季节	培养时间	示踪物	丰度比例（%）	生物量比例（%）	参考文献
南大洋罗斯海	水体	春季	12h	荧光标记细菌	8~42		Moorthi 等（2009）
	冰芯	春季	12h	荧光标记细菌	3~25		
乔治滩	表层	夏季、秋季	24h、36h	荧光标记细菌	12~17（最大38）		Sanders 等（2000）
马尾藻海	表层	夏季、秋季	24h、36h	荧光标记细菌	12~17（最大30）		Sanders 等（2000）
地中海西北部沿岸	表层	周年	40min	荧光标记细菌	18（<5μm 含色素鞭毛虫）		Unrein 等（2007）
	表层	周年	40min	荧光标记细菌	11（5~20μm 含色素鞭毛虫）		
智利中部沿岸上升流区	深层叶绿素最大层、温跃层、25m	周年	1h	荧光标记细菌	<5		Vargas 等（2012）
北冰洋	上混合层	9月	2h	荧光标记细菌和荧光微球	2~32		Sanders 和 Gast（2012）

微微型真核浮游生物（picoeukaryotes，粒径 0.2~2μm）也具有吞噬营养的能力。在北冰洋，1%~7%的微微型真核浮游生物能摄食粒径 0.6μm 的荧光微球，但无法摄食粒径 1~1.2μm 的荧光标记细菌（Sanders and Gast，2012）。

第二节 混合营养纤毛虫

有关混合营养纤毛虫的研究历史较短，海洋浮游无壳纤毛虫最初被认为是异养的，但是后来发现某些种类细胞内含有叶绿体和眼点。Norris（1967）在印度洋 78m 水深处采得的一个纤毛虫体内排列有金黄色的物体，他怀疑这是某种金藻（Chrysophyceae）的叶绿体，而不是整个藻体。Blackbourn 等（1973）用电镜观察了纤毛虫的细胞切片，发现了完整的叶绿体，从而确认浮游纤毛虫体内确实有来自藻类的叶绿体。Laval-Peuto 和 Febvre（1986）用透射电镜在附属曳尾虫（*Tontonia appendiculariformis*）中发现了数百种质体，并由此推断这种纤毛虫营混合营养。McManus 和 Fuhrman（1986）鉴定了球果螺体虫（*Laboea strobila*）的色素组成，发现其中包括叶绿素和藻红蛋白（phycoerythrin）。Jonsson（1987）和 Stoecker 等（1987）的研究证实寡毛类纤毛虫可以吸收无机碳。Stoecker 等（1987）及 Laval-Peuto 和 Rassoulzadegan（1988）用落射荧光显微镜方法观察纤毛虫是否含有外来色素体，发现纤毛虫体内的色素体有的单独存在，有的则处于食物泡中。Laval-Peuto 和 Rassoulzadegan（1988）推测纤毛虫可能通过摄食行为获得这些色素体并具有混合营养能力，谨慎起见他们建议将这些纤毛虫统称为含色素体的纤毛虫（plastidic ciliates）。在之后的研究中，有的沿用这一称呼（Stoecker et al.，1989；Stoecker，1991；Verity and Vernet，1992；Modigh，2001），有的则将其称为混合营养纤毛虫（mixotrophic ciliates）。也有研究把含叶绿体的纤毛虫等同于混合营养纤毛虫（Stoecker et al.，1994；Stoecker et al.，1996）。

如何区分混合营养纤毛虫与异养纤毛虫是实际研究中常遇到的难题。严格异养的纤毛虫可摄食浮游植物，因此在浮游植物被彻底消化前，异养纤毛虫体内也能观察到未降解的色素体，难以与混合营养纤毛虫进行区分。纤毛虫通过捕食获取的叶绿体通常排列在细胞周边的表膜下（Stoecker et al.，1988），因此，有研究仅把叶绿体遍布纤毛虫细胞的个体视作混合营养纤毛虫（Modigh，2001）。

一、混合营养纤毛虫的计数方法

（一）荧光显微镜计数

可通过倒置荧光显微镜观察纤毛虫样品，叶绿体在蓝光激发后呈红色荧光，从而区分有叶绿体的纤毛虫和不具叶绿体的严格异养纤毛虫。制备倒置荧光显微镜观察样品的方法有：①直接采集海水使用甲醛进行固定，终浓度2%~3%（V/V）；②若纤毛虫在水体中丰度较低，可将自然海水中的纤毛虫用10μm筛绢通过逆向过滤（reverse filtration）的方法浓缩500倍，甲醛固定（终浓度2%~3%）。将上述方法固定后的样品放置于黑暗环境中4℃保存，5周内分析完毕。取上述固定样品的分样100ml放入沉降杯，用倒置荧光显微镜检查计数。

除倒置荧光显微镜外，也可使用落射荧光显微镜观察。将采集的纤毛虫水样过滤到黑膜上，使用荧光染料DAPI染色，在落射荧光显微镜下分别用紫色和蓝色激发光观察。使用紫色激发光观察可见DAPI染色的虫体呈蓝色，再切换显微镜到蓝色激发光，若虫体内能观察到红色颗粒，即可认为纤毛虫体内含有叶绿体。

（二）光学显微镜计数

可通过光学显微镜观察鲁氏碘液（Lugol's iodine solution）固定样品中纤毛虫的形态，判断部分混合营养纤毛虫种类的丰度。例如，球果螺体虫（Laboea strobila）呈锥形，有4至5圈螺纹（图6-1a），曳尾虫（Tontonia sp.）则个体较大且具有突出的尾部（图6-1b）。这两类混合营养纤毛虫的形态极易辨认，可直接在光学显微镜下进行镜检计数。而其他混合营养纤毛虫（如急游虫属）通常个体较小，不能通过光学显微镜来计数。

图 6-1 光学显微镜下的球果螺体虫（a）和曳尾虫（b）（张武昌 摄）
样品经鲁氏碘液固定

二、混合营养纤毛虫的种类

目前已确认的寡毛类（oligotrich）混合营养纤毛虫种类有 13 种（表 6-2），其中螺体虫属（*Laboea*）仅有球果螺体虫 1 种，曳尾虫属（*Tontonia*）有 2 种，急游虫属（*Strombidium*）有 6 种，欧米虫属（*Omegastrombidium*）、拟曳尾虫属（*Paratontonia*）、伪曳尾虫属（*Pseudotontonia*）及拟盗虫属（*Strombidinopsis*）均仅有 1 种。

中缢虫属（*Mesodinium*）是常见的非寡毛类混合营养纤毛虫，在海洋和咸淡水中广泛存在，且易发生赤潮。在用鲁氏碘液固定的样品中，中缢虫呈现黑色，但周边纤毛明显，中缢的形态很容易辨认。目前已知的中缢虫有 6 个种，其中 *M. chamaeleon*、*M. coatsi*、*M. major*、*M. rubrum* 是混合营养，虫体内可观察到红色和/或绿色的质体，*M. pulex* 和 *M. pupula* 是异养。红色中缢虫 *M. rubrum* 体内的藻细胞可能处于与中缢虫共生的状态（Qiu et al.，2016；Johnson et al.，2017）。

三、混合营养纤毛虫的丰度和生物量比例

混合营养纤毛虫在不同研究海区所占丰度和生物量的比例见表 6-3。不同研究对混合营养纤毛虫的鉴定及计数方式有些差异，如 Stoecker 等（1987）对所有含色素体的纤毛虫进行计数，而 Martin 和 Montagnes（1993）则认为螺体虫属、曳尾虫属、具沟急游虫等是主要的混合营养纤毛虫。混合营养纤毛虫在不同海区的丰度和生物量所占比例多低于 70%，在近岸海区所占比例比大洋海区稍高，在地中海其所占比例可高达 93%（Pitta and Giannakourou，2000）。

表 6-2 已知的寡毛类混合营养纤毛虫

现种名	原种名	中文名	体长 (μm)	最大丰度 (ind./L)	水深 (m)	海区	参考文献
Laboea strobila	—	球果螺体虫	70~80		0	地中海 Villefranche-sur-Mer 湾, 大西洋 Falmouth 河口	Laval-Peuto 和 Rassoulzadegan (1988); Stoecker 等 (1988)
Omegastrombidium elegans	Strombidium elegans	优雅敞米虫	55		0	地中海 Villefranche-sur-Mer 湾	Laval-Peuto 和 Rassoulzadegan (1988)
Paratontonia gracillima	Tontonia gracillima	纤细拟曳尾虫	31~48	10	50	地中海 Villefranche-sur-Mer 湾, 爱琴海	Laval-Peuto 和 Rassoulzadegan (1988); Pitta 和 Giannakourou (2000)
Pseudotontonia simplicidens	Tontonia simplicidens	简单伪曳尾虫	30~40	10	75	爱琴海	Pitta 和 Giannakourou (2000)
Strombidinopsis batos	—	黑荇拟盗虫		40	10	爱琴海	Pitta 和 Giannakourou (2000)
Strombidium capitatum	—	具甲急游虫	52		0	地中海 Villefranche-sur-Mer 湾	Laval-Peuto 和 Rassoulzadegan (1988)
Strombidium conicum	—	锥形急游虫	50~75	490	1	爱琴海	Pitta 和 Giannakourou (2000)
Strombidium dalum	—	火炬急游虫	12~21	170	10	爱琴海	Pitta 和 Giannakourou (2000)
Strombidium stylifer	Strombidium oculatum	楔尾急游虫	50~60			大西洋长岛海峡	McManus 等 (2004)
Strombidium tintinnodes	—	丁丁急游虫	50~60			爱尔兰海	McManus 等 (2004)
Strombidium vestitum	Strombidium delicatissimum	浅裂急游虫	22~26	190	1	爱琴海, 地中海 Villefranche-sur-Mer 湾	Laval-Peuto 和 Rassoulzadegan (1988); Pitta 和 Giannakourou (2000)
Tontonia appendiculariformis	—	附属曳尾虫	80~105		0	地中海 Villefranche-sur-Mer 湾	Laval-Peuto 和 Rassoulzadegan (1988)
Tontonia ovalis	—	卵圆曳尾虫	45~53		0	地中海 Villefranche-sur-Mer 湾	Laval-Peuto 和 Rassoulzadegan (1988)

表 6-3　各海区混合营养纤毛虫的丰度及比例

海区	水深(m)	采样时间	混合营养纤毛虫类型	最大丰度(ind./L)	丰度比例(%)	生物量比例(%)	参考文献
近岸海区							
大西洋伍兹霍尔近岸	0	春季和夏季	含色素纤毛虫		47~51		Stoecker 等（1987）
	0	秋季和冬季			22		Stoecker 等（1987）
大西洋 Nantucket 海湾	0~9	7 月	含色素纤毛虫	8206	48~65		Lynn 等（1991）
加勒比海	5	1 年	螺体虫属	72	0.7~11.9	0.4~29.6	
			曳尾虫属	14	2.7~25.1	1.3~34.8	
北太平洋不列颠哥伦比亚峡湾	2	2 月	具甲急游虫	82	1.5		Martin 和 Montagnes（1993）
			球果螺体虫		0.25		
			纤细机曳尾虫		2.5		
北极海	0	夏季	含色素纤毛虫	497	58~65	59~63*	Putt（1990）
	50			65	14~24		
陆架/陆坡海区							
大西洋乔治滩	0~35	7 月	含色素纤毛虫		21~70		Stoecker 等（1989）
	真光层	早夏			39		
大洋区							
马尾藻海		晚春	含色素纤毛虫		37		Stoecker（1991）
墨西哥湾流区		晚春	含色素纤毛虫		25		Stoecker（1991）
亚北极太平洋	5~30	6 月和 9 月，次年 5 月和 8 月	含色素纤毛虫			30~50	Booth 等（1993）
北大西洋	0~20	春季	含色素寡毛类纤毛虫	3006	16~39		Stoecker 等（1994）
南大洋威德尔海和斯科舍海	0~85	6~8 月	含色素寡毛类纤毛虫	5	10		Gowing 和 Garrison（1992）
赤道太平洋		3~4 月		2	10		Stoecker 等（1996）
		10 月			<10		

续表

海区	水深 (m)	采样时间	混合营养纤毛虫类型	最大丰度 (ind./L)	丰度比例 (%)	生物量比例 (%)	参考文献
地中海						
加泰罗尼亚海	5	6月	螺体虫属和曳尾虫属		19	63	Dolan 和 Marrasé (1995)
	0~20				18	48	
	0~80			135	6	21	
爱琴海南部	0~20	3月	含色素体纂毛类纤毛虫	70~210	17~55	5~89	Pitta 和 Giannakourou (2000)
	20~100			20~140	7~33	0.1~39	
爱琴海北部受黑海水团影响的海区	0~20			40~1280	8~88	30~86	
	20~100			0~10	0~14	0~5	
爱琴海北部不受黑海水团影响的海区	0~20			30~260	28~87	18~93	
	20~100			0~30	0~20	0~15	
地中海 Villefranche-sur-Mer 湾	0	秋冬	含色素体		41*		Laval-Peuto 和 Rassoulzadegan (1988)
地中海西北部近岸	0	全年	含色素体			51*	Bernard 和 Rassoulzadegan (1994)
利古里亚海	0	5月	含色素体纤毛虫		46	39	Pérez 等 (2000)

*表示占无壳纤毛虫的比例

四、混合营养纤毛虫对叶绿素的贡献

将混合营养纤毛虫挑选出来测定其单个个体平均叶绿素含量,再结合丰度可估算混合营养纤毛虫对水体中总浮游生物叶绿素的贡献。在北欧海(Nordic Seas)7～8月,急游虫 *Strombidium* sp. A、*Strombidium* sp. B、球果螺体虫、曳尾虫的叶绿素含量分别为 52pg/cell、30pg/cell、82pg/cell 和 21pg/cell,混合营养纤毛虫的叶绿素含量占总叶绿素含量的比例小于15%,但在叶绿素浓度较低(<0.2μg/L)的站位,急游虫对总浮游生物叶绿素的贡献可达24%(Putt,1990)。在温带近岸海区,球果螺体虫的叶绿素含量约占总叶绿素含量的 2%～3%(McManus and Fuhrman,1986)。

五、混合营养纤毛虫代表——球果螺体虫

球果螺体虫是含色素纤毛虫中个体最大的种,体积约为 $1.1 \times 10^5 \mu m^3$,固定后体积约为 $8.1 \times 10^4 \mu m^3$(Stoecker et al.,1987；Stoecker et al.,1989),因具有螺纹形态特征使其很容易分辨。球果螺体虫主要分布在温带和极区海域。极地海区如加拿大东部北冰洋(Paranjape,1987)、冰岛和格陵兰岛附近海域(Putt,1990),温带海区有大西洋伍兹霍尔近岸(Stoecker et al.,1987)、大西洋乔治滩(Stoecker et al.,1988；Stoecker et al.,1989)、大西洋切萨皮克湾(Dolan,1991)、新西兰周边(James et al.,1996)、挪威峡湾(Verity and Vernet,1992)、地中海沿岸(Dolan and Marrasé,1995；Dolan et al.,1999；Modigh,2001)等海域,在我国近海也有分布。

在地中海的那不勒斯湾(Gulf of Naples),球果螺体虫最大丰度为 1.7×10^3 ind./L,平均丰度占到含色素纤毛虫的4.7%。地中海加泰罗尼亚海(Catalan Sea)球果螺体虫的最大丰度为135ind./L,出现在5m水层(Dolan and Marrasé,1995)。在大西洋切萨皮克湾,球果螺体虫最大丰度为 1.8×10^3 ind./L,出现在2m水层(Pitta and Giannakourou,2000)。在季节变化方面,球果螺体虫在春季丰度最大,其他季节丰度很低(Modigh,2001)。由于个体较大,所以球果螺体虫对含色素纤毛虫生物量的贡献相当可观,可达46%(Stoecker et al.,1987)。

球果螺体虫所处的水层可能与虫体寻找最佳光强进行光合作用有关。Dale 和 Dahl(1987)发现,球果螺体虫在水深5～10m进行昼夜迁移。Stoecker等(1989)发现,球果螺体虫在早上日出前会升到表层,但是在中午表层和次表层(10m)的丰度相当。

球果螺体虫体内的叶绿体有光合作用功能,但是不能再生,在球果螺体虫体内大约停留2周(Stoecker et al.,1988),因此需要不断获取新叶绿体。自然样品

的球果螺体虫体内叶绿素平均含量范围在 49～187pg/cell(McManus and Fuhrman，1986；Stoecker et al.，1987；Putt，1990）。实验室培养的球果螺体虫的叶绿素含量更高，在 100～500pg/cell（Stoecker et al.，1989）。

参 考 文 献

Arenovski A L, Lim E L, Caron D A. 1995. Mixotrophic nanoplankton in oligotrophic surface waters of the Sargasso Sea may employ phagotrophy to obtain major nutrients. Journal of Plankton Research, 17(4): 801-820.

Avrahami Y, Frada M J. 2020. Detection of phagotrophy in the marine phytoplankton group of the Coccolithophores(Calcihaptophycidae, Haptophyta) during nutrient-replete and phosphate-limited growth. Journal of Phycology, 56(4): 1103-1108.

Bernard C, Rassoulzadegan F. 1994. Seasonal variations of mixotrophic ciliates in the northwest Mediterranean Sea. Marine Ecology Progress Series, 108: 295-301.

Bird D F, Kalff J. 1986. Bacterial grazing by planktonic lake algae. Science, 231(4737): 493-495.

Blackbourn D J, Taylor F J R, Blackbourn J. 1973. Foreign organelle retention by ciliates. The Journal of Protozoology, 20(2): 286-288.

Booth B C, Lewin J, Postel J R. 1993. Temporal variation in the structure of autotrophic and heterotrophic communities in the subarctic Pacific. Progress in Oceanography, 32(1): 57-99.

Caron D A. 2000. Symbiosis and mixotrophy among pelagic microorganisms//Kirchman D L. Microbial Ecology of the Oceans. New York: Wiley-Liss: 495-523.

Chan Y F, Tsai A Y, Chiang K P, et al. 2009. Pigmented nanoflagellates grazing on Synechococcus: seasonal variations and effect of flagellate size in the coastal ecosystem of subtropical western Pacific. Microbial Ecology, 58(3): 548-557.

Christaki U, van Wambeke F, Dolan J R. 1999. Nanoflagellates(mixotrophs, heterotrophs and autotrophs) in the oligotrophic eastern Mediterranean: standing stocks, bacterivory and relationships with bacterial production. Marine Ecology Progress Series, 181: 297-307.

Czypionka T, Vargas C A, Silva N, et al. 2011. Importance of mixotrophic nanoplankton in Aysén Fjord(Southern Chile) during austral winter. Continental Shelf Research, 31(3): 216-224.

Dale T, Dahl E. 1987. Mass occurrence of planktonic oligotrichous ciliates in a bay in southern Norway. Journal of Plankton Research, 9(5): 871-879.

Davis P G, Sieburth J M. 1982. Differentiation of phototrophic and heterotrophic nanoplankton populations in marine waters by epifluorescence microscopy. Annales de l'Institut Oceanographique, Paris(Nouvelle Serie), 58(Suppl.): 249-260.

Davis P G, Sieburth J M. 1984. Differentiation and characterization of individual phototrophic and heterotrophic microflagellates by sequential epifluorescence and electron microscopy. Transactions of the American Microscopical Society, 103(3): 221-227.

Dolan J R. 1991. Guilds of ciliate microzooplankton in the Chesapeake Bay. Estuarine, Coastal and Shelf Science, 33(2): 137-152.

Dolan J R, Marrasé C. 1995. Planktonic ciliate distribution relative to a deep chlorophyll maximum: Catalan Sea, N.W. Mediterranean, June 1993. Deep Sea Research Part I: Oceanographic Research Papers, 42(11): 1965-1987.

Dolan J R, Vidussi F, Claustre H. 1999. Planktonic ciliates in the Mediterranean Sea: longitudinal

trends. Deep Sea Research Part I: Oceanographic Research Papers, 46(12): 2025-2039.
Eiler A. 2006. Evidence for the ubiquity of mixotrophic bacteria in the upper ocean: implications and consequences. Applied and Environmental Microbiology, 72(12): 7431-7437.
Estep K W, Davis P G, Keller M D, et al. 1986. How important are oceanic algal nanoflagellates in bacterivory? Limnology and Oceanography, 31(3): 646-650.
Flynn K J, Stoecker D K, Mitra A, et al. 2013. Misuse of the phytoplankton-zooplankton dichotomy: the need to assign organisms as mixotrophs within plankton functional types. Journal of Plankton Research, 35(1): 3-11.
Gowing M M, Garrison D L. 1992. Abundance and feeding ecology of larger protozooplankton in the ice edge zone of the Weddell and Scotia Seas during the austral winter. Deep Sea Research Part A. Oceanographic Research Papers, 39(5): 893-919.
Havskum H, Riemann B. 1996. Ecological importance of bacterivorous, pigmented flagellates (mixotrophs) in the Bay of Aarhus, Denmark. Marine Ecology Progress Series, 137: 251-263.
James M R, Hall J A, Barrett D P. 1996. Grazing by protozoa in marine coastal and oceanic ecosystems off New Zealand. New Zealand Journal of Marine and Freshwater Research, 30(3): 313-324.
Johnson M D, Lasek-Nesselquist E, Moeller H V, et al. 2017. *Mesodinium rubrum*: The symbiosis that wasn't. Proceedings of the National Academy of Sciences of the United States of America, 114(7): E1040-E1042.
Jonsson P R. 1987. Photosynthetic assimilation of inorganic carbon in marine oligotrich ciliates(Ciliophora, Oligotrichina). Marine Microbial Food Webs, 2: 55-67.
Kofoid C A, Swezy O. 1921. The Free-Living Unarmored Dinoflagellata. Memoirs of the University of California. Berkeley: University of California Press: 562.
Laval-Peuto M, Febvre M. 1986. On plastid symbiosis in *Tontonia appendiculariformis*(Ciliophora, Oligotrichina). Biosystems, 19(2): 137-158.
Laval-Peuto M, Rassoulzadegan F. 1988. Autofluorescence of marine planktonic Oligotrichina and other ciliates. Hydrobiologia, 159(1): 99-110.
Lewin J C. 1953. Heterotrophy in diatoms. Microbiology, 9(2): 305-313.
Lynn D H, Roff J C, Hopcroft R R. 1991. Annual abundance and biomass of aloricate ciliates in tropical neritic waters off Kingston, Jamaica. Marine Biology, 110(3): 437-448.
Martin A J, Montagnes D J S. 1993. Winter ciliates in a British Columbian fjord: six new species and an analysis of ciliate putative prey. Journal of Eukaryotic Microbiology, 40(5): 535-549.
McManus G B, Fuhrman J A. 1986. Bacterivory in seawater studied with the use of inert fluorescent particles. Limnology and Oceanography, 31(2): 420-426.
McManus G B, Zhang H, Lin S. 2004. Marine planktonic ciliates that prey on macroalgae and enslave their chloroplasts. Limnology and Oceanography, 49(1): 308-313.
Modigh M. 2001. Seasonal variations of photosynthetic ciliates at a Mediterranean coastal site. Aquatic Microbial Ecology, 23(2): 163-175.
Moorthi S, Caron D A, Gast R J, et al. 2009. Mixotrophy: a widespread and important ecological strategy for planktonic and sea-ice nanoflagellates in the Ross Sea, Antarctica. Aquatic Microbial Ecology, 54(3): 269-277.
Norris R E. 1967. Algal consortisms in marine plankton//Krishnamurthy V. Proceedings of the Seminar on Sea, Salt and Plants held at CSMCRI-Bhavnagar on December 20-23, 1965. Bhavhagar: Central Salt and Marine Chemicals Research Institute(India): 178-189.

Paranjape M A. 1987. The seasonal cycles and vertical distribution of tintinnines in Bedford Basin, Nova Scotia, Canada. Canadian Journal of Zoology, 65(1): 41-48.

Pérez M T, Dolan J R, Vidussi F, et al. 2000. Diel vertical distribution of planktonic ciliates within the surface layer of the NW Mediterranean(May 1995). Deep Sea Research Part I: Oceanographic Research Papers, 47(3): 479-503.

Pitta P, Giannakourou A. 2000. Planktonic ciliates in the oligotrophic Eastern Mediterranean: vertical, spatial distribution and mixotrophy. Marine Ecology Progress Series, 194: 269-282.

Porter K G. 1988. Phagotrophic phytoflagellates in microbial food webs. Hydrobiologia, 159(1): 89-97.

Porter K G, Sherr E B, Sherr B F, et al. 1985. Protozoa in planktonic food webs. The Journal of Protozoology, 32(3): 409-415.

Putt M. 1990. Abundance, chlorophyll content and photosynthetic rates of ciliates in the Nordic Seas during summer. Deep Sea Research Part A. Oceanographic Research Papers, 37(11): 1713-1731.

Qiu D, Huang L, Lin S. 2016. Cryptophyte farming by symbiotic ciliate host detected *in situ*. Proceedings of the National Academy of Sciences of the United States of America, 113(43): 12208-12213.

Sanders R W, Berninger U G, Lim E L, et al. 2000. Heterotrophic and mixotrophic nanoplankton predation on picoplankton in the Sargasso Sea and on Georges Bank. Marine Ecology Progress Series, 192: 103-118.

Sanders R W, Gast R J. 2012. Bacterivory by phototrophic picoplankton and nanoplankton in Arctic waters. FEMS Microbiology Ecology, 82(2): 242-253.

Sanders R W, Porter K G. 1988. Phagotrophic Phytoflagellates//Marshall K C. Advances in Microbial Ecology. Boston: Springer US: 167-192.

Stoecker D K. 1991. Mixotrophy in marine planktonic ciliates: physiological and ecological aspects of plastid-retention by oligotrichs//Reid P C, Turley C M, Burkill P H. Protozoa and Their Role in Marine Processes. Berlin: Springer Berlin Heidelberg: 161-179.

Stoecker D K. 1998. Conceptual models of mixotrophy in planktonic protists and some ecological and evolutionary implications. European Journal of Protistology, 34(3): 281-290.

Stoecker D K, Gustafson D E, Verity P G. 1996. Micro- and mesoprotozooplankton at 140 °W in the equatorial Pacific: heterotrophs and mixotrophs. Aquatic Microbial Ecology, 10(3): 273-282.

Stoecker D K, Michaels A E, Davis L H. 1987. Large proportion of marine planktonic ciliates found to contain functional chloroplasts. Nature, 326(6115): 790-792.

Stoecker D K, Sieracki M E, Verity P G, et al. 1994. Nanoplankton and protozoan microzooplankton during the JGOFS North Atlantic Bloom Experiment: 1989 and 1990. Journal of the Marine Biological Association of the United Kingdom, 74(2): 427-443.

Stoecker D K, Silver M W, Michaels A E, et al. 1988. Obligate mixotrophy in *Laboea strobila*, a ciliate which retains chloroplasts. Marine Biology, 99(3): 415-423.

Stoecker D K, Taniguchi A, Michaels A E. 1989. Abundance of autotrophic, mixotrophic and heterotrophic planktonic ciliates in shelf and slope waters. Marine Ecology Progress Series, 50: 241-254.

Unrein F, Massana R, Alonso-Sáez L, et al. 2007. Significant year-round effect of small mixotrophic flagellates on bacterioplankton in an oligotrophic coastal system. Limnology and Oceanography, 52(1): 456-469.

Vargas C A, Contreras P Y, Iriarte J L. 2012. Relative importance of phototrophic, heterotrophic, and mixotrophic nanoflagellates in the microbial food web of a river-influenced coastal upwelling area. Aquatic Microbial Ecology, 65(3): 233-248.

Verity P G, Vernet M. 1992. Microzooplankton grazing, pigments, and composition of plankton communities during late spring in two Norwegian fjords. Sarsia, 77(3-4): 263-274.

Zubkov M V, Tarran G A, Fuchs B M. 2004. Depth related amino acid uptake by *Prochlorococcus* cyanobacteria in the Southern Atlantic tropical gyre. FEMS Microbiology Ecology, 50(3): 153-161.

第七章 微型浮游动物和浮游病毒对细菌的下行控制

自然海水中细菌的表观生长率是细菌生长和被摄食（或被病毒裂解）这两种因素共同作用的结果。为研究微型浮游动物和浮游病毒对细菌的下行控制，可对细菌与摄食者进行共培养实验，通过人为调整控制饵料与摄食者的接触概率，改变细菌的表观生长率以估算细菌被摄食（或裂解）的情况，也可通过统计细菌被病毒侵染的比例间接估算死亡率。本章将对这些方法和获得的结果进行介绍。

第一节 微型浮游动物对细菌的下行控制

一、分粒级培养法

顾名思义，分粒级培养法（size-fractionation incubation）就是将海水中的大颗粒去除后进行培养的方法。首先，将海水用一定孔径的滤膜过滤，以去除比细菌粒级大的摄食者，其次，对过滤后的海水进行培养，通过培养前后细菌丰度的变化可估计细菌的生长率。若同时培养过滤海水和原始海水，两个培养体系之间细菌生长率的差异即为细菌的被摄食率。

早在1955年即有研究使用"海水培养法"（seawater culture technique），尝试通过监测自然海水培养前后细菌的数量变化来估计水体中细菌的生长情况（Ivanov, 1955），但这一研究未考虑微型浮游动物对细菌的摄食，而且当时也没有准确计数细菌的方法。到20世纪70年代初，学者开始意识到海水中存在细菌的摄食者，因此在估计细菌生长的时候，先采取过滤较大颗粒的方法去除细菌的摄食者（Gak et al., 1972），这是真正意义上使用海水分粒级培养法估计得到的细菌生长率，但此时也无准确计数海洋细菌的方法。随着准确计数海水细菌丰度方法的成熟（Hobbie et al., 1977），Fuhrman 和 Azam（1980）率先用 3μm 滤膜分粒级培养法研究了海洋浮游细菌的生长，随后对不同海区浮游细菌生长率的估算研究逐渐展开（表7-1）。

表7-1 海水分粒级培养法获得的自然海区细菌生长率

海区	滤膜孔径（μm）	培养时间（h）	生长率（/d）	参考文献
北大西洋佐治亚近岸	3	40	0.792	Newell 和 Christian（1981）
北大西洋佐治亚近岸	3	5、10	0.72	Newell 和 Fallon（1982）
北大西洋 Skidaway 盐沼	3	19	0.00~2.88	Christian 等（1982）

续表

海区	滤膜孔径 (μm)	培养时间 (h)	生长率 (/d)	参考文献
亚热带东南太平洋（47°S）	3	>40	0.00~4.32	Hanson 等（1983）
北太平洋加利福尼亚近岸	3	—	0.48~2.64	Fuhrman 和 Azam（1980）
北太平洋北加利福尼亚陆架区	3	24	−1.2~0.72	Ferguson 等（1984）

注：—表示无数据，下同

不同孔径滤膜对细菌摄食者的滤过效率是分粒级培养法面临的主要问题。有研究比较了孔径为 1μm、3μm、8μm、12μm、37μm、120μm 和 270μm 的滤膜的滤过效果，建议选用孔径 1μm 的滤膜过滤海水进行分粒级培养（Wright and Coffin，1984）。但也有研究发现，即便使用孔径为 0.6μm 的滤膜也会有一些鞭毛虫穿过（Fuhrman and McManus，1984；Cynar et al.，1985）。这意味着滤膜无法有效地将捕食者与细菌分离，该方法获得的细菌生长率并不准确，因此，之后使用滤膜过滤和分粒级培养法测定海洋细菌生长率的研究就很少了。

尽管海水分粒级培养法不适用于估计细菌的生长率，但在不同粒级微型浮游动物对细菌的下行控制研究中带来了很多新认识：首先，分粒级培养实验的结果表明小粒级的微型浮游动物是细菌的主要摄食者（Gak et al.，1972；Hagström et al.，1988）；其次，多级过滤培养的结果证明微型浮游动物存在多个营养级（Rassoulzadegan and Sheldon，1986；Wikner and Hagström，1988；Chen et al.，2009）；最后，还有研究揭示微型浮游动物倾向选择摄食正在分裂的较大的细菌（Sherr et al.，1992）。

在自然海区，细菌和鞭毛虫的丰度均存在短时周期性波动，变化的周期为 10~20d，且二者的周期存在一定的相位差，即摄食者鞭毛虫的生长落后于饵料生物细菌。海水分粒级培养实验可以在实验室内再现细菌和鞭毛虫的这种周期波动。Andersen 和 Fenchel（1985）使用孔径 8μm 的滤膜过滤自然海水，在大于 350h 培养的过程中重现了细菌和鞭毛虫的丰度周期性波动，且细菌和鞭毛虫的摄食关系符合洛特卡-沃尔泰拉捕食模型（Lotka-Volterra predator-prey model）。

二、生物抑制剂培养法

在海水中添加选择性生物抑制剂进行培养（biological inhibitors incubation），不会改变海水中原有细菌和微型浮游动物的组成，但可以改变它们的功能。通过选择性抑制细菌的生长或微型浮游动物的摄食，可改变细菌的表观生长率，从而估计细菌的生长率和被摄食率。添加真核生物抑制剂能够限制真核摄食者的活动，使微型浮游动物的摄食率为零，因此可估计细菌的生长率。添加原核生物抑制剂能够限制细菌生长,使细菌的生长率为零,故而可估计微型浮游动物的摄食率。

理想的选择性生物抑制剂应对目标生物有较好的抑制作用,同时自身不能作为底物直接促进细菌生长,也不能杀死或促使浮游植物、微型浮游动物释放体内有机物质而间接促进细菌的生长,也不应限制微型浮游动物排泄从而间接抑制细菌的生长。

多项研究对真核生物抑制剂和原核生物抑制剂的效果进行了探索,研究者在实验室内用纯培养或从海水中分离获得的原生动物,通过添加不同浓度的抑制剂和不添加抑制剂作为对照的方法进行研究(Newell et al.,1983;Sherr et al.,1986;Taylor and Pace,1987)。已得到检验的真核生物抑制剂有环己酰亚胺、双硫胺甲酰、秋水仙酰胺、秋水仙碱、中性红和灰黄霉素等,它们对目标海洋生物的抑制效果不同,其中,双硫胺甲酰和中性红对海洋鞭毛虫和纤毛虫都有很好的抑制作用,而其他抑制剂不能同时对两类生物有很好的抑制(表 7-2)。得到检验的原核生物抑制剂有吡硫头孢菌素、氯霉素、青霉素和万古霉素,在受试浓度下,吡硫头孢菌素对海洋细菌生长没有影响,而氯霉素、青霉素和万古霉素在较高浓度下可以抑制海洋细菌生长(表 7-2)。目前尚无检测真核生物抑制剂能否作为细菌生长的底物的报道。

表 7-2 常见生物抑制剂对海洋细菌和微型浮游动物的作用

抑制剂类型	中文名称	英文名称	测试浓度(mg/L)	作用机理	对海洋细菌	对海洋鞭毛虫	对海洋纤毛虫
真核生物抑制剂	放线酮(环己酰亚胺)	cycloheximide	100~200	抑制蛋白质合成	无作用	完全抑制	部分抑制
	福美双(双硫胺甲酰)	thiram	1~100	抑制蛋白质合成	部分抑制	完全抑制	完全抑制
	秋水仙酰胺	demecolcine	0.1~1.0	抑制微管微丝聚合	无作用	无作用	
	秋水仙碱	colchicine	20~400	抑制微管微丝聚合	无作用	无作用	部分抑制
	中性红	neutral red	0.1~100	刺激自噬作用并抑制内吞作用		完全抑制	完全抑制
	灰黄霉素	griseofulvin	10~350	抑制微管微丝聚合		部分抑制	完全抑制
原核生物抑制剂	头孢匹林	cephapirin	1~100	抑制细胞壁合成	无作用	无作用	
	氯霉素	chloramphenicol	1~100	抑制蛋白质合成	抑制率为 90%	无作用	
	青霉素	penicillin	1~100	抑制细胞壁合成	高浓度时有影响	部分抑制	无作用
	万古霉素	vancomycin	100~200	抑制细胞壁合成	高浓度时有影响	无作用	

与不加入抑制剂的单种浮游植物培养对照相比,加入真核生物抑制剂后的浮游植物生长率有不同程度的降低,如硅藻对添加双硫胺甲酰、环己酰亚胺和中性红较为敏感(Taylor and Pace,1987)。通常来说,浮游植物死亡会导致水体中溶解有机物增加从而提高细菌的生长率,但添加生物抑制剂后,微型浮游动物却会

减少营养盐释放，使细菌的生长率偏低（Sherr et al.，1986）。

由于尚未发现哪种选择性抑制剂能抑制全部目标生物同时不具有副作用，因此生物抑制剂培养法的应用也不广泛。仅有的几个研究中，红海细菌的生长率为 0.014~0.097/h，微型浮游动物对细菌的摄食率为 0.010~0.108/h（Weisse，1989）；在北大西洋 Vineyard 海湾，细菌的生长率为 0~0.47/d（Caron et al.，1991）；在智利北部上升流区，细菌的生长率和被摄食率的范围都是 0~0.07/h（Cuevas and Morales，2006）。

三、稀释培养法

两种生物间摄食关系的研究通常涉及如下几个参数。

1. 内禀生长率

内禀生长率（intrinsic growth rate，μ，/d）指浮游生物种群在理想条件下，即最适环境条件且不存在摄食者时，能够达到的生长速率。通常假定饵料生物 A 以指数形式增长，培养开始时（t_0 时刻）饵料的丰度为 A_0，培养一段时间（t）后，饵料丰度变为 A_t，$A_t=A_0 \times e^{\mu t}$，则内禀生长率 $\mu=(1/t) \times \ln(A_t/A_0)$。

2. 摄食率

摄食率（grazing rate，g，/d）指在含有摄食者和饵料生物的培养体系中，摄食者造成的饵料表观生长率的降低。摄食率是一个率的概念，不是摄食饵料的绝对数量，微型浮游动物摄食细菌的绝对数量需要用细菌的丰度进行换算。

3. 表观生长率

在含有摄食者和饵料生物的培养体系中，饵料生物的丰度仍以指数形式增长，但摄食者将造成饵料生物丰度比没有摄食者的对照组降低，因此经培养时间 t 后的饵料丰度为 $A_t=A_0 \times e^{(\mu-g)t}$，此时，饵料生物表现出的生长率即为表观生长率（apparent growth rate，k，/d），$k=\mu-g=(1/t) \times \ln(A_t/A_0)$。

研究细菌生长率和微型浮游动物摄食率的稀释培养法（dilution incubation）有 3 个理论假设：第一，细菌的生长依赖水体中的溶解有机物，生长率不会受细菌及摄食者丰度变化的影响；第二，细菌的死亡率与摄食者的丰度成正比；第三，细菌初始丰度（B_0）和 t 时刻丰度（B_t）符合指数公式：

$$B_t=B_0 \times e^{(\mu-g)t} \tag{7-1}$$

式中，μ 为细菌的内禀生长率（以下简称生长率），g 为微型浮游动物摄食率。因此，在自然海水中，细菌的表观生长率（k_n）可表示为

$$k_n=(1/t) \times \ln(B_t/B_0)=\mu-g \tag{7-2}$$

而在稀释度为 d 的海水中,细菌的表观生长率(k_d)为

$$k_d = \mu - dg \tag{7-3}$$

测量各稀释度培养后单位时间内细菌丰度的变化,可获得细菌的表观生长率。通过一系列的细菌表观生长率 k_d 值与对应稀释度 d 值建立一元线性回归模型,所得的截距与斜率的绝对值分别对应细菌生长率(μ)和微型浮游动物摄食率(g)。

为测定细菌的生长率,首先需要去除摄食者的影响。最初的实验策略是先用孔径 0.2μm 的滤膜过滤,得到的过滤海水(filtered seawater,FSW)被认为是去除了细菌和微型浮游动物。使用 FSW 对含有细菌和微型浮游动物的自然海水(whole seawater,WSW)进行稀释,可达到减少自然海水中摄食者丰度的目的。对稀释后的混合海水进行一定时间的培养,根据单位时间内细菌丰度的变化可获得细菌的生长率。有研究为了最大可能地去除摄食者,会使用 0.6~1.0μm 滤膜对 WSW 进行预过滤,再加入 FSW 进行稀释培养,但这一预处理仍难以保证完全去除摄食者。这种单一稀释度的海水稀释培养法重复性较差,无法获得准确的海洋细菌生长率,因此使用较少(表 7-3)。

表 7-3 单一稀释度海水稀释培养法获得的自然海区细菌生长率

海区	稀释度	培养时间(h)	生长率(/d)	参考文献
波罗的海基尔湾	—	—	0.44(0.24)[*]	Meyer-Reil(1977)
北大西洋佐治亚近岸	1∶1	40	0.528	Newell 和 Christian(1981)
北大西洋纽约湾	1∶10	40	1.44~2.64	Kirchman 等(1982)
北大西洋 Great Sippewisset 沼泽	1∶10	20	3.84	
北大西洋北卡罗来纳陆架区	1∶10	34	1.2~2.16	Ferguson 等(1984)
北大西洋埃塞克斯河口至近岸区	1∶10	24	2.16	Wright 和 Coffin(1984)[**]
北大西洋 Vineyard 海湾	1∶20	48、72	0.24~5.76	
北大西洋纽芬兰大浅滩 1500m 深层水	1∶20	48、72	−0.72~9.6	Höfle(1984)[***]
北大西洋墨西哥湾流暖核环	1∶10~1∶2	6~12	−1.92~5.14(平均 1.94)	Ducklow 和 Hill(1985)
北太平洋斯克利普斯近岸	1∶50~1∶5	80	1.92	Ammerman 等(1984)[****]

[*]表示该研究仅有 40%的细菌生长,高值表示活性较高的个体的生长率,低值表示总体的生长率;[**]表示该研究使用 1μm 的滤膜制备过滤海水;[***]表示该研究使用孔径 0.05μm 的滤膜制备过滤海水,且自然海水在稀释前经孔径 0.8μm 滤膜过滤去除摄食者;[****]表示该研究使用的自然海水在稀释前经孔径 1.0μm 或 0.6μm 滤膜过滤去除摄食者

多稀释度海水稀释培养法同样使用 0.2μm 滤膜过滤制备 FSW,随后将 FSW 按多个梯度比例与 WSW 混合,WSW 在混合海水培养体系中的体积比例即为稀释度(d),如采用 100%FSW、75%FSW、50%FSW、25%FSW 和 0%FSW 这 5 个

梯度比例，对应的 WSW 稀释度分别为 0、0.25、0.5、0.75 和 1。尽管多稀释度培养法也无法完全去除摄食者，但对稀释后的混合海水分别进行培养，通过对多个稀释度分别得到的数据进行回归分析，可更为准确地估算细菌生长率（k）和微型浮游动物摄食率（g），因此应用更为广泛。

从极地到赤道海区，使用多稀释度海水稀释培养实验获得的细菌生长率和被摄食率范围分别在 0.04~3.12/d 与 0~5.52/d（表 7-4）。总的来说，在细菌生长率低的情况下，微型浮游动物的摄食与细菌的生长基本是平衡的，但在细菌生长率高的时候，微型浮游动物的摄食速率高于细菌的生长率（图 7-1），此时对细菌丰度的下行控制可能还源于其他因素，如病毒对细菌的杀死作用。

表 7-4 多稀释度海水稀释培养法获得的自然海区细菌生长率和被摄食率

海区	培养时间（h）	生长率（/d）	被摄食率（/d）	参考文献
黄海南部	24	0.54~0.6	0.35~0.44	Zhao 等（2020）
东北太平洋夏威夷近岸	24	1.2~1.9	0.5~1.1	Landry 等（1984）
东南太平洋塔卡波托环礁	24	约 0.9	平均 0.25	Sakka 等（2000）
墨西哥湾密西西比河河口	24	0.8	0.5	Jochem（2003）
红海北部	36	0.61~0.72	0.73~0.89	Sommer 等（2002）
红海亚喀巴湾		0.86~1.3	0.75~1.06	
地中海西北部	24	表层：0.88±0.43（其中 HNA 细菌 1.18±0.60，LNA 细菌 0.47±0.28）；DCM：0.71±0.23（其中 HNA 细菌 0.36±0.23，LNA 细菌 0.90±0.46）	表层：0.75±0.23（其中 HNA 细菌 1.02±0.31，LNA 细菌 0.37±0.19）；DCM：0.58±0.29（其中 HNA 细菌 0.26±0.17，LNA 细菌 0.77±0.50）	Scharek 和 Latasa（2007）
凯尔特海和北海	>20~30	0.48~3.12	0.48~5.52	Geider（1989）
亚北极东北太平洋	48	0.16~0.57	0.32~0.43	Rivkin 等（1999）
北冰洋雷索卢特湾	24、48	3~5 月：<0.2；6~9 月：0.6~1.0	0.15~0.88	Anderson 和 Rivkin（2001）
南大洋麦克默多海峡	24、48	9~11 月：0.4~0.65；11 月至次年 1 月：0.15~0.3	0.05~0.73	Anderson 和 Rivkin（2001）
南大洋杰拉许海峡	48	0.5~0.8	0	Bird 和 Karl（1999）
南大洋东部（30°~80°E）	24	0.4~2.6	0.3~2.3	Pearce 等（2010）
南大洋亚南极海区	24	0.14~0.87	0.2~1.03	Pearce 等（2011）
南大洋南极半岛西部海区	72	0.04~0.82（其中 HNA 细菌 0.17~0.95，LNA 细菌 0.03~0.17）	0.08~0.38（其中 HNA 细菌 0.08~0.43，LNA 细菌 0.06~0.20）	Garzio 等（2013）

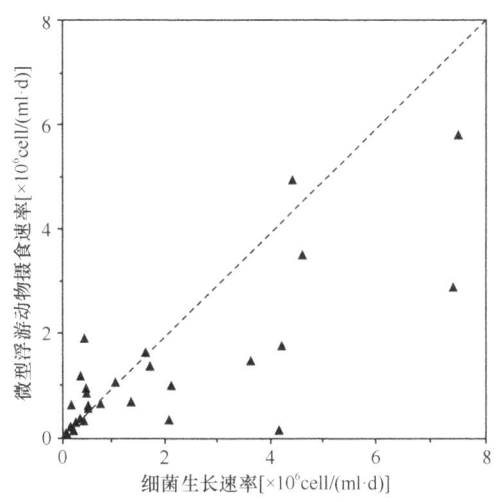

图 7-1　自然水体中细菌生长速率与微型浮游动物摄食速率的关系[改绘自 Strom（2000）]

微型浮游动物对细菌的摄食压力与水华有一定关系。在北冰洋和南大洋海区，当有水华发生时，微型浮游动物摄食消耗了几乎全部的细菌生产，而在非水华期微型浮游动物对细菌的摄食率可以忽略不计（Anderson and Rivkin，2001）。

使用流式细胞术计数时，可根据细胞内核酸含量的高低，将海洋细菌分为高核酸含量（HNA）细菌和低核酸含量（LNA）细菌（详见本书第三章）。稀释培养法与流式细胞术相结合，可以分别估算 HNA 细菌与 LNA 细菌的生长率及被摄食率。研究发现，不同水层 HNA 细菌和 LNA 细菌的生长率与被摄食率有差异，如在地中海西北部和南极半岛西部海区的表层水体，HNA 细菌的生长率和被摄食率都高于 LNA 细菌；而在深层叶绿素最大层（DCM）的趋势恰好相反，LNA 细菌的生长率和被摄食率都高于 HNA 细菌（Scharek and Latasa，2007；Garzio et al.，2013）。

稀释培养法能同时测定微型浮游动物对细菌、蓝细菌和浮游植物的摄食，因此可用于比较微型浮游动物对不同饵料类群的摄食选择性。在太平洋塔卡波托环礁，微型浮游动物对蓝细菌的摄食率高于细菌（Sakka et al.，2000）；在夏威夷近岸海区，微型浮游动物对细菌的摄食率高于蓝细菌（Landry et al.，1984）；在密西西比河羽状流区（Mississippi River plume），微型浮游动物对微微型浮游植物的摄食率高于细菌（Jochem，2003）；在寡营养的红海北部海区和亚喀巴湾，微型浮游动物对细菌和超微型浮游植物（ultraphytoplankton）的摄食率都很高，但对蓝细菌的摄食率极低（Sommer et al.，2002）；在黄海南部的寡营养海区，微型浮游动物对蓝细菌的摄食率远高于细菌和微微型浮游植物（Zhao et al.，2020）。

第二节 浮游病毒对细菌的下行控制

海洋浮游病毒主要通过侵染和裂解宿主实现对细菌的下行控制。由于从病毒感染到细菌裂解的时间长短不一，且难以直接观察病毒对细菌的杀死过程，因此研究由病毒介导的细菌死亡率（virus-induced bacterial mortality，VBM）主要采用间接方法来估算。

一、根据细菌被病毒感染的比例估算 VBM

浮游病毒是海洋细菌死亡的重要贡献者。在侵染后期海洋细菌即将被病毒裂解杀死前，病毒在细菌内大量合成复制。使用电子显微镜观察细菌的完整细胞或细胞切片，若细菌体内含有成熟的病毒衣壳，则可判定细菌被感染，这些被感染的细菌最终将被病毒裂解导致死亡。对自然水样中的细菌进行电镜观察和统计，可获得细菌群落的可见细胞感染率（frequency of visibly infected cell，FVIC）。在不同的海洋环境中，可见 0.2%~4.3%的浮游细菌被病毒感染（Wommack and Colwell，2000）。

完整的病毒衣壳只有在侵染后期才大量合成并被观察到，因此能观察到的被感染细菌只占实际被感染细菌总数的一小部分。需考虑从最初的侵染到病毒衣壳在被感染细胞内首次出现之间的潜伏期长度，通过特定转换系数将 FVIC 值换算为浮游细菌群落的总感染率（frequency of infected cell，FIC）。最初的研究采用转换系数 10（Proctor and Fuhrman，1990），然而单一转换系数隐含着所有病毒均具有相同的侵染发育模式的假设，这一假设用于自然海水样品显然是不切实际的。后续的研究多建议使用一定范围的转换系数，如 3.7~7.14（Proctor et al.，1993），或 4.34~10.78（平均值 7.11）（Weinbauer et al.，2002）。若采用转换系数 7，在上述不同的海洋环境中约有 1.4%~30%的细菌被病毒感染。考虑到 FVIC 值的范围和转换系数的不准确性，病毒对某一特定海区 FIC 的估算值范围可超过一个数量级。

FIC 数据代表的是病毒对浮游细菌群落的总体感染水平，并不能直接等同为 VBM，还需考虑病毒感染的潜伏期长度与宿主世代时间（generation time，即宿主繁殖一代所需时间）之间的关系。假设潜伏期与宿主世代时间长度相同，那么在一个稳定的浮游细菌群落中，每一个细菌分裂为二，其中一个子细胞被病毒感染死亡，另一个子细胞存活并继续分裂形成下一代。这意味着在稳态系统中，若病毒杀死了 50%的细菌个体，将造成 100%的群落死亡率，这就是"系数二规则"（factor-of-two rule）（Binder，1999），因此估算 VBM 只需将 FIC 值加倍即可。

根据细菌被病毒感染的比例，研究发现在海洋环境中病毒裂解是导致细菌死亡的重要因素，贡献了 2%~60% 的细菌死亡率（Wommack and Colwell，2000）。环境因素对病毒裂解细菌有一定影响，如在亚得里亚海中营养站位，VBM 为 3.5%~24%，在高营养站位，VBM 则升高至 7.0%~64.3%（Weinbauer et al.，1993）。在波罗的海，有氧的表层水体 VBM 为 12.6%~17.9%，中层低氧水体为 11.6%~19.4%，缺氧深层水体为 27.4%~50.6%，个别样品甚至可高达 71%。在地中海的缺氧海盆，VBM 超过 90%，甚至接近 100%，这意味着此处细菌的死亡几乎全部是病毒侵染裂解导致的（Corinaldesi et al.，2007）。

"系数二规则"的内在隐含假设是系统中不存在微型浮游动物摄食，所有被感染的细菌都被病毒裂解杀死，这一假设显然不符合自然海水样品的情形。Binder (1999) 使用数学模型测试了"系数二规则"的假设，该模型考虑了感染细菌的被摄食，以及潜伏期与宿主世代时间之间的关系，提出 VBM 可由 FVIC 值估算：

$$\text{VBM} \cong \text{FVIC} / \left[\gamma \ln(2)(1 - \varepsilon - \text{FVIC}) \right] \tag{7-4}$$

式中，γ 为潜伏期与宿主世代时间长度之比，ε 为潜伏期中无法观察到病毒颗粒的时长与总潜伏期时长之比。若假设被感染和未被感染的细菌的被捕食率相等、生长率相等，潜伏期与宿主世代时间相等（$\gamma=1$），VBM 也可由 FIC 值估算：

$$\text{VBM} \cong \left(\text{FIC} + 0.6\text{FIC}^2 \right) / (1 - 1.2\text{FIC}) \tag{7-5}$$

根据已发表的 FVIC/FIC 数据重新计算，Binder（1999）发现，在各海洋环境中，病毒介导了 6%~45% 的细菌死亡，而"系数二规则"系统高估了 6%~41% 的 VBM，平均高估了约 23%。

二、根据病毒生产速率和裂解量估算 VBM

病毒的生产速率由细菌的裂解死亡速度（单位时间内裂解死亡细菌的数量）与裂解量（burst size，每个细菌裂解释放出的病毒数量）决定，可通过病毒的生产速率与裂解量估算 VBM。

（一）病毒生产速率

病毒生产速率可由直接法和间接法检测，其中直接法包括同位素标记示踪法和荧光标记病毒（fluorescently labeled virus，FLV）示踪法，间接法则是由病毒失活速率估算生产速率。

同位素标记示踪法是在海水样品中添加氚标记胸腺嘧啶（^{3}H-thymidine，^{3}H-TdR），海洋异养细菌可以吸收游离的 ^{3}H-TdR，但自养生物不能或只能极少吸收利用游离的 ^{3}H-TdR。病毒侵染后在细菌体内利用宿主细胞的遗传机制进行自我复制和繁殖，将细菌吸收的 ^{3}H-TdR 整合到病毒基因组中。当细菌被裂解时，^{3}H-TdR

标记病毒被释放到培养海水中。测量培养海水的放射性活度值，再结合游离病毒颗粒 ^{3}H-TdR 含量转换系数，可估计病毒生产速率（Steward et al.，1992a；Steward et al.，1992b）。

FLV 示踪法是从环境样本中浓缩收集病毒，用荧光染色剂（如 SYBR Green I 等）进行荧光标记。洗去多余的染色剂后，将标记病毒以示踪剂水平，即环境病毒浓度的 10%添加回自然水样。经一定的培养时间，取样对标记病毒和总病毒分别进行计数。标记病毒和未标记病毒的失活（viral decay）速率相同，但新产生的病毒都是未标记的。随培养时间增长，标记病毒的比例会逐渐减少，从而可估算出病毒的生产速率和清除率（Noble and Fuhrman，2000）。

间接法估算病毒生产速率基于如下假设：通常自然海水中的病毒丰度在短时间内保持稳定，这意味着失活导致的病毒死亡速率和细菌裂解带来的新病毒生产速率应大抵平衡，测定病毒的失活速度，即可间接获得病毒的生产率。为测定病毒失活速率，可使用过滤法去除自然海水中所有的宿主，仅保留病毒，也可以对自然海水加入氰化物（cyanide）以抑制细菌呼吸，从而阻止新病毒的复制，经上述处理的海水中不会产生新的病毒，而已有的病毒会逐渐失活。进行定时采样监测病毒失活的速率，即可获知病毒的生产速率（Heldal and Bratbak，1991）。

（二）裂解量

病毒裂解量的测定主要通过使用电子显微镜观察，统计侵染后期单个细菌中存在的病毒颗粒数量。这种观察方法也有一定的不确定性，因为病毒在细菌体内的分布不是均匀的，可能在细胞某一区域集中重叠排列，不管是观察完整细胞还是观察细胞切片，都难以保证统计到细胞内部所有的病毒颗粒。此外，若观察统计的是某样品中所有被感染细菌，获得的裂解量应为最小估计值，因为细胞内的病毒数量仍可能增加；而若仅选取完全充满病毒的细菌来观察统计裂解量，获得的则是最大估计值，因为有些细菌可能在完全充满病毒之前就被裂解。考虑到裂解量值对估计 VBM 的巨大影响，许多研究会选择选取一定范围的裂解量数值。

链霉素（streptomycin）可破坏细菌细胞的完整性从而向水体释放病毒，使病毒的计数更容易（Heldal and Bratbak，1991）。这种裂解方式被称为"外源裂解"（lysis from without），与病毒介导的"内源裂解"（lysis from within）相对应。用这种方法获得的裂解量也应为最小估计值，因为并非所有的感染细菌都会被链霉素裂解，而且链霉素可能会提前停止病毒的生成。

根据已有报道，海洋浮游细菌的病毒裂解量在 20~300，平均值约 24（Wommack and Colwell，2000），也有研究使用的裂解量为 50（Heldal and Bratbak，1991；Suttle and Chen，1992；Steward et al.，1996）。实验室培养细菌的病毒裂解量高于自然样品，平均值约 185，范围为 6~600（Børsheim，1993）。宿主的大小

通常与裂解量呈显著正相关，如在亚得里亚海，个体较大的杆状细菌的病毒裂解量平均为 51，（范围 12～124，n=53），而个体较小的球形细菌的病毒裂解量平均仅 28（范围 6～61，n=53）（Weinbauer and Peduzzi，1994）。此外，营养条件、宿主生产力也是影响裂解量的重要因素，富营养水体中浮游细菌的病毒裂解量明显高于中营养水体（Weinbauer et al.，1993），随宿主生产力提高病毒裂解量也随之升高（Middelboe，2000）。

由于裂解量估算存在一定不确定性，对于同一样品由其推算的 VBM 范围也较大，如在亚得里亚海表层 VBM 范围为 39.6%～212.2%/d，22m 层 VBM 范围为 19.9%～157.2%/d（Weinbauer et al.，1995）。总的来说，根据病毒失活速率和裂解量估算，在各海洋环境中每天病毒介导了 9%～100%的细菌死亡（Wommack and Colwell，2000）。

三、改变群落法估算 VBM

病毒对细菌的感染和致死作用与二者的接触率有关，因此人工调整病毒与细菌的相对浓度，可改变细菌的感染率和致死率。根据病毒浓度和细菌致死率变化的关系即可估算 VBM。

（一）增加病毒浓度

将自然海水用孔径为 0.22μm 的滤膜过滤，再将滤液用 30kD 的切向流系统过滤，获得浓缩病毒。使用 0.6μm 滤膜过滤自然海水去除摄食者，随后加入浓缩病毒进行培养（实验组），同时也培养 0.6μm 滤膜过滤的自然海水（对照组）。实验组提高了病毒与宿主的接触率，意味着细菌被感染和裂解的概率升高，使用这一方法，Noble 等（1999）发现，对照组的 FIC 为 5%～10%，实验组的 FIC 提高至 10%～35%。比较对照组和实验组细菌丰度的相对变化，其差值即视为被病毒裂解杀死的细菌，可由此估算 VBM（Noble et al.，1999）。但细菌被病毒裂解后会向培养体系释放营养物质，对未感染细菌的生长有促进作用，这部分生长可能相当可观，会对 VBM 的估算带来影响，因此该方法的应用并不广泛。

（二）降低病毒浓度——稀释培养法

稀释培养法也可以用来评估病毒对细菌的杀死作用。与细菌的多稀释度培养类似，病毒稀释培养法是使用不含病毒的海水对自然海水进行一系列梯度的稀释，病毒减少降低了细菌被感染和裂解的概率，对多个稀释度培养分别得到的细菌丰度数据进行回归分析，可获得细菌的死亡率。通过培养病毒稀释海水和原海水，获得的是病毒与微型浮游动物介导的细菌死亡，同时进行常规的细菌稀释培养实验，获得微型浮游动物介导的细菌死亡，二者之差即为 VBM。

通过应用稀释培养法，有研究发现，在西北太平洋北海道近岸海区 VBM 为 0.53~0.98/d，远高于微型浮游动物对细菌的摄食率（0.05~0.13/d）（Taira et al., 2009）。在东海近岸表层，夏季鞭毛虫摄食是导致细菌死亡的主要原因，而秋冬季病毒是介导细菌死亡的主要因素（Tsai et al., 2013），在 135m 层病毒是介导细菌死亡的主要因素（VBM 为 0.013~0.016/h），高于鞭毛虫对细菌的摄食率（0~0.004/h）（Tsai et al., 2016）。在热带西太平洋深海（2000m）病毒裂解是细菌死亡的主要原因，病毒对细菌丰度和群落结构有显著的下行控制（Zhang et al., 2020）。

对不同方法估算的 VBM 进行比较并评估其利弊的研究较少。Fuhrman 和 Noble（1995）使用电镜观察细菌切片法和同位素标记示踪法分别估算了病毒生产，Steward 等（1996）使用全细胞透射电镜观察法和同位素标记示踪法分别估算了病毒生产，均发现不同方法得出接近的 VBM 值。Fischer 和 Velimirov（2002）用病毒失活速率估算的 VBM 平均为 56%，而用全细胞透射电镜观察法估算的 VBM 平均为 63%。另外，有报道提出，与电镜观察细菌切片法（Weinbauer and Höfle, 1998）、病毒失活速率估算法（Guixa-Boixereu et al., 1999）、同位素标记示踪法（Fuhrman and Noble, 1995）或改变群落法（Middelboe et al., 2002）相比，全细胞透射电镜观察法低估了 VBM。

参 考 文 献

Ammerman J, Fuhrman J, Hagström Å, et al. 1984. Bacterioplankton growth in seawater: I. Growth kinetics and cellular characteristics in seawater cultures. Marine Ecology Progress Series, 18(1-2): 31-39.

Andersen P, Fenchel T. 1985. Bacterivory by microheterotrophic flagellates in seawater samples. Limnology and Oceanography, 30(1): 198-202.

Anderson M R, Rivkin R B. 2001. Seasonal patterns in grazing mortality of bacterioplankton in polar oceans: a bipolar comparison. Aquatic Microbial Ecology, 25(2): 195-206.

Binder B. 1999. Reconsidering the relationship between virally induced bacterial mortality and frequency of infected cells. Aquatic Microbial Ecology, 18: 207-215.

Bird D, Karl D. 1999. Uncoupling of bacteria and phytoplankton during the austral spring bloom in Gerlache Strait, Antarctic Peninsula. Aquatic Microbial Ecology, 19(1): 13-27.

Børsheim K Y. 1993. Native marine bacteriophages. FEMS Microbiology Letters, 102(3): 141-159.

Caron D A, Lim E L, Miceli G, et al. 1991. Grazing and utilization of chroococcoid cyanobacteria and heterotrophic bacteria by protozoa in laboratory cultures and a coastal plankton community. Marine Ecology Progress Series, 76(3): 205-217.

Chen B, Liu H, Wang Z. 2009. Trophic interactions within the microbial food web in the South China Sea revealed by size-fractionation method. Journal of Experimental Marine Biology and Ecology, 368(1): 59-66.

Christian R R, Hanson R B, Newell S Y. 1982. Comparison of methods for measurement of bacterial growth rates in mixed batch cultures. Applied and Environmental Microbiology, 43(5): 1160-1165.

Corinaldesi C, Dell'Anno A, Danovaro R. 2007. Viral infection plays a key role in extracellular DNA dynamics in marine anoxic systems. Limnology and Oceanography, 52(2): 508-516.

Cuevas L A, Morales C E. 2006. Nanoheterotroph grazing on bacteria and cyanobacteria in oxic and suboxic waters in coastal upwelling areas off northern Chile. Journal of Plankton Research, 28(4): 385-397.

Cynar F J, Estep K W, Sieburth J M. 1985. The detection and characterization of bacteria-sized protists in "protist-free" filtrates and their potential impact on experimental marine ecology. Microbial Ecology, 11(4): 281-288.

Ducklow H W, Hill S M. 1985. The growth of heterotrophic bacteria in the surface waters of warm core rings. Limnology and Oceanography, 30(2): 239-259.

Ferguson R L, Buckley E N, Palumbo A V. 1984. Response of marine bacterioplankton to differential filtration and confinement. Applied and Environmental Microbiology, 47(1): 49-55.

Fischer U R, Velimirov B. 2002. High control of bacterial production by viruses in a eutrophic oxbow lake. Aquatic Microbial Ecology, 27: 1-12.

Fuhrman J A, Azam F. 1980. Bacterioplankton secondary production estimates for coastal waters of British Columbia, Antarctica, and California. Applied and Environmental Microbiology, 39(6): 1085-1095.

Fuhrman J A, McManus G B. 1984. Do bacteria-sized marine eukaryotes consume significant bacterial production? Science, 224(4654): 1257-1260.

Fuhrman J A, Noble R T. 1995. Viruses and protists cause similar bacterial mortality in coastal seawater. Limnology and Oceanography, 40(7): 1236-1242.

Gak D, Romanova E, Romanenko V, et al. 1972. Estimation of changes in number of bacteria in the isolated water samples//Sorokin Y I, Kadota H. Microbial Production and Decomposition in Fresh Waters. Oxford: Blackwell Scientific Publications: 78-82.

Garzio L M, Steinberg D K, Erickson M, et al. 2013. Microzooplankton grazing along the Western Antarctic Peninsula. Aquatic Microbial Ecology, 70(3): 215-232.

Geider R J. 1989. Use of radiolabeled tracers in dilution grazing experiments to estimate bacterial growth and loss rates. Microbial Ecology, 17(1): 77-87.

Guixa-Boixereu N, Lysnes K, Pedros-Alio C. 1999. Viral lysis and bacterivory during a phytoplankton bloom in a coastal water microcosm. Applied and Environmental Microbiology, 65(5): 1949-1958.

Hagström Å, Azam F, Andersson A, et al. 1988. Microbial loop in an oligotrophic pelagic marine ecosystem: possible roles of cyanobacteria and nanoflagellates in the organic fluxes. Marine Ecology Progress Series, 49: 171-178.

Hanson R B, Shafer D, Ryan T, et al. 1983. Bacterioplankton in Antarctic Ocean waters during late austral winter: abundance, frequency of dividing cells, and estimates of production. Applied and Environmental Microbiology, 45(5): 1622-1632.

Heldal M, Bratbak G. 1991. Production and decay of viruses in aquatic environments. Marine Ecology Progress Series, 72(3): 205-212.

Hobbie J E, Daley R J, Jasper S. 1977. Use of nuclepore filters for counting bacteria by fluorescence microscopy. Applied and Environmental Microbiology, 33(5): 1225-1228.

Höfle M G. 1984. Degradation of putrescine and cadaverine in seawater cultures by marine bacteria. Applied and Environmental Microbiology, 47(4): 843-849.

Ivanov M V. 1955. Method for determination of production of bacterial biomass in water bodies. Mikrobiologiya, 29(1): 79-89.

Jochem F J. 2003. Photo- and heterotrophic pico- and nanoplankton in the Mississippi River plume: distribution and grazing activity. Journal of Plankton Research, 25(10): 1201-1214.

Kirchman D L, Ducklow H W, Mitchell R. 1982. Estimates of bacterial growth from changes in uptake rates and biomass. Applied and Environmental Microbiology, 44(6): 1296-1307.

Landry M R, Haas L W, Fagerness V L. 1984. Dynamics of microbial plankton communities: experiments in Kaneohe Bay, Hawaii. Marine Ecology Progress Series, 16: 127-133.

Meyer-Reil L A. 1977. Bacterial growth rates and biomass production//Rheinheimer G. Microbial Ecology of a Brackish Water Environment. Berlin: Springer Berlin Heidelberg: 223-236.

Middelboe M. 2000. Bacterial growth rate and marine virus-host dynamics. Microbial Ecology, 40(2): 114-124.

Middelboe M, Nielsen T G, Bjornsen P K. 2002. Viral and bacterial production in the North Water: *in situ* measurements, batch-culture experiments and characterization and distribution of a virus-host system. Deep Sea Research Part II: Topical Studies in Oceanography, 49(22-23): 5063-5079.

Newell S Y, Christian R R. 1981. Frequency of dividing cells as an estimator of bacterial productivity. Applied and Environmental Microbiology, 42(1): 23-31.

Newell S Y, Fallon R D. 1982. Bacterial productivity in the water column and sediments of the Georgia(USA) coastal zone: estimates via direct counting and parallel measurement of thymidine incorporation. Microbial Ecology, 8(1): 33-46.

Newell S Y, Sherr B F, Sherr E B, et al. 1983. Bacterial response to presence of eukaryote inhibitors in water from a coastal marine environment. Marine Environmental Research, 10(3): 147-157.

Noble R T, Fuhrman J A. 2000. Rapid virus production and removal as measured with fluorescently labeled viruses as tracers. Applied and Environmental Microbiology, 66(9): 3790-3797.

Noble R T, Middelboe M, Fuhrman J A. 1999. Effects of viral enrichment on the mortality and growth of heterotrophic bacterioplankton. Aquatic Microbial Ecology, 18: 1-13.

Pearce I, Davidson A T, Thomson P G, et al. 2010. Marine microbial ecology off East Antarctica(30 - 80°E): rates of bacterial and phytoplankton growth and grazing by heterotrophic protists. Deep Sea Research Part II: Topical Studies in Oceanography, 57(9): 849-862.

Pearce I, Davidson A T, Thomson P G, et al. 2011. Marine microbial ecology in the sub-Antarctic Zone: rates of bacterial and phytoplankton growth and grazing by heterotrophic protists. Deep Sea Research Part II: Topical Studies in Oceanography, 58(21): 2248-2259.

Proctor L M, Fuhrman J A. 1990. Viral mortality of marine bacteria and cyanobacteria. Nature, 343(6253): 60-62.

Proctor L M, Okubo A, Fuhrman J A. 1993. Calibrating estimates of phage-induced mortality in marine bacteria: ultrastructural studies of marine bacteriophage development from one-step growth experiments. Microbial Ecology, 25(2): 161-182.

Rassoulzadegan F, Sheldon R. 1986. Predator-prey interactions of nanozooplankton and bacteria in an oligotrophic marine environment. Limnology and Oceanography, 31(5): 1010-1021.

Rivkin R B, Putland J N, Anderson M R, et al. 1999. Microzooplankton bacterivory and herbivory in the NE subarctic Pacific. Deep Sea Research Part II: Topical Studies in Oceanography, 46(11):

2579-2618.

Sakka A, Legendre L, Gosselin M, et al. 2000. Structure of the oligotrophic planktonic food web under low grazing of heterotrophic bacteria: Takapoto Atoll, French Polynesia. Marine Ecology Progress Series, 197: 1-17.

Scharek R, Latasa M. 2007. Growth, grazing and carbon flux of high and low nucleic acid bacteria differ in surface and deep chlorophyll maximum layers in the NW Mediterranean Sea. Aquatic Microbial Ecology, 46(2): 153-161.

Sherr B F, Sherr E B, Andrew T L, et al. 1986. Trophic interactions between heterotrophic Protozoa and bacterioplankton in estuarine water analyzed with selective metabolic inhibitors. Marine Ecology Progress Series, 32: 169-179.

Sherr B F, Sherr E B, Mcdaniel J. 1992. Effect of protistan grazing on the frequency of dividing cells in bacterioplankton assemblages. Applied and Environmental Microbiology, 58(8): 2381-2385.

Sommer U, Berninger U G, Böttger-Schnack R, et al. 2002. Grazing during early spring in the Gulf of Aqaba and the northern Red Sea. Marine Ecology Progress Series, 239: 251-261.

Steward G F, Smith D C, Azam F. 1996. Abundance and production of bacteria and viruses in the Bering and Chukchi Seas. Marine Ecology Progress Series, 131: 287-300.

Steward G F, Wikner J, Cochlan W P, et al. 1992a. Estimation of virus production in the sea: II. field results. Marine Microbial Food Webs, 6(2): 79-90.

Steward G F, Wikner J, Smith D C, et al. 1992b. Estimation of virus production in the sea: I. method developement. Marine Microbial Food Webs, 6(2): 57-78.

Strom S L. 2000. Bacterivory: interactions between bacteria and their grazers//Kirchman D L. Microbial Ecology of the Oceans. New York: Wiley-Liss: 351-386.

Suttle C A, Chen F. 1992. Mechanisms and rates of decay of marine viruses in seawater. Applied and Environmental Microbiology, 58(11): 3721-3729.

Taira Y, Uchimiya M, Kudo I. 2009. Simultaneous estimation of viral lysis and protozoan grazing on bacterial mortality using a modified virus-dilution method. Marine Ecology Progress Series, 379: 23-32.

Taylor G T, Pace M L. 1987. Validity of eukaryote inhibitors for assessing production and grazing mortality of marine bacterioplankton. Applied and Environmental Microbiology, 53(1): 119-128.

Tsai A Y, Gong G C, Chao C F. 2016. Contribution of viral lysis and nanoflagellate grazing to bacterial mortality at surface waters and deeper depths in the coastal ecosystem of subtropical Western Pacific. Estuaries and Coasts, 39(5): 1357-1366.

Tsai A Y, Gong G C, Hung J. 2013. Seasonal variations of virus- and nanoflagellate-mediated mortality of heterotrophic bacteria in the coastal ecosystem of subtropical western Pacific. Biogeosciences, 10(5): 3055-3065.

Weinbauer M G, Fuks D, Peduzzi P. 1993. Distribution of viruses and dissolved DNA along a coastal trophic gradient in the northern Adriatic Sea. Applied and Environmental Microbiology, 59(12): 4074-4082.

Weinbauer M G, Fuks D, Puskaric S, et al. 1995. Diel, seasonal, and depth-related variability of viruses and dissolved DNA in the northern Adriatic Sea. Microbial Ecology, 30(1): 25-41.

Weinbauer M G, Höfle M G. 1998. Significance of viral lysis and flagellate grazing as factors controlling bacterioplankton production in a eutrophic lake. Appl Environ Microbiol, 64(2): 431-438.

Weinbauer M G, Peduzzi P. 1994. Frequency, size and distribution of bacteriophages in different marine bacterial morphotypes. Marine Ecology Progress Series, 108(1-2): 11-20.

Weinbauer M G, Winter C, Höfle M G. 2002. Reconsidering transmission electron microscopy based estimates of viral infection of bacterioplankton using conversion factors derived from natural communities. Aquatic Microbial Ecology, 27: 103-110.

Weisse T. 1989. The microbial loop in the Red Sea. dynamics of pelagic bacteria and heterotrophic nanoflagellates. Marine Ecology Progress Series, 55: 241-250.

Wikner J, Hagström Å. 1988. Evidence for a tightly coupled nanoplanktonic predator-prey link regulating the bacterivores in the marine environment. Marine Ecology Progress Series, 50(1): 137-145.

Wommack K E, Colwell R R. 2000. Virioplankton: viruses in aquatic ecosystems. Microbiology and Molecular Biology Reviews, 64(1): 69-114.

Wright R T, Coffin R B. 1984. Measuring microzooplankton grazing on planktonic marine bacteria by its impact on bacterial production. Microbial Ecology, 10(2): 137-149.

Zhang R, Li Y, Yan W, et al. 2020. Viral control of biomass and diversity of bacterioplankton in the deep sea. Communications Biology, 3(1): 256.

Zhao Y, Dong Y, Li H, et al. 2020. Grazing by microzooplankton and copepods on the microbial food web in spring in the southern Yellow Sea, China. Marine Life Science & Technology, 2: 442-455.

第八章　微型浮游动物对自养微食物网生物的摄食

自养生物是微食物网的重要组成部分，主要包括原核的聚球藻属（*Synechococcus*）蓝细菌、原绿球藻属（*Prochlorococcus*）蓝细菌，以及微微型真核浮游生物（picoeukaryotes）和微型浮游生物（nanoplankton）等。在多数寡营养海区和中营养海区中，浮游生物的初级生产主要是由 pico 级（0.2~2μm）自养生物，即聚球藻、原绿球藻和微微型自养浮游生物所贡献的。

自养微食物网生物繁殖快，其中 pico 级自养生物每天可分裂一次，然而它们的丰度在几天甚至几年内保持基本恒定，表明这些生物的死亡率和生长率形成了动态平衡。被摄食、病毒侵染裂解和沉降是自养微食物网生物消亡的主要原因。微型浮游动物对自养微食物网生物的摄食是初级生产向较高营养级转化的关键步骤，因此受到研究者的重视。本章将以聚球藻和微型浮游植物为例，介绍微型浮游动物对自养微食物网生物的摄食情况。

第一节　微型浮游动物对聚球藻的摄食

一、摄食关系

研究微型浮游动物是否摄食聚球藻，所使用的方法与摄食细菌的研究类似。聚球藻体内含有叶绿素 a 和藻胆素，因此在荧光显微镜蓝色激发光下呈现明亮的橙色。可通过观察摄食者体内是否有橙色荧光，直接确定微型浮游动物是否摄食聚球藻。也有研究使用荧光染料对聚球藻染色再进行喂食培养，常用的染料包括 DTAF[5-(4,6-二氯三嗪基)氨基荧光素，5-(4,6-dichlorotriazin-2-yl)amino] 和 DAPI(4′,6-二脒基-2-苯基吲哚，4′,6-diamidino-2-phenylindole)。考虑到很多聚球藻可以被摄食但难以被消化，因此体内饵料颗粒法是确定摄食关系的最佳方法。此外也有研究用直径 1μm 的荧光微球作为聚球藻饵料的替代物来喂食微型浮游动物，但由于对微型浮游动物接触荧光微球的反应了解很少，尚不能确定这种方法反映真实状况的程度。将聚球藻与摄食者混合培养，间隔一定时间周期采样，监测摄食者体内饵料颗粒的增多情况，最初一段时间摄食者体内的饵料数量呈线性增长，经一定时间后摄食者消化和排出饵料，其体内的聚球藻数量会保持恒定，线性增长期摄食者体内聚球藻数量增加的斜率即为摄食速率[ingestion rate，prey/（grazer•h）]。

另一个研究方法是将摄食者与饵料聚球藻混合培养，监测摄食者的生长情况，若摄食者能存活和生长，则证明其利用了饵料聚球藻。同一种摄食者和不同的饵料混合培养，摄食者表现出不同的生长率，可以指示不同饵料对于摄食者的营养价值。根据实验组（含摄食者和聚球藻）与对照组（只有聚球藻）在培养过程中聚球藻丰度的差异，可以估算摄食速率。

Johnson 等（1982）最早在实验室内研究了微型浮游动物与聚球藻的摄食关系，他用聚球藻菌株分别与纤毛虫 Uronema sp.和鞭毛虫 Actinomonas sp.进行共培养，摄食者在培养体系中存活了将近 1 年。Caron 等（1991）用聚球藻菌株 WH7803、WH8012 和 WH8101 喂养在马萨诸塞近岸海水中分离的鞭毛虫 Paraphysomonas sp.和两种纤毛虫（scuticociliate 和 hymenostome），发现这 3 株聚球藻都能支持上述 3 种摄食者的生长，但喂食聚球藻时摄食者的生长率低于喂食异养细菌的情况。喂食聚球藻时 Paraphysomonas sp.的生长率相当于喂食异养细菌生长率的 74%，hymenostome 为 48%，而 scuticociliate 仅为 20%。

Zwirglmaier 等（2009）用 37 株聚球藻喂养两种鞭毛虫，发现鞭毛虫 Paraphysomonas imperforata 可摄食并利用 5 株聚球藻，可摄食但不消化利用 11 株聚球藻；鞭毛虫 Pteridomonas danica 可摄食并利用 11 株聚球藻，有 9 株聚球藻可被摄食但不被利用。这表明，鞭毛虫对聚球藻的摄食是有选择性的，可能与饵料细胞表面特性如疏水性相关。

Jeong 等（2005）发现 17 株培养和 1 株野外采集的赤潮甲藻均可摄食聚球藻。大个体的可形成赤潮的甲藻摄食速率更快，当聚球藻丰度在 $1.1×10^6$～$2.3×10^6$cell/ml 时，甲藻的粒径（5.2～38.2μm，等效球直径）与摄食速率[1.0～64.2prey/（grazer·h）]呈正相关。赤潮甲藻 Karenia brevis 也能够摄食聚球藻（Glibert et al.，2009）。不同种甲藻的摄食速率不同，Prorocentrum donghaiense 在聚球藻丰度为 $1.1×10^6$cell/ml 时摄食速率达到饱和，最大摄食速率为 7.7prey/（grazer·h）；Prorocentrum micans 在聚球藻丰度为 $1.4×10^6$cell/ml 时摄食速率达到饱和，最大摄食速率为 38.2prey/（grazer·h）（Jeong et al.，2005）。同一种甲藻对不同聚球藻菌株的摄食速率也有差异，甲藻 Oxyrrhis marina 对聚球藻菌株 CC9605 和 CC9902 的摄食速率分别为 56prey/(grazer·h)和 71prey/(grazer·h)，远高于对菌株 WH8102 的摄食速率[1.7prey/（grazer·h）]（Apple et al.，2011）。

在自然海区样品中也有微型浮游动物摄食聚球藻的证据，如自然海区纤毛虫的食物泡中可观察到聚球藻（Campbell and Carpenter，1986；Sherr et al.，1986；Bernard and Rassoulzadegan，1993），在地中海 Villefranche 湾聚球藻是砂壳纤毛虫 Rhabdonella spiralis 的主要饵料，是砂壳纤毛虫 Salpingella acuminata 的唯一饵料（Bernard and Rassoulzadegan，1993）。在法国 Villefranche-sur-Mer 近岸的鞭毛虫体内（Laval，1971）、Chesapeake 湾的原生动物体内（Perkins et al.，1981）也能

观察到聚球藻。

在地中海的 Villefranche 湾，每个异养鞭毛虫体内的聚球藻数量为 0.06~0.34 个，夜间样品平均为 0.11 个，白天样品平均为 0.19 个，但差异不显著（Dolan and Šimek，1999）。海水中正在进行复制分裂的聚球藻比例高时，异养鞭毛虫体内平均聚球藻数量降低，可能是由于异养鞭毛虫只能摄食粒径较小的聚球藻，正在分裂的聚球藻体积变大，无法被摄食。此外，异养鞭毛虫对聚球藻的摄食有昼夜节律，摄食速率在晚上达到最大，白天减小，傍晚最小，这与聚球藻的分裂节律是一致的（Dolan and Šimek，1999；Christaki et al.，2001）。

聚球藻对部分微型浮游动物来说是很难消化的饵料。在培养实验中常发现聚球藻被微型浮游动物摄食，但是摄食者生长率不高或出现死亡的现象（Caron et al.，1991；Christaki et al.，1998；Zwirglmaier et al.，2009；Apple et al.，2011）。这种现象在中型浮游动物摄食研究中也有发现，例如，Stukel 等（2013）在中型浮游动物海上现场培养实验的粪便颗粒中发现有单个或成团的聚球藻保存完好，在现场采集到的中型浮游动物的肠道中也检测到了聚球藻的特征色素藻红蛋白（Phycoerythrin，PE）。在部分大型浮游动物的粪便中也发现了聚球藻，如飞马哲水蚤（Johnson et al.，1982）、海樽类（Gorsky et al.，1999）和翼足类（Silver and Bruland，1981）的粪便，可能是这些大型浮游动物摄食了体内含有未消化聚球藻的微型浮游动物。

二、摄食率和摄食压力

要研究微型浮游动物对聚球藻的摄食率，除前面介绍的体内颗粒增加法和水体颗粒减少法外，更多研究采用人为调整控制饵料与摄食者接触概率的策略，具体方法与第七章微型浮游动物和浮游病毒对细菌的下行控制的研究相同，也包括分粒级培养法、稀释培养法、生物抑制剂培养法。使用这些方法不仅能够获得聚球藻的生长率（growth rate，/d）、浮游动物摄食率（grazing rate，/d），还可以结合摄食者和聚球藻的丰度评估摄食压力（grazing pressure，%）。

（一）摄食率

1. 分粒级培养法

海水分粒级培养使用的对照组为自然海水，实验组的海水则需经重力过滤，使用孔径为 1μm 的滤膜去除所有摄食者，培养 24h 后实验组中聚球藻的生长率即为内禀生长率，减去对照组中的表观生长率即可获得微型浮游动物的摄食率。用这一方法在夏威夷近岸海域获得的聚球藻内禀生长率为 1.42/d，微型浮游动物摄食率为 0.39/d（Landry et al.，1984）。

Caron 等（1991）将自然海水分别用孔径为 20μm（实验组）和 200μm（对照组）的滤膜过滤并培养 24h，发现实验组和对照组中聚球藻的丰度相差不大，部分实验组的聚球藻丰度低于对照组，说明聚球藻的摄食者主要是小于 20μm 的浮游生物。Guillou 等（2001）指出在 2μm 或者 3μm 分级过滤的海水中仍有更小的摄食者存在。Ning 和 Vaulot（1992）则认为在英吉利海峡 70%聚球藻的摄食压力来自小于 2μm 的微型浮游动物。

Chiang 等（2014）在东海南部进行了海水分粒级培养，实验组用孔径 2μm 滤膜过滤，对照组用孔径 10μm 滤膜过滤。实验组获得的聚球藻的表观生长率（该论文作者认为是内禀生长率）减去对照组的表观生长率，即为粒级为 2~10μm 微型浮游动物的摄食率，为 0.30~1.33/d，且摄食率随内禀生长率升高而升高，尽管这一升高并不显著。

多级过滤的方法也被用于评估不同粒级的浮游生物对聚球藻的摄食情况。有学者使用孔径为 1μm、5μm 和 10μm 的 Nuclepore® 核孔滤膜过滤培养海水，其中 1μm 孔径滤膜过滤后的滤液中，所有的纤毛虫和 95%的异养鞭毛虫都被过滤掉了，5μm 孔径滤膜过滤后的滤液中所有的纤毛虫被过滤掉了，10μm 孔径滤膜过滤后的滤液中，87%的纤毛虫被过滤掉了（Kudoh et al.，1990）。1μm 孔径滤膜过滤后的滤液中聚球藻的生长率被认为接近内禀生长率，为 0.1/h，5μm 孔径滤膜过滤的海水中聚球藻的表观生长率为 0.064/h，而 10μm 孔径滤膜过滤的海水中聚球藻的表观生长率为 0.053/h，不经任何过滤的自然海水中聚球藻的表观生长率为 0.004/h。这一结果表明，聚球藻有很高的内禀生长率，但在自然海水中被微型浮游动物的摄食抵消，其中聚球藻死亡率的 2/3 由纤毛虫的摄食导致。假设聚球藻细胞的氮含量为 10fgN/cell，以该研究中聚球藻丰度 5×10^3~15×10^3cell/ml 来估算，聚球藻可以利用海水中无机氮的 10%来维持其较高的生长率，表明聚球藻在寡营养的海水中具有竞争优势（Kudoh et al.，1990）。

Tsai 等（2009）分别使用 2μm、5μm、10μm 和 20μm 孔径的滤膜过滤自然海水，以研究混合营养鞭毛虫对聚球藻摄食的级联效应。他们发现，级联效应的发生需要满足下列条件：①聚球藻丰度要大于鞭毛虫摄食的阈值浓度，即 6×10^4cell/ml；②夜间发生，可能是因为这一时段的摄食速率高；③<5μm 与>5μm 的含色素鞭毛虫的丰度比值介于 1∶1 和 2∶1 之间。

总的来说，尽管各研究所使用的具体过滤粒级不同，但是得出的结果是一致的，即聚球藻的主要摄食者个体微小，绝大部分小于 20μm，其中粒径小于 5μm 的微型浮游动物是聚球藻最重要的摄食者。

2. 稀释培养法

采用稀释培养法研究微型浮游动物对聚球藻摄食的原理可参考本书第七章第

一节。已有的研究表明微型浮游动物对聚球藻的摄食率大多低于 0.9/d，最大约 4.30/d（表 8-1）。

表 8-1 稀释培养法获得的聚球藻生长率和微型浮游动物摄食率

海区	聚球藻生长率（/d）	微型浮游动物摄食率（/d）	参考文献
东北太平洋夏威夷近岸	1.98	0.14	Landry 等（1984）
亚北极东北太平洋	0.12~0.73	0.19~0.42	Rivkin 等（1999）
日本 Uwa 海	0~1.39	0~1.54	Hirose 等（2008）
黄海南部	−0.34~3.08	1.17~4.30	Zhao 等（2020）
南海北部	−0.04~2.39	0.25~2.44	Chen 和 Liu（2010）
赤道中太平洋	0.57~0.61	0.52~1.08	Landry 等（1995a）；Landry 等（1995b）
南加利福尼亚湾	0.52~0.86	0.14~0.39	Worden 等（2004）
西北大西洋	0.52~0.94	0.06~0.83	Campbell 和 Carpenter（1986）
东北大西洋	0.44	0.33	Quevedo 和 Anadon（2001）
英吉利海峡西部	0.13~0.50	0.22~0.67	Kimmance 等（2007）
马尾藻海西北部	0.54~0.87	0.20~0.33	Lessard 和 Murrell（1998）
马尾藻海和加利福尼亚海流区	0.37~0.67	0.13~0.51	Worden 和 Binder（2003）
墨西哥湾密西西比河口		0.2~3	Jochem（2003）
印度洋澳大利亚西部海区	−0.12~1.34	0~0.45	Paterson 等（2007）
阿拉伯海西部	0.18~0.90	0.06~0.73	Reckermann 和 Veldhuis（1997）

在稀释培养体系中添加营养盐可促进聚球藻的生长率增长，但微型浮游动物对聚球藻的摄食率也随之提高。添加营养盐后使用流式细胞仪检测聚球藻，可发现细胞的红色荧光信号（代表细胞叶绿素含量）和前向散射光信号（代表细胞大小）都升高了，Worden 和 Binder（2003）认为，这些指标代表着聚球藻生理状况出现改善，细胞变得更有营养或更加适口，进而导致微型浮游动物的摄食行为也出现了变化，增加了对这些饵料的摄食压力。另外由于饵料藻营养丰富使得微型浮游动物繁殖速度加快，所以摄食压力的升高还有可能来自微型浮游动物丰度的增加。

结合分粒级过滤和稀释培养法，有研究发现去除＞10μm 和＞3μm 的微型浮游动物后，pico 级生物的内禀生长率和被摄食率都增加了（Reckermann and Veldhuis，1997）。内禀生长率的增加可能是由于去除＞10μm 的微型浮游动物后，营养盐的再生速率提高（Ferrier and Rassoulzadegan，1991；Glibert et al.，1992）。被摄食率增加可能是因为微型浮游动物中存在营养级联作用，较大个体的微型浮游动物被去除后，小个体的微型浮游动物失去了摄食者的抑制，开始更多地摄食 pico 级生物进行生长。Chen 和 Liu（2010）使用分粒级稀释培养，去除＞5μm 的微型浮游动物后，也发现 pico 级生物被摄食率增加。

3. 生物抑制剂培养法

如第七章所介绍，生物抑制剂分为原核生物抑制剂和真核生物抑制剂。常用的聚球藻抑制剂包括氨苄青霉素（ampicillin）和卡那霉素（kanamycin），它们能抑制原核细胞的分裂，但不会引起细胞自溶和死亡，且对真核摄食者几乎没有影响；常用的真核摄食者抑制剂包括放线酮（cycloheximide）、秋水仙碱（colchicine）等。

添加生物抑制剂培养估计微型浮游动物对聚球藻的摄食率有 3 种实验方法。第一种方法是在培养组海水中加入原核生物抑制剂，聚球藻不再生长，原生动物摄食使得聚球藻丰度降低。在对照组中同时加入原核和真核生物抑制剂，聚球藻不再生长，原生动物也不能摄食，所以聚球藻丰度会保持不变。比较培养组与对照组的聚球藻丰度差异，可估计微型浮游动物对聚球藻的摄食率（Campbell and Carpenter，1986；刘镇盛，1990；Ning and Vaulot，1992；Ning and Vaulot，1996）。

第二种方法仅在培养组中加入真核生物抑制剂，对照组为不加抑制剂的自然海水。培养组的结果可获得聚球藻的内禀生长率，而对照组得出表观生长率，内禀生长率减去表观生长率即为微型浮游动物摄食率（Caron et al.，1991）。

第三种方法为仅在培养组中加入原核生物抑制剂，对照组为不加抑制剂的自然海水。培养一段时间后，对照组得出聚球藻的表观生长率，而培养组得出微型浮游动物的摄食率（Liu et al.，1995；Chang et al.，2003；蔡昱明等，2006；乐凤凤等，2011）。

真核生物抑制剂使真核生物停止摄食，但也减少了营养盐的再生，所以为了保证聚球藻生长所需的营养盐，添加真核生物抑制剂的同时需要在培养体系中添加营养盐。而且为了使抑制剂发生作用，应在加入抑制剂 1h 后再开始取样作为初始样品。

使用生物抑制剂方法获得的微型浮游动物对聚球藻的摄食率范围为 0.04～1.06/d（表 8-2）。微型浮游动物的摄食率通常低于聚球藻的生长率，如在夏威夷 ALOHA 观测站，聚球藻的生长率达 1.0/d，微型浮游动物的摄食率为聚球藻生长率的 43%～87%（Liu et al.，1995）。

表 8-2　生物抑制剂培养法获得的聚球藻生长率和微型浮游动物摄食率

海区	聚球藻生长率（/d）	微型浮游动物摄食率（/d）	参考文献
西北大西洋		0.33～0.79	Campbell 和 Carpenter（1986）
北大西洋 Vineyard 海湾	0	0.35～0.63*	Caron 等（1991）
地中海罗斯科夫近岸	0.18～0.66	0.09～0.98	刘镇盛（1990）
英吉利海峡	0.25～0.72（平均 0.51）	0.21～0.64（平均 0.44）	Ning 和 Vaulot（1992）；Ning 和 Vaulot（1996）

续表

海区	聚球藻生长率（/d）	微型浮游动物摄食率（/d）	参考文献
东北太平洋夏威夷ALOHA观测站	0.17~1.06	0.09~0.47	Liu 等（1995）
东海	0.20~0.68	0.07~0.40	Chang 等（2003）
南海	0.11~1.18	0.11~0.76	蔡昱明等（2006）
东海	0.1~2.47（平均0.61）	0.04~1.06（平均0.23）	乐凤凤等（2011）

*表示根据原文数据计算得出

我国学者刘镇盛在 1990 年率先用选择抑制剂技术估计了英吉利海峡近岸表层水微型浮游动物对聚球藻的摄食压力，发现异养微型浮游生物摄食率与聚球藻生长率呈动态平衡（刘镇盛，1990）。Chang 等（2003）发现，在东海微型浮游动物对聚球藻的摄食率有季节差异，随水温的升高摄食率上升，但是没有明显的空间变化。黄海聚球藻生长率和微型浮游动物摄食率的分布格局是近岸（>0.3/d）高于黄海中部（<0.3/d），生长率和摄食率呈显著线性正相关，但其关系不是1：1，而是略低于1：1（乐凤凤等，2011）。微型浮游动物的摄食率有昼夜差异，在英吉利海峡微型浮游动物白天的摄食率（0.61/d）大于夜间（0.11/d）（Ning and Vaulot，1992）。

在部分海域由于聚球藻饵料的丰度过低，微型浮游动物停止摄食，用稀释培养的方法难以准确测定微型浮游动物的摄食率。这种情况下，用生物抑制剂培养法能够较好地评估摄食率。例如，在西北大西洋远岸站位，稀释培养得出的摄食率仅为 0.06/d，几乎可以忽略不计，但用生物抑制剂培养法得到的微型浮游动物的摄食率为 0.33/d（Campbell and Carpenter，1986）。

（二）摄食压力

微型浮游动物的清滤速率[clearance rate，F，nl/（grazer·h）]与摄食率（g，/d）的关系可用公式 $F=g/D_p$ 表示，式中 D_p 为摄食者的丰度（Landry，1981）。根据这一公式，可由稀释培养获得的摄食率估算主要摄食者的清滤速率。而根据摄食率、清滤速率，结合自然海水中摄食者与饵料生物的丰度，可以评估自然海区摄食者对饵料生物的摄食压力（grazing pressure，%）。例如，在韩国马山湾（Masan Bay），聚球藻饵料丰度为 5.5×10^2~1.61×10^4cell/ml，摄食者甲藻 *Prorocentrum donghaiense* 丰度为 1.71×10^3~5.5×10^4cell/ml，摄食率为 0.1~3.6/h，意味着每小时 *P. donghaiense* 可去除聚球藻现存量的 11%~98%；在济州岛海域 *P. donghaiense*（丰度 12~328cell/ml）对聚球藻（丰度 7.01×10^4~2.03×10^5cell/ml）的摄食率为 0.0011~0.014/h，表明 *P. donghaiense* 每小时可去除聚球藻现存量的 0.1%~1.5%（Jeong et al.，2005）。

多数研究发现，聚球藻的生长率与微型浮游动物摄食率相耦合，即生长率升高时摄食率也增加。一般来说，在寡营养大洋微型浮游动物的摄食与聚球藻的生产基本相抵，而在近岸海区聚球藻生长大于摄食的损失。但也有例外，如在北大西洋的寡营养暖涡，微型浮游动物对聚球藻生产的摄食压力仅为 37%~52%（Campbell and Carpenter，1986）。这一现象可能是聚球藻对营养盐迅速响应造成的，在寡营养海区的表层水中，无机氮浓度 nmol 量级的升高即可使聚球藻迅速达到最大生长率（Glover et al.，1988），饵料丰度迅速升高使摄食压力降低。

第二节　微型浮游动物对浮游植物的摄食

一、常用研究方法

为评估微型浮游动物在海洋生态系统中的重要性，最初的研究试图用摄食者和饵料的营养关系间接估计微型浮游动物的摄食量。例如，有学者提出假设，认为原生动物每天摄食饵料的体积是其本身体积的 50%，后生动物（如桡足类幼虫）则为 10%（Lohmann，1908），还有研究者提出原生动物与桡足类幼虫每天摄食饵料的生物量分别为其本身生物量的 3.1 倍和 30%（Beers and Stewart，1971）。根据这类假设，结合摄食者与饵料的丰度，即可估算微型浮游动物的摄食率和摄食压力。此后一些间接估计或直接测量微型浮游动物摄食影响的方法被提出和应用（表 8-3）。

表 8-3　微型浮游动物摄食浮游植物的常用研究方法

	方法	优点	缺点
间接方法	1. 根据摄食者和饵料的营养关系估算	无须摄食培养实验	不精确
	2. 根据实验室培养数据估算野外自然水体的情况	无须摄食培养实验	可能无法反映现场情况
	3. 根据其他海区的数据估算本海区摄食率	无须摄食培养实验	可能无法反映现场情况
直接方法	1. 摄食标记物		
	（1）惰性颗粒标记	定量，直接显示吞噬速度	选择性摄食对结果有影响
	（2）放射性物质标记	定量，灵敏度高	放射性物质有其他的流通渠道，影响标记作用
	2. 生物抑制剂培养	定量	部分抑制剂的专一性差
	3. 分粒级培养	定量，使用自然群体	操作步骤较多，对生物的影响大，摄食者和饵料生物并非完全分离
	4. 稀释培养	定量，可同时获得浮游植物的生长率与微型浮游动物的摄食率，对自然群体的干扰较少	可能改变自然群体，很难确定摄食阈值效应是否存在

	方法	优点	缺点
直接方法	5. 色素收支	定量，没有对自然群体的干扰	叶绿素转化为脱镁叶绿酸的比例尚不确定

由表 8-3 可以看出，尽管可以使用的研究方法多样，但是目前还没有哪种方法能够简单、精确地估计微型浮游动物的摄食率。海洋研究科学委员会（Scientific Committee on Oceanic Research，SCOR）推荐了 2 种研究微型浮游动物摄食的方法，包括利用摄食者和饵料的营养关系的间接估算法，以及稀释培养法。

利用微型浮游动物的现存量估计摄食压力的方法主要基于如下 4 个假设：①鞭毛虫摄食异养细菌和 pico 级自养生物；②纤毛虫和甲藻摄食 nano 级自养生物；③每一类群的微型浮游动物摄食率保持不变；④摄食的速度随温度而改变。其中微型浮游动物各个类群的清滤速率[ml/(grazer·h)]可参考文献获得（Lessard and Swift，1985；Sherr et al.，1986；Fenchel，1987）。因此，这个方法只需测定下列参数，即微型浮游动物各主要类群如鞭毛虫、纤毛虫和甲藻的丰度和生物量，饵料各主要类群包括 pico 级和 nano 级自养生物及异养细菌的丰度和生物量。通过清滤速率乘以饵料生物量就可以分别得到鞭毛虫、纤毛虫和甲藻的摄食速率，各个类群的摄食速率相加即为总的微型浮游动物摄食速率。由于是使用文献的清滤速率数据来间接估算摄食压力，这个方法的准确度相对不足，大概为–60%～+300%。

稀释培养法在微型浮游动物对浮游植物摄食压力研究中的应用更为广泛。针对浮游植物的稀释培养有如下理论假设：①浮游植物丰度的改变不会影响其生长速率；②微型浮游动物摄食不存在阈值效应，即摄食者的清滤速率不会因饵料浓度的改变而改变，浮游动物在饵料浓度高时不会因达到自身的需要而停止摄食，在饵料浓度低时也不会因摄食不到饵料而减少摄食努力或停止摄食；③浮游植物初始丰度（P_0）和 t 时刻丰度（P_t）符合公式 $P_t = P_0 e^{(\mu-g)t}$，其中，μ 为浮游植物的内禀生长率（以下简称生长率），g 为微型浮游动物摄食率。根据这些假设，稀释度（过滤海水体积与总体积的比值）为 d 的培养体系中浮游植物的浓度符合公式 $P_t = P_0 e^{(\mu-dg)t}$。培养不同稀释度的海水，检测培养前后浮游植物的丰度 P_t 和 P_0，可得到浮游植物生长率（μ）和微型浮游动物摄食率（g）。

由浮游植物生长率和微型浮游动物摄食率，可进一步计算微型浮游动物对浮游植物现存量的摄食压力（grazing pressure on phytoplankton initial stock，P_i）、对初级生产力的摄食压力（grazing pressure on primary production，P_p）、浮游植物的加倍时间（time of doubling，T_d）、每天的加倍数（n）等参数，使用的计算公式如下。

$$P_i = (1-e^g) \times 100\% \tag{8-1}$$

$$P_\mathrm{p} = \left[\mathrm{e}^\mu - \mathrm{e}^{(\mu-g)} \right] / (\mathrm{e}^\mu - 1) \times 100\% \qquad (8\text{-}2)$$

$$T_\mathrm{d} = \ln 2 / \mu \qquad (8\text{-}3)$$

$$n = \mu / \ln 2 \qquad (8\text{-}4)$$

除丰度（ind./ml）外，还可以用生物量（μg C/L）、叶绿素 a 或其他特征色素的浓度（μg/L）来表示浮游植物的现存量，从而计算各种摄食压力参数。叶绿素 a 浓度检测最简单易行，所以多数稀释培养研究使用浮游植物的叶绿素浓度，而非碳生物量作为指标进行度量。这样得到的摄食压力，严格来讲应是微型浮游动物对叶绿素现存量和叶绿素生产力的摄食压力，尽管在实际研究中学者基本都将其等同于对浮游植物现存量和初级生产力的摄食压力。

叶绿素 a 浓度反映的是整个浮游植物群体，而非个别浮游植物类群的变化。将稀释培养与高效液相色谱技术（high performance liquid chromatography，HPLC）结合，通过 HPLC 分析培养前后海水中各种特征色素的浓度，可反映不同类群浮游植物丰度的变化。

二、摄食压力

值得注意的是，稀释培养获得的微型浮游动物对初级生产力的摄食压力 P_p，其概念与其他方法获得的摄食压力略有差异。如图 8-1 所示，P_S 为培养前浮游植物的现存量，P 为过滤去除微型浮游动物条件下的初级生产力，$P = P_\mathrm{S} \times (\mathrm{e}^\mu - 1)$。$P_1$ 为含有微型浮游动物摄食者条件下的初级生产力，$P_1 = P_\mathrm{S} \times \left[\mathrm{e}^{(\mu-g)} - 1 \right]$。$P_2$ 为微型浮游动物的摄食量，$P_2 = P - P_1 = P_\mathrm{S} \times \left[\mathrm{e}^\mu - \mathrm{e}^{(\mu-g)} \right]$。因此，微型浮游动物对初级生产力的摄食压力 P_p 为 P_2 与 P 的比值。由于 P 是假设没有微型浮游动物摄食时可能达到的最大初级生产力，故而 P_p 实际上代表的是微型浮游动物对潜在最大初级生产力的摄食压力。用其他方法研究微型浮游动物对初级生产力的摄食压力，在测定初级生产力时未将微型浮游动物排除，因此得到的是微型浮游动物摄食对图 8-1 中 P_1 的摄食压力。

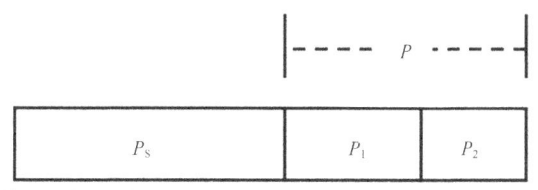

图 8-1 稀释培养法估算微型浮游动物对初级生产力的摄食压力图示（说明见正文）

尽管各种研究方法均存在一定不足，学者还是对微型动物对浮游植物的摄食压力作出了一些估计。使用实验室或野外调查数据估算研究发现微型浮游动物（或其中某一类群）可摄食初级生产力的 4%～100%（表 8-4）。使用稀释培养法进行研究的报道更多，这些研究表明，微型浮游动物可摄食浮游植物现存量的 0%～79%，摄食初级生产力的 0%～271%（表 8-5）。

表 8-4 不同方法获得的微型浮游动物对浮游植物的摄食压力（除稀释培养法外）

海区	摄食者	时间	方法	对初级生产力的摄食压力（%）	参考文献
美国长岛湾	微型浮游动物	周年	根据野外调查数据估算	43	Riley（1956）
美国长岛湾	砂壳纤毛虫	周年	根据实验室培养数据估算	27	Capriulo 和 Carpenter（1983）
美国南加利福尼亚湾	砂壳纤毛虫	周年	根据实验室培养数据估算	4～20	Heinbokel 和 Beers（1979）
太平洋秘鲁上升流海区	纤毛虫	6月	根据实验室培养数据估算	5～24	Beers 等（1971）
太平洋加利福尼亚洋流区	微型浮游动物	4～9月	根据实验室培养数据估算	7～52	Beers 和 Stewart（1970）
加拿大萨尼奇湾	纤毛虫和桡足类幼虫	冬季	根据实验室培养数据估算	30	Takahashi 和 Hoskins（1978）
日本 Akkeshi 湾	微型浮游动物	周年	根据实验室培养数据估算	10	Taguchi（1976）
英国南安普敦湾	砂壳纤毛虫	周年	根据实验室培养数据估算	60	Burkill（1982）
瑞典 Gullmar 峡湾	砂壳纤毛虫和轮虫	周年	根据实验室培养数据估算	100	Hernroth（1983）
美国长岛湾	砂壳纤毛虫	6～11月	分粒级培养	12～21	Capriulo 和 Carpenter（1980）
美国纳拉甘西特湾	砂壳纤毛虫	周年	分粒级培养	62	Verity（1986）

表 8-5 稀释培养法获得的微型浮游动物对浮游植物的摄食压力

海区	浮游植物生长率（/d）	微型浮游动物摄食率（/d）	对浮游植物现存量的摄食压力（%）	对初级生产力的摄食压力（%）	参考文献
渤海	0.23～0.73	0.43～0.69	34～50	85～100	Wang 等（1998）
挪威奥斯陆峡湾	0.4～1.6	0.02～1.08			Andersen 等（1991）
东北大西洋				39～115	Burkill 等（1993a）
别林斯高晋海				21～271	Burkill 等（1995）
印度洋西北部				31～71	Burkill 等（1993b）
英国凯尔特湾		0.4～1.0	13～65		Burkill 等（1987）
热带太平洋	0.7	0.5		75	Chavez 等（1991）
比斯开湾 Mundaka 河口			43～51	0～203	Cotano 等（1998）
大西洋南部	0.06～1.87	0～0.58	0～44	0～60	Froneman 和 Perissinotto（1996）

续表

海区	浮游植物生长率（/d）	微型浮游动物摄食率（/d）	对浮游植物现存量的摄食压力（%）	对初级生产力的摄食压力（%）	参考文献
南大西洋副热带辐合带	0.07~1.32	0~0.66	14~48	45~81	Froneman 等（1996b）
拉扎列夫海	0.019~0.080	0.012~0.052	1.3~7	45~97	Froneman 等（1996a）
美国罗德河		0.2~2.0	17~79	45~104	Gallegos（1988）
加拿大哈利法克斯港			38	0~100	Gifford（1988）
日本广岛湾	0.26~1.88	0.2~1.39	15.3~75.2		Kamiyama（1994）
美国华盛顿州近岸	0.455~0.628	0.065~0.278	6~24	17~52	Landry 和 Hassett（1982）
阿拉伯海	1.1	0.6			Landry 等（1998）
热带太平洋	0.20~1.00	0.21~0.72		55~83	Landry 等（1995a）
夏威夷湾	1.2~2.0	0.1~1.1	29~37		Landry（1984）
北冰洋加拿大沿岸	0.06~0.34	0.02~0.17	8~15	40~114	Paranjape（1987）
阿拉伯海	0.52~1.12	0.2~1.19	38	67	Reckermann 和 Veldhuis（1997）
亚北极太平洋	0~0.8	0~0.6		40~50	Strom 和 Welschmeyer（1991）
南大洋	0.1~0.4	0~0.3			Tsuda 和 Kawaguchi（1997）
圣劳伦斯湾西部	0.41~1.09	0.34~0.55	29~42	54~125	Tamigneaux 等（1997）
挪威峡湾				50~100	Verity 和 Vernet（1992）
北大西洋				37~100	Verity 等（1993）
赤道太平洋	0.4~1.1	0.2~1.0		70~123	Verity 等（1996）
美国蒙特雷湾	0.53~1.30	0.23~0.79	21~55		Waterhouse 和 Welschmeyer（1995）
美国 Fourleague 湾	0.46~2.14	0.32~2.11			Dagg（1995）
美国莫比尔湾	0.70~1.62	0.57~1.10			Lehrter 等（1999）

在自然海区，微型浮游动物和中、大型浮游动物（>200μm）均摄食浮游植物，但中、大型浮游动物对初级生产力的摄食压力通常在 10%~20%，大大低于微型浮游动物对初级生产力的摄食压力。在自然海区进行培养实验测得的初级生产力通常很高，但是浮游植物的生物量却没有明显的变化，微型浮游动物摄食被认为是重要原因之一

第三节 微型浮游动物的 $\delta^{15}N$ 营养级

稳定碳、氮同位素比值分析是研究生态系统物质流动和食物链营养关系的有效方法。生物圈中轻、重同位素在沿食物链传递过程中会产生分馏（fractionation）

效应，较重的同位素会滞留富集，因此生物体内的稳定性同位素比值可作为自然标记，用来示踪营养物质在生态系统中的流动。碳稳定同位素比值（$\delta^{13}C$）在营养级间变化很小，仅有 0.1‰~0.4‰，常用来分析消费者食物来源。氮稳定同位素比值（$\delta^{15}N$）的营养级间富集因子（trophic enrichment factor，TEF）较高（3‰~5‰，平均 3.4‰），表明氮同位素在营养级间存在明显的分馏，可用于确定消费者在食物网中的营养级（trophic position，TP）。营养级计算公式为

$$TP = \frac{\delta^{15}N_{consumer} - \delta^{15}N_{baseline}}{TEF} + \lambda \tag{8-5}$$

式中，$\delta^{15}N_{consumer}$ 为消费者的氮稳定同位素比值，$\delta^{15}N_{baseline}$ 为基线生物的氮稳定同位素比值，λ 为基线生物的营养级（即 $\lambda=1$ 时，基线生物为生态系统的初级生产者，$\lambda=2$ 时，基线生物为初级消费者）。当消费者营养级大于 2 时，营养级一般为非整数。

对于海洋浮游动物的营养级研究，还有一种改进的稳定同位素比值方法，称为氨基酸类化合物特异性同位素分析（compound specific isotopic analyses of amino acids，CSIA-AA）。海洋浮游动物体内的氨基酸可分为两类：源氨基酸（source amino acid，主要为苯丙氨酸，Phe）的 $\delta^{15}N$ 与食物网基线值相近，而营养氨基酸（trophic amino acids，trAA）的 $\delta^{15}N$ 随营养级升高而不断富集（McClelland and Montoya，2002；McClelland et al.，2003）。根据浮游动物体内 Phe 和 trAA 的 $\delta^{15}N$ 的差异即可计算营养级，计算公式为

$$TP_{trAA} = \frac{(\delta^{15}N_{trAA} - \delta^{15}N_{Phe} - \beta_{trAA})}{TEF} + 1 \tag{8-6}$$

式中，β_{trAA} 是初级生产者的营养氨基酸和源氨基酸之间的同位素差异（Chikaraishi et al.，2009）。

目前对浮游动物 $\delta^{15}N$ 的营养级研究主要集中在大、中型浮游动物，对微型浮游动物营养级的理解非常滞后。在自然样品中，微型浮游动物与饵料的粒级重叠混杂，很难严格分离微型浮游动物以测定其 $\delta^{15}N$。此外，后生动物主要从其饵料中吸收氨基酸，而微型浮游动物自身能合成多种氨基酸，所以即使能够获取自然样品中不含饵料的微型浮游动物个体，用其整个细胞的 $\delta^{15}N$ 也难以准确计算微型浮游动物的营养级。CSIA-AA 法可针对性选择微型浮游动物适用的营养氨基酸。

谷氨酸是 CSIA-AA 法最常用的营养氨基酸选择，大、中型浮游动物对谷氨酸都有很好的富集作用（Chikaraishi et al.，2009）。但对于微型浮游动物，丙氨酸是更合适的营养氨基酸种类。有研究发现，纤毛虫 *Favella* sp.摄食甲藻 *Heterocapsa triquetra* 时（Décima et al.，2017），以及鞭毛虫 *Oxyrrhis marina* 摄食绿藻 *Dunaliella tertiolecta* 时（Gutiérrez-Rodríguez et al.，2014），对丙氨酸都有较强的富集作用，

但对谷氨酸的营养富集却很少。因此，若使用谷氨酸进行 CSIA-AA 会显示微型浮游动物的 $\delta^{15}N$ 营养级缺失，从而严重低估微型浮游动物在海洋食物网中的生态功能（Gutiérrez-Rodríguez et al., 2014）。若选择丙氨酸进行 CSIA-AA，则发现在海洋食物网中，微型浮游动物占据了浮游植物和中型浮游动物之间的一个完整的营养级（Landry and Décima, 2017）。

参 考 文 献

蔡昱明, 宁修仁, 刘诚刚. 2006. 南海北部海域 Synechococcus 和 Prochlorococcus 生长率和被摄食消亡率——变化范围及其与环境因子的关系. 生态学报, 26(7): 2237-2246.

乐凤凤, 刘诚刚, 郝锵, 等. 2011. 2009 年春季南黄海聚球藻生长率、被摄食消亡率及其与环境因子的关系. 海洋学研究, 29(1): 34-41.

刘镇盛. 1990. 用选择抑制剂技术评价近岸表层水中异养微型浮游生物对聚球藻的摄食压力. 东海海洋, 8(3): 49-54.

Andersen T, Schartau A, Paasche E. 1991. Quantifying external and internal nitrogen and phosphorus pools, as well as nitrogen and phosphorus supplied through remineralization, in coastal marine plankton by means of a dilution technique. Marine Ecology Progress Series, 69(1-2): 67-80.

Apple J, Strom S L, Palenik B, et al. 2011. Variability in protist grazing and growth on different marine Synechococcus isolates. Applied and Environmental Microbiology, 77(9): 3074-3084.

Beers J R, Stevenson M R, Eppley R W, et al. 1971. Plankton populations and upwelling off the coast of Peru, June 1969. Fishery Bulletin, 69(4): 859-876.

Beers J R, Stewart G L. 1970. The ecology of the plankton off La Jolla, California in the period April through September, 1967. Part VI. Numerical abundance and estimated biomass of microzooplankton. Bulletin of the Scripps Institution of Oceanography, 17: 67-87.

Beers J R, Stewart G L. 1971. Micro-zooplankters in the plankton communities of the upper waters of the eastern tropical Pacific. Deep Sea Research and Oceanographic Abstracts, 18(9): 861-883.

Bernard C, Rassoulzadegan F. 1993. The role of picoplankton(cyanobacteria and plastidic picoflagellates) in the diet of tintinnids. Journal of Plankton Research, 15(4): 361-373.

Burkill P H. 1982. Ciliates and other microplankton components of a nearshore food web: standing stocks and production processes. Annales de l'Institut Oceanographique, Paris(Nouvelle Serie), 58(Suppl.): 335-350.

Burkill P H, Edwards E S, John A W G, et al. 1993a. Microzooplankton and their herbivorous activity in the northeastern Atlantic Ocean. Deep Sea Research Part II: Topical Studies in Oceanography, 40(1-2): 479-493.

Burkill P H, Edwards E S, Sleight M A. 1995. Microzooplankton and their role in controlling phytoplankton growth in the marginal ice zone of the Bellingshausen Sea. Deep Sea Research Part II: Topical Studies in Oceanography, 42(4-5): 1277-1290.

Burkill P H, Leakey R J G, Owens N J P, et al. 1993b. Synechococcus and its importance to the microbial foodweb of the northwestern Indian Ocean. Deep Sea Research Part II: Topical Studies in Oceanography, 40(3): 773-782.

Burkill P H, Mantoura R F C, Llewellyn C A, et al. 1987. Microzooplankton grazing and selectivity of phytoplankton in coastal waters. Marine Biology, 93(4): 581-590.

Campbell L, Carpenter E J. 1986. Estimating the grazing pressure of heterotrophic nanoplankton on

Synechococcus spp. using the sea water dilution and selective inhibitor techniques. Marine Ecology Progress Series, 33(2): 121-129.

Capriulo G M, Carpenter E J. 1980. Grazing by 35 to 202 μm micro-zooplankton in Long Island Sound. Marine Biology, 56(4): 319-326.

Capriulo G M, Carpenter E J. 1983. Abundance, species composition and feeding impact of tintinnid micro-zooplankton in central Long Island Sound. Marine Ecology Progress Series, 10: 277-288.

Caron D A, Lim E L, Miceli G, et al. 1991. Grazing and utilization of chroococcoid cyanobacteria and heterotrophic bacteria by protozoa in laboratory cultures and a coastal plankton community. Marine Ecology Progress Series, 76(3): 205-217.

Chang J, Lin K H, Chen K M, et al. 2003. *Synechococcus* growth and mortality rates in the East China Sea: range of variations and correlation with environmental factors. Deep Sea Research Part II: Topical Studies in Oceanography, 50(6-7): 1265-1278.

Chavez F P, Buck K R, Coale K H, et al. 1991. Growth rates, grazing, sinking, and iron limitation of equatorial Pacific phytoplankton. Limnology and Oceanography, 36(8): 1816-1833.

Chen B, Liu H. 2010. Trophic linkages between grazers and ultraplankton within the microbial food web in subtropical coastal waters. Marine Ecology Progress Series, 407: 43-53.

Chiang K P, Tsai A Y, Tsai P J, et al. 2014. The influence of nanoflagellates on the spatial variety of picoplankton and the carbon flow of the microbial food web in the oligotrophic subtropical pelagic continental shelf ecosystem. Continental Shelf Research, 80: 57-66.

Chikaraishi Y, Ogawa N O, Kashiyama Y, et al. 2009. Determination of aquatic food-web structure based on compound-specific nitrogen isotopic composition of amino acids. Limnology and Oceanography: Methods, 7(11): 740-750.

Christaki U, Dolan J R, Pelegri S, et al. 1998. Consumption of picoplankton-size particles by marine ciliates: effects of physiological state of the ciliate and particle quality. Limnology and Oceanography, 43(3): 458-464.

Christaki U, Giannakourou A, van Wambeke F, et al. 2001. Nanoflagellate predation on auto- and heterotrophic picoplankton in the oligotrophic Mediterranean Sea. Journal of Plankton Research, 23(11): 1297-1310.

Cotano U, Uriarte I, Villate F. 1998. Herbivory of nanozooplankton in polyhaline and euhaline zones of a small temperate estuarine system(Estuary of Mundaka): seasonal variations. Journal of Experimental Marine Biology and Ecology, 227(2): 265-279.

Dagg M J. 1995. Ingestion of phytoplankton by the micro- and mesozooplankton communities in a productive subtropical estuary. Journal of Plankton Research, 17(4): 845-857.

Décima M, Landry M R, Bradley C J, et al. 2017. Alanine $\delta^{15}N$ trophic fractionation in heterotrophic protists. Limnology and Oceanography, 62(5): 2308-2322.

Dolan J R, Šimek K. 1999. Diel periodicity in *Synechococcus* populations and grazing by heterotrophic nanoflagellates: analysis of food vacuole contents. Limnology and Oceanography, 44(6): 1565-1570.

Fenchel T M. 1987. Ecology of Protozoa. The Biology of Free-Living Phagotrophic Protists. Brock Springer Series in Contemporary BioScience. Berlin: Springer-Verlag Berlin Heidelberg: 197.

Ferrier C, Rassoulzadegan F. 1991. Density-dependent effects of protozoans on specific growth rates in pico- and nanoplanktonic assemblages. Limnology and Oceanography, 36(4): 657-669.

Froneman P W, Perissinotto R. 1996. Microzooplankton grazing and protozooplankton community structure in the South Atlantic and in the Atlantic sector of the Southern Ocean. Deep Sea

Research Part I: Oceanographic Research Papers, 43(5): 703-721.
Froneman P W, Perissinotto R, McQuaid C D. 1996a. Dynamics of microplankton communities at the ice-edge zone of the Lazarev Sea during a summer drogue study. Journal of Plankton Research, 18(8): 1455-1470.
Froneman P W, Perissinotto R, McQuaid C D. 1996b. Seasonal variations in microzooplankton grazing in the region of the Subtropical Convergence. Marine Biology, 126(3): 433-442.
Gallegos C L. 1988. Microzooplankton grazing on phytoplankton in the Rhode River, Maryland: nonlinear feeding kinetics. Marine Ecology Progress Series, 57: 23-33.
Gifford D J. 1988. Impact of grazing by microzooplankton in the Northwest Arm of Halifax Harbour, Nova Scotia. Marine Ecology Progress Series, 47: 249-258.
Glibert P M, Burkholder J M, Kana T M, et al. 2009. Grazing by *Karenia brevis* on *Synechococcus* enhances its growth rate and may help to sustain blooms. Aquatic Microbial Ecology, 55(1): 17-30.
Glibert P M, Miller C A, Garside C, et al. 1992. NH_4^+ regeneration and grazing: interdependent processes in size-fractionated $^{15}NH_4^+$ experiments. Marine Ecology Progress Series, 82: 65-74.
Glover H E, Prézelin B B, Campbell L, et al. 1988. A nitrate-dependent *Synechococcus* bloom in surface Sargasso Sea water. Nature, 331(6152): 161-163.
Gorsky G, Chrétiennot-Dinet M J, Blanchot J, et al. 1999. Picoplankton and nanoplankton aggregation by appendicularians: fecal pellet contents of *Megalocercus huxleyi* in the equatorial Pacific. Journal of Geophysical Research, 104: 3381-3390.
Guillou L, Jacquet S, Chrétiennot-Dinet M J, et al. 2001. Grazing impact of two small heterotrophic flagellates on *Prochlorococcus* and *Synechococcus*. Aquatic Microbial Ecology, 26(2): 201-207.
Gutiérrez-Rodríguez A, Décima M, Popp B N, et al. 2014. Isotopic invisibility of protozoan trophic steps in marine food webs. Limnology and Oceanography, 59(5): 1590-1598.
Heinbokel J F, Beers J R. 1979. Studies on the functional role of tintinnids in the Southern California Bight. III. Grazing impact of natural assemblages. Marine Biology, 52(1): 23-32.
Hernroth L. 1983. Marine pelagic rotifers and tintinnids-important trophic links in the spring plankton community of the Gullmar Fjord, Sweden. Journal of Plankton Research, 5(6): 835-846.
Hirose M, Katano T, Nakano S I. 2008. Growth and grazing mortality rates of *Prochlorococcus*, *Synechococcus* and eukaryotic picophytoplankton in a bay of the Uwa Sea, Japan. Journal of Plankton Research, 30(3): 241-250.
Jeong H J, Park J Y, Nho J H, et al. 2005. Feeding by red-tide dinoflagellates on the cyanobacterium *Synechococcus*. Aquatic Microbial Ecology, 41(2): 131-143.
Jochem F J. 2003. Photo- and heterotrophic pico- and nanoplankton in the Mississippi River plume: distribution and grazing activity. Journal of Plankton Research, 25(10): 1201-1214.
Johnson P W, Xu H S, Sieburth J M. 1982. The utilization of chroococcoid cyanobacteria by protozooplankers but not by calanoid copepods. Annales de l'Institut Oceanographique, Paris(Nouvelle Serie), 58(Suppl.): 297-308.
Kamiyama T. 1994. The impact of grazing by microzooplankton in northern Hiroshima Bay, the Seto Inland Seam, Japan. Marine Biology, 119(1): 77-88.
Kimmance S A, Wilson W H, Archer S D. 2007. Modified dilution technique to estimate viral versus grazing mortality of phytoplankton: limitations associated with method sensitivity in natural waters. Aquatic Microbial Ecology, 49: 207-222.
Kudoh S, Kanda J, Takahashi M. 1990. Specific growth rates and grazing mortality of chroococcoid

cyanobacteria *Synechococcus* spp. in pelagic surface waters in the sea. Journal of Experimental Marine Biology and Ecology, 142(3): 201-212.

Landry M R. 1981. Switching between herbivory and carnivory by the planktonic marine copepod *Calanus pacificus*. Marine Biology, 65(1): 77-82.

Landry M R, Brown S L, Campbell L, et al. 1998. Spatial patterns in phytoplankton growth and microzooplankton grazing in the Arabian Sea during monsoon forcing. Deep Sea Research Part II: Topical Studies in Oceanography, 45(10-11): 2353-2368.

Landry M R, Constantinou J, Kirshtein J. 1995a. Microzooplankton grazing in the central equatorial Pacific during February and August, 1992. Deep Sea Research Part II: Topical Studies in Oceanography, 42(2-3): 657-671.

Landry M R, Décima M R. 2017. Protistan microzooplankton and the trophic position of tuna: quantifying the trophic link between micro- and mesozooplankton in marine foodwebs. ICES Journal of Marine Science, 74(7): 1885-1892.

Landry M R, Haas L W, Fagerness V L. 1984. Dynamics of microbial plankton communities: experiments in Kaneohe Bay, Hawaii. Marine Ecology Progress Series, 16: 127-133.

Landry M R, Hassett R P. 1982. Estimating the grazing impact of marine micro-zooplankton. Marine Biology, 67(3): 283-288.

Landry M R, Kirshtein J, Constantinou J. 1995b. A refined dilution technique for measuring the community grazing impact of microzooplankton, with experimental tests in the central equatorial Pacific. Marine Ecology Progress Series, 120: 53-63.

Laval M. 1971. Ultrastructure et mode de nutrition du choanoflagelle *Salpingoeca pelagica* sp. nov. Comparaison avec les choanocytes des spongiaires. Protistologica, 7: 325-336.

Lehrter J C, Pennock J, Mcmanus G B. 1999. Microzooplankton grazing and nitrogen excretion across a surface estuarine-coastal interface. Estuaries, 22(1): 113-125.

Lessard E J, Murrell M C. 1998. Microzooplankton herbivory and phytoplankton growth in the northwestern Sargasso Sea. Aquatic Microbial Ecology, 16(2): 173-188.

Lessard E J, Swift E. 1985. Species-specific grazing rates of heterotrophic dinoflagellates in oceanic waters, measured with a dual-label radioisotope technique. Marine Biology, 87(3): 289-296.

Liu H, Campbell L, Landry M R. 1995. Growth and mortality rates of *Prochlorococcus* and *Synechococcus* measured with a selective inhibitor technique. Marine Ecology Progress Series, 116: 277-287.

Lohmann H. 1908. Untersuchungen zur feststellung des vollständigen gehaltes des meeres an plankton. Wissenschaftliche Meeresuntersuchungen Abt Kiel, Neue Folge, 10: 129-370.

McClelland J W, Holl C M, Montoya J P. 2003. Relating low $\delta^{15}N$ values of zooplankton to N_2-fixation in the tropical North Atlantic: insights provided by stable isotope ratios of amino acids. Deep Sea Research Part I: Oceanographic Research Papers, 50(7): 849-861.

McClelland J W, Montoya J P. 2002. Trophic relationships and the nitrogen isotopic composition of amino acids in plankton. Ecology, 83(8): 2173-2180.

Ning X, Vaulot D. 1992. Estimating *Synechococcus* spp. growth rates and grazing pressure by heterotrophic nanoplankton in the English Channel and the Celtic Sea. Acta Oceanologica Sinica, 11(2): 255-273.

Ning X R, Vaulot D. 1996. Simultaneous estimates of *Synechococcus* spp. growth and grazing mortality rates in the English Channel. Chinese Journal of Oceanology and Limnology, 14(1): 8-16.

Paranjape M A. 1987. Grazing by microzooplankton in the eastern Canadian Arctic in summer 1983. Marine Ecology Progress Series, 40: 239-246.

Paterson H, Knott B, Waite A M. 2007. Microzooplankton community structure and grazing on phytoplankton, in an eddy pair in the Indian Ocean off Western Australia. Deep Sea Research Part II: Topical Studies in Oceanography, 54(8): 1076-1093.

Perkins F O, Haas L W, Phillips D E, et al. 1981. Ultrastructure of a marine *Synechococcus* possessing spinae. Canadian Journal of Microbiology, 27(3): 318-329.

Quevedo M, Anadon R. 2001. Protist control of phytoplankton growth in the subtropical north-east Atlantic. Marine Ecology Progress Series, 221: 29-38.

Reckermann M, Veldhuis M J. 1997. Trophic interactions between picophytoplankton and micro-and nanozooplankton in the western Arabian Sea during the NE monsoon 1993. Aquatic Microbial Ecology, 12(3): 263-273.

Riley G A. 1956. Oceanography of Long Island Sound, 1952-1954. IX. Production and utilization of organic matter. Bulletin of the Bingham Oceanographic Collection, 15: 324-341.

Rivkin R B, Putland J N, Anderson M R, et al. 1999. Microzooplankton bacterivory and herbivory in the NE subarctic Pacific. Deep Sea Research Part II: Topical Studies in Oceanography, 46(11): 2579-2618.

Sherr B F, Sherr E B, Andrew T L, et al. 1986. Trophic interactions between heterotrophic protozoa and bacterioplankton in estuarine water analyzed with selective metabolic inhibitors. Marine Ecology Progress Series, 32: 169-179.

Silver M W, Bruland K W. 1981. Differential feeding and fecal pellet composition of salps and pteropods, and the possible origin of the deep-water flora and olive-green "cells". Marine Biology, 62(4): 263-273.

Strom S, Welschmeyer N A. 1991. Pigment-specific rates of phytoplankton growth and microzooplankton grazing in the open subarctic Pacific Ocean. Limnology and Oceanography, 36(1): 50-63.

Stukel M R, Décima M, Selph K E, et al. 2013. The role of *Synechococcus* in vertical flux in the Costa Rica upwelling dome. Progress in Oceanography, 112-113: 49-59.

Taguchi S. 1976. Microzooplankton and seston in Akkeshi Bay, Japan. Hydrobiologia, 50(3): 195-204.

Takahashi M, Hoskins K D. 1978. Winter condition of marine plankton populations in Saanich Inlet, B.C., Canada. II. Micro-zooplankton. Journal of Experimental Marine Biology and Ecology, 32(1): 27-37.

Tamigneaux E, Mingelbier M, Klein B, et al. 1997. Grazing by protists and seasonal changes in the size structure of protozooplankton and phytoplankton in a temperate nearshore environment(western Gulf of St. Lawrence, Canada). Marine Ecology Progress Series, 146: 231-247.

Tsai A Y, Chin W M, Chiang K P. 2009. Diel patterns of grazing by pigmented nanoflagellates on *Synechococcus* spp. in the coastal ecosystem of subtropical western Pacific. Hydrobiologia, 636(1): 249-256.

Tsuda A, Kawaguchi S. 1997. Microzooplankton grazing in the surface water of the Southern Ocean during an austral summer. Polar Biology, 18(4): 240-245.

Verity P G. 1986. Grazing of phototrophic nanoplankton by microzooplankton in Narragansett Bay. Marine Ecology Progress Series, 29: 105-115.

Verity P G, Stoecker D K, Sieracki M E, et al. 1993. Grazing, growth and mortality of microzooplankton during the 1989 North Atlantic spring bloom at 47°N, 18°W. Deep Sea Research Part I: Oceanographic Research Papers, 40(9): 1793-1814.

Verity P G, Stoecker D K, Sieracki M E, et al. 1996. Microzooplankton grazing of primary production at 140°W in the equatorial Pacific. Deep Sea Research Part II: Topical Studies in Oceanography, 43(4): 1227-1255.

Verity P G, Vernet M. 1992. Microzooplankton grazing, pigments, and composition of plankton communities during late spring in two Norwegian fjords. Sarsia, 77(3-4): 263-274.

Wang R, Li C, Wang K, et al. 1998. Feeding activities of zooplankton in the Bohai Sea. Fisheries Oceanography, 7(3-4): 265-271.

Waterhouse T Y, Welschmeyer N A. 1995. Taxon-specific analysis of microzooplankton grazing rates and phytoplankton growth rates. Limnology and Oceanography, 40(4): 827-834.

Worden A Z, Binder B J. 2003. Application of dilution experiments for measuring growth and mortality rates among *Prochlorococcus* and *Synechococcus* populations in oligotrophic environments. Aquatic Microbial Ecology, 30(2): 159-174.

Worden A Z, Nolan J K, Palenik B. 2004. Assessing the dynamics and ecology of marine picophytoplankton: the importance of the eukaryotic component. Limnology and Oceanography, 49(1): 168-179.

Zhao Y, Dong Y, Li H, et al. 2020. Grazing by microzooplankton and copepods on the microbial food web in spring in the southern Yellow Sea, China. Marine Life Science & Technology, 2: 442-455.

Zwirglmaier K, Spence E, Zubkov M V, et al. 2009. Differential grazing of two heterotrophic nanoflagellates on marine *Synechococcus* strains. Environmental Microbiology, 11(7): 1767-1776.

第九章　海洋浮游纤毛虫的摄食

海洋浮游纤毛虫（以下简称纤毛虫）根据肉体外是否具有壳体，可分为砂壳纤毛虫（tintinnid）和无壳纤毛虫（aloricate ciliate）两大类。纤毛虫主要隶属于旋毛纲（Sprirotrichea）下的环毛亚纲（Choreotrichia）和寡毛亚纲（Oligotrichia），另外也包括叶口纲（Litostomatea）下的中缢虫，以及一些寡膜纲（Oligohymenophora）下浮游生活的尾丝虫。纤毛虫是海洋微食物网的重要组成部分，是连接微食物网与经典食物链的重要环节。研究纤毛虫的摄食，包括食性、对不同饵料的摄食强度和选择性、影响摄食的因素及摄食后的同化和排遗等内容，对了解纤毛虫以至微食物网在海洋浮游生态系统物质循环和能量流动中的作用有重要意义。

第一节　纤毛虫的食性

研究纤毛虫的摄食首先要确定其食性，即饵料组成。研究纤毛虫食性的方法主要有两种：第一种是体内饵料颗粒法，检查纤毛虫体内因摄食而形成的食物泡，鉴定饵料的种类组成；第二种是维持生长法，在实验室内用不同的饵料来喂养纤毛虫，依据纤毛虫的生长状况来判断纤毛虫对饵料的摄食情况。

体内饵料颗粒法最先应用于纤毛虫食性研究。浮游纤毛虫研究的早期，在海上采集的砂壳纤毛虫体内的食物泡中发现有小的砂壳纤毛虫（Daday，1887；Blackbourn，1974）、颗石藻和无壳纤毛虫（Entz，1909）、细菌、硅藻、放射虫和金胞藻类（Campbell，1926，1927），以及甲藻（Beers and Stewart，1967）。在砂壳纤毛虫的食物泡中很少看到硅藻，可能是因为硅藻太大，无法被砂壳纤毛虫摄食（Blackbourn，1974）。体内饵料颗粒法在实验室内也得到了应用，Gold（1968）首次尝试在实验室用混合饵料喂养砂壳纤毛虫 *Tintinnopsis* sp.，混合饵料中包括藻类 *Rhodomonas lens*、*Isochrysis galbana*、*Platymonas tetrathele*、*Saccharomyces cerevisiae*、*Diaphanoeca grandis* 和细菌，经培养后观察 *Tintinnopsis* sp.的食物泡，发现 *Tintinnopsis* sp.可摄食这些藻类和细菌。在室内实验中，砂壳纤毛虫 *Favella* sp.在饵料缺乏时会摄食自己的粪粒（Stoecker，1984）。

体内饵料颗粒法可以得到纤毛虫的饵料组成，但是该方法存在局限性，无法说明被摄食饵料对纤毛虫的营养价值。可将体内饵料颗粒法与维持生长法相结合，先观察纤毛虫体内存在哪些类型的饵料颗粒，再使用相同或相近的饵料喂食纤毛虫，检验不同种类的饵料对于纤毛虫的营养价值。用不同饵料培养纤毛虫时，纤毛虫的生长情况可以作为饵料对纤毛虫是否具有营养价值的指标。

Gold（1968，1969，1973）最先在实验室内用含色素鞭毛虫作为饵料成功培养砂壳纤毛虫，虽然他没有进行纤毛虫摄食观察，但这些培养实验已经可以说明砂壳纤毛虫摄食自养鞭毛虫并维持生长。也有培养实验认为鞭毛虫并不是纤毛虫的适宜饵料，如 Jonsson（1986）发现无壳纤毛虫 *Strombidium vestitum*、*S. reticulatum* 和 *Lohmanniella spiralis* 虽然能够摄食一些异养和混合营养鞭毛虫，但是却不生长。

在自然海区有研究观察到砂壳纤毛虫偏好摄食鞭毛虫。Modigh 等（2003）检查自然海区采集的砂壳纤毛虫体内的食物泡时发现其中含有聚球藻，但更主要的饵料是 nano 级鞭毛虫。Modigh 等（2003）认为 nano 级鞭毛虫的丰度较低，而砂壳纤毛虫若偏好以 nano 级鞭毛虫为食，必然饵料受限，导致丰度不会太高，这可能是自然海区砂壳纤毛虫丰度不高的原因之一。

多数研究表明，硅藻基本不能作为饵料维持纤毛虫的生长。Blackbourn（1974）发现，砂壳纤毛虫 *Favella serrata* 对硅藻不摄食或很少摄食。Johansen（1976）报道，砂壳纤毛虫在只有硅藻作为饵料的情况下不能培养成功。Jonsson（1986）在实验室中观察到无壳纤毛虫 *Strombidium* 属的部分种可以摄食硅藻，但是并不生长。Capriulo 和 Carpenter（1980）在自然海区进行培养实验，发现包括浮游纤毛虫在内的微型浮游动物可以摄食较小的硅藻，但是有长须的硅藻（*Thalassiosira* sp. 或 *Chaetoceros* sp.）不易被摄食，这可能是因为硅藻的 β-几丁质长须（β-chitin thread）使其不易被浮游动物摄食（Gifford et al.，1981）。Verity 和 Villareal（1986）也得到了类似的实验结果，他们使用带长须的硅藻和不带长须的硅藻分别喂养砂壳纤毛虫 *Tintinnopsis acuminata* 和 *T. vasculum*，两种砂壳纤毛虫在没有长须的硅藻作为饵料时生长很快，在有长须的硅藻作为饵料时几乎不生长，较小的 *T. vasculum* 还出现了个体的死亡。

不同种类的甲藻对于纤毛虫营养价值不同。Stoecker 等（1981）发现，并不是所有的甲藻都是砂壳纤毛虫 *Favella ehrenbergii* 的适宜饵料，甲藻 *Gonyaulax tamarensis*、*G. polyedra*、*Heterocapsa* sp.能很好地支持 *F. ehrenbergii* 的生长；甲藻 *Prorocentrum mariaelebouriae* 虽然被摄食，但 *F. ehrenbergii* 的生长率却较低；甲藻 *Amphidinium carterae* 会产生类似胆碱的物质，不被 *F. ehrenbergii* 摄食。通过对比喂食各种饵料对纤毛虫生长的影响，研究发现砂壳纤毛虫 *F. campanula* 和 *F. ehrenbergii* 必须要有甲藻作为饵料才能存活（Gold，1969；Stoecker et al.，1981），而甲藻 *Heterocapsa triquetra* 是培养砂壳纤毛虫 *F. taraikaensis* 的理想饵料（Kamiyama et al.，2005）。

Verity 和 Villareal（1986）在实验室内用海洋聚球藻培养两种海洋浮游砂壳纤毛虫（*Tintinnopsis acuminate* 和 *T. vasculum*）都没有成功，可能是这两种砂壳纤毛虫不摄食聚球藻，也可能是由于聚球藻营养不够，无法维持砂壳纤毛虫的生长。但也有研究表明，在自然海区中聚球藻是纤毛虫的重要饵料之一。Sherr 等（1986）用荧光显微镜在自然海区纤毛虫体内的食物泡中发现了聚球藻。Pérez 等（1996）发现在自然海水中添加聚球藻，纤毛虫的生长得到促进，而在自然海水中添加纤毛虫，聚球藻的生长得到明显抑制。Bernard 和 Rassoulzadegan（1993）在法国的 Villefranche 湾检查砂壳纤毛虫体内的食物泡时发现，聚球藻是砂壳纤毛虫 *Rhabdonella spiralis* 的主要饵料，是砂壳纤毛虫 *Salpingella acuminata* 的唯一饵料。Pitta 等（2001）用荧光显微镜计数地中海纤毛虫体内的聚球藻时发现，纤毛虫体内的聚球藻最多可达 14 个，平均为 0.94 个，其中无壳纤毛虫体内平均有 0.28 个聚球藻。

除了上述方法之外，还有研究用其他方法来确认纤毛虫的饵料组成，如通过监测饵料的减少来证明纤毛虫对饵料的摄食。Gold（1969）将砂壳纤毛虫 *Favella campanula* 和甲藻 *Glenodinium foliaceum* 或 *Peridinium trochoideum* 混合（丰度比例 1∶14），经过一晚的摄食后加入 ^{14}C 标记的 $NaHCO_3$，光合作用 1h，然后测定甲藻固定的 ^{14}C，发现实验组甲藻的光合作用比对照组平均低了 38%，因此可以证明纤毛虫摄食了甲藻。

第二节　纤毛虫的摄食强度和对饵料的选择性

如第五章所介绍，衡量纤毛虫的摄食强度的参数主要包括摄食速率和清滤速率等，主要研究方法包括体内颗粒增加法或水体颗粒减少法。其中，体内颗粒增加法主要用于研究无壳纤毛虫和透明壳砂壳纤毛虫的摄食，黏着壳砂壳纤毛虫由于其壳上有黏着颗粒，不易观察体内的饵料颗粒，所以黏着壳砂壳纤毛虫摄食的研究较少。

自然饵料进入纤毛虫体内后不易被辨认，所以有人使用各种颗粒来替代饵料研究摄食，这些饵料替代颗粒包括活性炭（Spittler，1973）、淀粉（Spittler，1973；Kivi and Setälä，1995）、刚果红染色的酵母菌（Spittler，1973）等。后来随着荧光标记技术的发展，更多的研究开始使用荧光标记颗粒喂食纤毛虫，使用荧光显微镜观察即可统计纤毛虫摄入体内的颗粒数量。这些颗粒可以是荧光微球（Jonsson，1986），也可以是荧光标记细菌（Sherr et al.，1987）或荧光标记藻类（Sherr et al.，1991）。由于纤毛虫摄食过程可能存在机械感应或化学感应，有些替代饵料获得的实验结果可能无法反映真实的摄食情况。另外，多数替代饵料不能运动，也可能导致被捕获的概率降低（Bernard and Rassoulzadegan，1990）。

除了这两种方法外，还有一些间接估算方法。例如，对于无壳纤毛虫 *Strombidium sulcatum*，其游泳速度约为 600μm/s，用于摄食的围口小膜的面积约为 1200μm², 由此可估算 *S. sulcatum* 的清滤速率为 2.5μl/(grazer·h)(Fenchel and Jonsson, 1988)。

一、摄食强度

迄今为止, 纤毛虫对替代饵料颗粒、纤毛虫、甲藻、鞭毛虫、聚球藻和原绿球藻及细菌的摄食研究均有展开。

个体较大的纤毛虫可以摄食较小的纤毛虫, 纤毛虫对纤毛虫的摄食速率范围在 0.15～4.5prey/(grazer·h), 清滤速率在 12.2～129.86μl/(grazer·h)。纤毛虫对甲藻摄食的研究最广泛, 摄食速率通常在 0.1～500prey/(grazer·h), 清滤速率范围为 0.3～110μl/(grazer·h)。纤毛虫对鞭毛虫的摄食研究报道也较多, 摄食速率与清滤速率的范围分别为 0.095～51prey/(grazer·h) 和 0.02～17.4μl/(grazer·h)。纤毛虫对聚球藻和原绿球藻的摄食速率, 实验室培养实验的结果与自然海区的结果差异较大, 自然海区的摄食速率范围是 0.13～0.41prey/(grazer·h), 而实验室内得出的摄食速率则很高, 范围通常在 18～920prey/(grazer·h)。纤毛虫对聚球藻和原绿球藻的清滤速率较小, 范围是 44～515nl/(grazer·h)。详细的纤毛虫摄食速率和清滤速率数据可参见张武昌等(2016)。

纤毛虫摄食细菌研究使用的饵料有两种, 一种是天然细菌(也包括用同位素标记的细菌), 另一种是荧光标记细菌(fluorescence labeled bacteria, FLB)。Sherr 等 (1987) 使用荧光染料 DTAF[5-(4,6-二氯三嗪基)氨基荧光素, 5-(4,6-dichlorotriazin-2-yl) amino]对细菌进行染色得到 FLB, 用 FLB 喂食纤毛虫以研究其对细菌的摄食情况。实验证明 FLB 对原生动物无毒性, 纤毛虫能消化 FLB 并生长, 生长状况和摄食未染色的同株细菌没有差异, 表明纤毛虫对 FLB 与未标记细菌不存在选择性摄食。纤毛虫对细菌的摄食速率为 3.2～630prey/(grazer·h), 清滤速率常见范围在 20～300nl/(grazer·h), 但在马尾藻海分离的一株未定种无壳纤毛虫对细菌具有极高的清滤速率, 达 213 000nl/(grazer·h)(Lessard and Swift, 1985)。

另外, 还有研究利用细菌的生长率和总生长效率(gross growth efficiency)的数据推算纤毛虫对细菌的摄食速率, 提出纤毛虫 *Strombidium sulcatum* 对细菌的摄食速率为 4.3×10^3～237.6×10^3 prey/(grazer·h), 清滤速率为 602～73100nl/(grazer·h)(Rivier et al., 1985), 但该研究的摄食速率结果远高于其他文献, 这种方法的可靠性还需进一步验证。

考虑到不同纤毛虫及饵料种类均有较大的粒径范围, 比较不同纤毛虫种类的摄食强度通常使用特定碳生物量摄食速率(carbon specific ingestion rate)和特定体积清滤速率(body-volume specific clearance rate)这两个参数, 前者为摄食者在单位时间内摄食的饵料生物量与摄食者自身生物量的比值，后者为摄食者在单位时间内清滤的体积与摄食者自身体积的比值。

不同种类纤毛虫和饵料的特定碳生物量摄食速率较为接近，如无壳纤毛虫 *Strombidium* sp.的特定碳生物量摄食速率为 0.05～0.19/h（Rose et al., 2013），砂壳纤毛虫 *Favella taraikaensis* 的特定碳生物量摄食速率为 0.05～0.12/h，最大为 0.15/h（Kamiyama et al., 2005）。但不同种类纤毛虫对饵料的特定体积清滤速率有较大差异，范围可相差 1～3 个数量级，如无壳纤毛虫 *Strombidium sulcatum* 对 0.49μm 荧光微球的特定体积清滤速率是 $6×10^2$～$1.6×10^3$/h，对 1.03μm 荧光微球的特定体积清滤速率为 $0.8×10^4$～$1.2×10^4$/h，对原绿球藻的特定体积清滤速率为 $2.9×10^3$/h，对聚球藻的特定体积清滤速率为 $3.29×10^4$/h（Christaki et al., 1998），*Strombidium* sp.对微藻 *Isochrysis galbana* 和 *Nannochloropsis* sp.的最大特定体积清滤速率分别为 $5.2×10^5$/h 和 $1.8×10^5$/h（Chen et al., 2010）。

二、对饵料的选择性

纤毛虫对不同的饵料有一定选择性，这可能源自其对饵料物理性质（大小、形状、质地）或化学性质（如化学成分）的辨别和喜好。

（一）饵料的粒级

小个体的纤毛虫种类通常仅能摄食小粒级的饵料，较大个体的纤毛虫种类摄食饵料的粒级则通常较广，但随着纤毛虫本身粒级的增大，其摄食小粒级饵料的比例降低。通过检查自然海水中的纤毛虫体内的饵料颗粒，Rassoulzadegan 等（1988）发现，粒级小于 30μm 的纤毛虫的饵料有 72% 是 pico 级（0.2～2μm）颗粒，28% 是 nano 级（2～20μm）；粒级为 30～50μm 纤毛虫的饵料由 30% 的 pico 级和 70% 的 nano 级颗粒组成；粒级大于 50μm 的纤毛虫的饵料则由 5% 的 pico 级和 95% 的 nano 级颗粒组成。这种选择性在无壳纤毛虫和砂壳纤毛虫摄食中都存在，如在地中海 88% 的小个体（＜30μm）无壳纤毛虫和 80% 的小个体砂壳纤毛虫仅摄食小于 3μm 的饵料，只有 12% 的无壳纤毛虫和 20% 的砂壳纤毛虫可以摄食大于 3μm 的饵料；而对于大个体（30～50μm）纤毛虫，无壳纤毛虫和砂壳纤毛虫分别约有 39% 和 46% 的个体可以摄食大于 3μm 的饵料（Pitta et al., 2001）。

纤毛虫所能捕捉的饵料的粒级受其围口部（peristome）结构限制。图 9-1 以无壳纤毛虫 *Lohmanniella spiralis* 为例展示纤毛虫的围口部结构，多根纤毛融合成一片围口小膜（membranelle），多片围口小膜按一定间距排列在围口部周边。邻近两片围口小膜间的空隙距离决定了纤毛虫能够捕获的最小饵料，只有饵料粒径大于小膜间隙时才能成功被小膜捕获，若饵料粒径小于小膜间隙，将会随水流从小膜之间流失。例如，*L. spiralis* 小膜基部的间隙约为 2μm，它可以摄食粒径 2.11μm 的微球，但无法摄食粒径 1.11μm 的微球；无壳纤毛虫 *Strombidium reticulatum* 的小膜间隙更小（约 1.2μm），它可以成功摄食粒径 1.11μm 的微球，但对其的清滤速率远低于粒级 2.11~5.7μm 的微球（Jonsson，1986）。

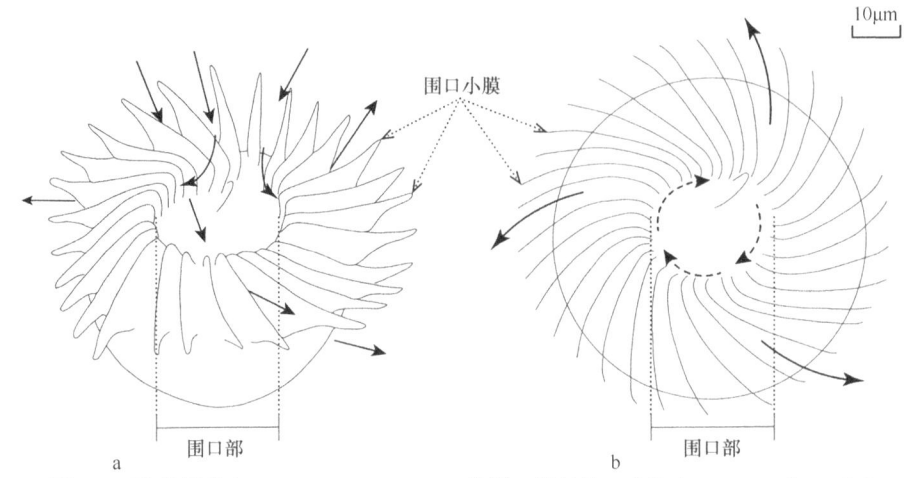

图 9-1　无壳纤毛虫 *Lohmanniella spiralis* 的围口部结构 [改绘自 Jonsson（1986）]
a. 侧面观；b. 顶面观，实线箭头表示摄食时由小膜摆动产生的水流方向，虚线箭头表示围口部内部的水流方向

纤毛虫所能摄食饵料的粒径上限由虫体围口部直径决定。即使能被围口小膜截留，超过围口部直径的饵料通常也不被纤毛虫摄入。对于砂壳纤毛虫来说，饵料粒径还需小于壳开口直径。在实验室培养实验中，无壳纤毛虫和砂壳纤毛虫能摄食的最大饵料颗粒直径与围口部直径（或壳开口直径）之比通常在 40%~45%。例如，无壳纤毛虫 *Lohmanniella spiralis*、*Strombidium reticulatum* 和 *S. vestitum* 能够摄食的最大饵料粒径分别为 14.4μm、7.9μm 和 6.4μm，均为其围口部直径的 40%（Jonsson，1986）。Spittler（1973）用直径为 3~60μm 的淀粉颗粒投喂砂壳纤毛虫，观察到砂壳纤毛虫摄食的最大颗粒直径约为壳开口直径的 41.2%~45%。Heinbokel（1978b）发现，砂壳纤毛虫能够摄食的最大淀粉颗粒为其围口部直径的 43%，且清滤速率与壳开口直径成正比。

也有部分室内培养实验发现纤毛虫可摄食的颗粒较大。例如，砂壳纤毛虫 *Favella ehrenbergii* 和 *F. taraikaensis* 摄食饵料的最大粒径分别为壳开口直径的 59% 与 78%（Kamiyama and Arima，2001）。砂壳纤毛虫 *Tintinnopsis* sp. 可摄食横截面直径超过纤毛虫壳开口直径 50% 的硅藻饵料（Verity and Villareal，1986）。砂壳纤毛虫 *Stenosemella ventricosa* 口径为 32~36μm，其主要的饵料粒级为 3~12μm，但它也能摄食粒径超过 20μm 的颗粒，最大可达 26μm（Rassoulzadegan and Etienne，1981）。寡毛类纤毛虫能摄食与自己一样大的颗粒，在饵料缺乏时甚至可以同类相食（Gifford，1985）。还有研究使用野外收集的砂壳纤毛虫与饵料进行培养，发现部分砂壳纤毛虫甚至可以摄食超过其壳开口直径尺寸的饵料（Capriulo，1982）。自然饵料和人工饵料的培养实验结果不同，表明纤毛虫的化感作用可能在饵料选择过程中起重要作用，当饵料可口时，即使粒径超过最适范围也可以被摄食（Kamiyama and Arima，2001）。

研究表明，纤毛虫的摄食强度与饵料颗粒大小有关，通常饵料颗粒越大，纤毛虫对饵料的摄食速率和清滤速率就越高（Capriulo，1982）。例如，无壳纤毛虫 *Strombidium sulcatum* 对 0.51μm 颗粒的清滤速率仅为 2.83μm 颗粒的 5%（Fenchel and Jonsson，1988）。另一个研究发现，在饵料浓度相同的情况下，*S. sulcatum* 对聚球藻（粒径 0.98μm）的清滤速率为 0.58μl/（grazer·h），对微球藻（*Nannochloris* sp.，粒径 2.50μm）的清滤速率为 6.38μl/（grazer·h），可相差 1 个数量级（Bernard and Rassoulzadegan，1990）。

（二）饵料的化学性质

纤毛虫对饵料化学性质选择性的研究较少，也比较困难，主要难点在于寻找形状和运动能力相同（或相近）的饵料颗粒作为对照。有研究使用相同粒径的无涂层的荧光微球（plain fluorescent microsphere，pMS）和有羧酸酯涂层的微球（carboxylate microsphere，cMS）进行对照喂食培养，发现纤毛虫 *Strombidium sulcatum* 对 pMS 的清滤速率显著高于 cMS，表明纤毛虫在摄食时对饵料的化学性质有一定的选择性（Christaki et al.，1998）。

另一个研究使用微藻 *Nannochloropsis* sp.和 *Isochrysis galbana* 作为饵料培养一株未定种纤毛虫 *Strobilidium* sp.，这两种单胞藻饵料的粒径和 C、N 含量相近，但 *Strobilidium* sp.对 *Nannochloropsis* sp.的最大清滤速率远低于 *I. galbana*，证明纤毛虫对饵料的化学性质有选择性（Chen et al.，2010）。该研究还发现，在以 *Nannochloropsis* sp.为饵料时，*Strobilidium* sp.有更高的生长速率和总生长效率。这可能是由于在饵料质量不理想的情况下，摄食者可能需要提高摄食速率以最大限度地吸收必需营养物质，从而导致总生长效率降低。此外，*Nannochloropsis* sp.运动能力很差，而 *I. galbana* 则有很强的运动能力，也可能影响摄食速率。

第三节 影响纤毛虫摄食的因素

一、饵料浓度

饵料浓度对纤毛虫摄食的影响研究较多。通常在饵料浓度较低的条件下，随饵料浓度增加纤毛虫的清滤速率和摄食速率也增加，但当饵料浓度升高到一定阈值时，清滤速率和摄食速率达到峰值，经一定平台期之后则随饵料浓度的继续增加而降低。

砂壳纤毛虫 Favella azorica 摄食甲藻 Heterocapsa triquetra 和 H. circularisquama 的培养实验中，当饵料浓度在 400cell/ml 左右时，F. azorica 对 H. triquetra 和 H. circularisquama 的清滤速率达到最大值，分别为 18.2μl/（grazer·h）和 27.5μl/（grazer·h）；当饵料浓度超过 400cell/ml 时清滤速率则迅速降低；当饵料浓度升高至约 $7×10^3$cell/ml 时，F. azorica 对 H. triquetra 和 H. circularisquama 的清滤速率分别只有 0.9μl/（grazer·h）和 4.1μl/（grazer·h），比峰值分别下降了 95%和 85%；当饵料浓度高于 10^4cell/ml 时，已无法检测到清滤速率（Kamiyama，1997）。无壳纤毛虫 Strobilidium sp.摄食饵料藻 Nannochloropsis sp.时，清滤速率随着饵料浓度的增加而升高，饵料浓度为 1000μgC/L 时清滤速率达到最大，饵料浓度升高到 2000μgC/L 时清滤速率降低（Chen et al.，2010）。

拟铃虫属（Tintinnopsis）砂壳纤毛虫的摄食速率随着饵料浓度的增加而增加，在饵料浓度为 80μgC/L 时达到峰值，之后随饵料浓度升高摄食速率保持不变，直至饵料浓度超过 500μgC/L 时 Tintinnopsis sp.的摄食速率下降。但与摄食速率的变化趋势不同，Tintinnopsis sp.的清滤速率反而与饵料浓度呈负相关，在低饵料浓度（15～50μgC/L）条件下清滤速率最高（Verity，1985）。

也有研究发现，纤毛虫的摄食速率到达平台期后保持稳定，并不随饵料浓度升高而降低。例如，砂壳纤毛虫 Amphorides quadrilineata 在饵料藻 Isochrysis galbana 浓度为 100μgC/L 时达到最大摄食速率，此后随饵料浓度增加摄食速率始终保持恒定，甚至饵料浓度达到 600μgC/L 时仍未出现下降（Jakobsen et al.，2001）。砂壳纤毛虫 Favella taraikaensis 的摄食速率先随饵料甲藻 Alexandrium tamarense 浓度的升高而升高，当 A. tamarense 的浓度超过 100cell/ml 后，F. taraikaensis 的摄食速率维持恒定，饵料浓度为 800cell/ml 时仍未出现下降（Kamiyama et al.，2005）。

二、温度

多数研究表明温度升高将促使纤毛虫摄食速率增加。在 15～26℃条件下，寡毛类纤毛虫 Lohmanniella spiralis 的摄食速率随温度的升高而升高（Rassoulzadegan，1982）。在 5～25℃，温度每升高 10℃，砂壳纤毛虫 Tintinnopsis sp.的清滤速率与摄食速率分别升高 1.5～1.7 倍和 2.5～2.9 倍（Verity，1985）。砂壳纤毛虫 Favella sp. 的摄食速率在 8～12℃时较低，12～16℃时迅速升高，之后维持较高水平直至 21℃出现下降（Aelion and Chisholm，1985）。

也有研究未发现温度对砂壳纤毛虫摄食速率有影响。例如，在培养温度 8～20℃、饵料浓度 80～800cell/ml 的条件下，Favella sp.的清滤速率未表现出与温度的明显相关性（Stoecker and Guillard，1982）。Capriulo（1982）研究了 7 种从野外收集的砂壳纤毛虫对自然悬浮颗粒的摄食速率和清滤速率，认为纤毛虫的摄食与温度无关。

三、纤毛虫的生理状态

纤毛虫的生理状态可划分为 4 种时期：饥饿期、指数生长期（饵料不受限制）、静止期（饵料受限）和分裂滞后期（摄食很多饵料，但是还没有分裂），不同时期的纤毛虫摄食强度有差异。

饥饿期和指数生长期纤毛虫的摄食强度一般较高。Taniguchi 和 Kawakami（1985）发现，经过少于 45min 的饥饿处理后，砂壳纤毛虫的摄食速率升高。Rassoulzadegan 和 Etienne（1981）也发现经过一天的饥饿处理后，砂壳纤毛虫 *Favella ehrenbergii* 的摄食速率升高。纤毛虫 *Strombidium sulcatum* 和 *Uronema* sp. 在指数生长期的清滤速率均高于静止期，其中，*Uronema* sp.在指数生长期的清滤速率是静止期的 4~5 倍（Christaki et al.，1998）。

四、光照

光照对纤毛虫摄食强度的影响不同研究结论不同。有研究认为光照促使纤毛虫摄食强度增加。例如，Rassoulzadegan（1978）发现在有光照的条件下砂壳纤毛虫 *Favella ehrenbergii* 的摄食速率较高，Stoecker 和 Guillard（1982）发现砂壳纤毛虫 *Favella* sp.在有光照条件下清滤速率增加，Strom（2001）发现纤毛虫 *Coxliella* sp.和 *Strombidinopsis acuminatum* 在中等光照下的摄食速率是黑暗时的 2~7 倍。

也有研究认为，光照降低了纤毛虫的摄食强度，如使用荧光标记藻类喂养纤毛虫 *Lohmanniella* sp.，在黑暗状态下摄食速率为 0.4prey/(grazer·min)，在 115μE/(m^2·s) 光照强度下，摄食速率降为 0.07prey/(grazer·min)，而且当光照突然关闭或打开时摄食速率立即表现出变化（Chen and Chang，1999）。

另外，有一些研究认为光照条件对纤毛虫摄食没有影响。Heinbokel（1978a）的砂壳纤毛虫培养实验未观察到摄食行为有明显的昼夜节律。砂壳纤毛虫 *Tintinnopsis parvala* 和 *T. cylindrica* 摄食微藻 *Pavlova lutheri* 也没有明显的昼夜节律（Blackbourn，1974）。

五、扰动

在微观尺度上，黏滞力对流场的影响大于惯性力，流场中的扰动会被黏滞力平滑掉，流体分层流动，互不混合，称为层流。纤毛虫及其饵料就生活在这样的微观尺度上，层流剪切作用于纤毛虫与饵料，使二者之间产生相对运动，可能影响摄食者与饵料的接触概率和捕食成功率。

已有的研究发现，水流扰动对纤毛虫清滤速率的影响因种而异。砂壳纤毛虫 *Helicostomella* sp.在扰动水体中摄食的清滤速率只有静止水体的 42%，无壳纤毛虫 *Strombidium sulcatum* 在扰动条件下清滤速率降低，但扰动对砂壳纤毛虫 *Favella* sp.的清滤速率没有明显影响（Shimeta et al., 1995; Dolan et al., 2003）。

第四节 同化和排遗

同化（assimilation）是指摄食者对所摄入饵料的消化吸收作用。砂壳纤毛虫 *Favella ehrenbergii* 和 *Stenosemella ventricosa* 每天的同化量分别相当于自身碳生物量的 70%与 43%（Rassoulzadegan，1978；Rassoulzadegan and Etienne，1981）。

并非所有被摄入的饵料都能被消化吸收，未被同化的食物颗粒会被纤毛虫排出，称为排遗（egestion）。纤毛虫同化的饵料与摄食的总饵料之比为同化率。Stoecker（1984）用甲藻 *Heterocapsa triquetra* 喂养砂壳纤毛虫 *Favella* sp.，观察砂壳纤毛虫的粪粒，发现里面最多可以有 4 个饵料藻的残余，粒径最大为 19μm×32μm，粪粒的体积是其摄食饵料体积的 21%~22%，因为 *Favella* sp.摄食甲藻时将细胞整个吞下，所以估计纤毛虫的同化率为 78%~79%。Rassoulzadegan（1978）估计砂壳纤毛虫 *Favella ehrenbergii* 的同化率为 67%。

有研究分析了纤毛虫从摄入饵料到发生排遗的时间长度。对于无壳纤毛虫 *Strombidium sulcatum*，当以粒径为 2.5μm 的甲藻 *Nannochloris* sp.为受试饵料时，纤毛虫出现排遗的时间最早，发生在摄食后(13.7±2.2)min；以细菌、微藻 *Emiliania huxleyi* 和 *Isochrysis galbana* 为饵料时，发生排遗的时间较为接近，为 20~21min；以聚球藻为饵料时，纤毛虫发生排遗最慢，需要（23.1±6.0）min（Bernard and Rassoulzadegan，1990）。砂壳纤毛虫 *Favella* sp.的排遗过程主要发生在摄食开始 12h 以后，并在 24h 内完成（Rassoulzadegan et al., 1988）。

还有研究用半消化时间来评估纤毛虫对饵料的同化和排遗速度。对于无壳纤毛虫 *Strombidium sulcatum* 来说，微藻 *Isochrysis galbana* 的半消化时间为 126min，聚球藻的半消化时间为 44~86min（Dolan and Šimek，1997）。Kopylov 和 Tumantseva（1987）测得砂壳纤毛虫 *Eutintinnus fraknoii* 和 *Epiplocycloides reticulata* 摄食藻类的半消化时间约为（60±10）min。

参 考 文 献

张武昌, 陈雪, 李海波, 等. 2016. 海洋浮游纤毛虫摄食研究综述. 海洋与湖沼, 47(1): 276-289.

Aelion C M, Chisholm S W. 1985. Effect of temperature on growth and ingestion rates of *Favella* sp. Journal of Plankton Research, 7(6): 821-830.

Beers J R, Stewart G L. 1967. Micro-zooplankton in the euphotic zone at five locations across the California current. Journal of the Fisheries Research Board of Canada, 24(10): 2053-2068.

Bernard C, Rassoulzadegan F. 1990. Bacteria or microflagellates as a major food source for marine ciliates: possible implications for the microzooplankton. Marine Ecology Progress Series, 64: 147-155.

Bernard C, Rassoulzadegan F. 1993. The role of picoplankton(cyanobacteria and plastidic picoflagellates) in the diet of tintinnids. Journal of Plankton Research, 15(4): 361-373.

Blackbourn D J. 1974. The feeding biology of tintinnid protozoa and some other inshore microzooplankton. Vancouver: University of British Columbia.

Campbell A S. 1926. The cytology of *Tintinnopsis nucula*(FOL) Laackmann with an account of its neuromotor apparatus, division and a new intranuclear parasite. University of California Publications in Zoology, 29: 179-236.

Campbell A S. 1927. Studies on the marine ciliate *Favella*(Jörgensen), with special regard to the neuromotor apparatus and its role in the formation of the lorica. University of California Publications in Zoology, 29: 429-452.

Capriulo G M. 1982. Feeding of field collected tintinnid micro-zooplankton on natural food. Marine Biology, 71(1): 73-86.

Capriulo G M, Carpenter E J. 1980. Grazing by 35 to 202 μm micro-zooplankton in Long Island Sound. Marine Biology, 56(4): 319-326.

Chen B, Liu H, Lau M T S. 2010. Grazing and growth responses of a marine oligotrichous ciliate fed with two nanoplankton: does food quality matter for micrograzers? Aquatic Ecology, 44(1): 113-119.

Chen K, Chang J. 1999. Influence of light intensity on the ingestion rate of a marine ciliate, *Lohmanniella* sp. Journal of Plankton Research, 21(9): 1791-1798.

Christaki U, Dolan J R, Pelegri S, et al. 1998. Consumption of picoplankton-size particles by marine ciliates: effects of physiological state of the ciliate and particle quality. Limnology and Oceanography, 43(3): 458-464.

Daday E V. 1887. Monographie der familie der tintinnodeen. Mittheilungen aus der Zoologischen Station zu Neapel, 7: 473-591.

Dolan J R, Sall N, Metcalfe A, et al. 2003. Effects of turbulence on the feeding and growth of a marine oligotrich ciliate. Aquatic Microbial Ecology, 31(2): 183-192.

Dolan J R, Šimek K. 1997. Processing of ingested matter in *Strombidium sulcatum*, a marine ciliate(Oligotrichida). Limnology and Oceanography, 42(2): 393-397.

Entz G J. 1909. Studien über organisation und biologie der tintinniden. Archiv für Protistenkunde, 15: 93-226.

Fenchel T, Jonsson P R. 1988. The functional biology of *Strombidium sulcatum*, a marine oligotrich ciliate(Ciliophora, Oligotrichina). Marine Ecology Progress Series, 48: 1-15.

Gifford D J. 1985. Laboratory culture of marine planktonic oligotrichs(Ciliophora, Oligotrichida). Marine Ecology Progress Series, 23: 257-267.

Gifford D J, Bohrer R N, Boyd C M. 1981. Spines on diatoms: do copepods care? Limnology and Oceanography, 26(6): 1057-1061.

Gold K. 1968. Some observations on the biology of *Tintinnopsis* sp. Journal of Eukaryotic

Microbiology, 15(1): 193-194.

Gold K. 1969. Tintinnida: feeding experiments and lorica development. The Journal of Protozoology, 16(3): 507-509.

Gold K. 1973. Methods for growing tintinnida in continuous culture. Integrative and Comparative Biology, 13(1): 203-208.

Heinbokel J F. 1978a. Studies on the functional role of tintinnids in the Southern California Bight.I. Grazing and growth rates in laboratory cultures. Marine Biology, 47(2): 177-189.

Heinbokel J F. 1978b. Studies on the functional role of tintinnids in the Southern California Bight. II. Grazing rates of field populations. Marine Biology, 47(2): 191-197.

Jakobsen H H, Hyatt C J, Buskey E J. 2001. Growth and grazing on the 'Texas brown tide' alga *Aureoumbra lagunensis* by the tintinnid *Amphorides quadrilineata*. Aquatic Microbial Ecology, 23(3): 245-252.

Johansen P L. 1976. A study of tintinnids and other protozoa in eastern Canadian waters with special reference to tintinnid feeding, nitrogen excretion and reproduction rates. Nova Scotia: Dalhousie University.

Jonsson P R. 1986. Particle size selection, feeding rates and growth dynamics of marine planktonic oligotrichous ciliates(Ciliophora: Oligotrichina). Marine Ecology Progress Series, 33: 265-277.

Kamiyama T. 1997. Growth and grazing responses of tintinnid ciliates feeding on the toxic dinoflagellate *Heterocapsa circularisquama*. Marine Biology, 128(3): 509-515.

Kamiyama T, Arima S. 2001. Feeding characteristics of two tintinnid ciliate species on phytoplankton including harmful species: effects of prey size on ingestion rates and selectivity. Journal of Experimental Marine Biology and Ecology, 257(2): 281-296.

Kamiyama T, Tsujino M, Matsuyama Y, et al. 2005. Growth and grazing rates of the tintinnid ciliate *Favella taraikaensis* on the toxic dinoflagellate *Alexandrium tamarense*. Marine Biology, 147(4): 989-997.

Kivi K, Setälä O. 1995. Simultaneous measurement of food particle selection and clearance rates of planktonic oligotrich ciliates(Ciliophora: Oligotrichina). Marine Ecology Progress Series, 119: 125-137.

Kopylov A I, Tumantseva N I. 1987. Analysis of the contents of tintinnid food vacuoles and evaluation of their contribution to the consumption of phytoplankton production off the Peru coast. Oceanology, 27: 343-347.

Lessard E J, Swift E. 1985. Species-specific grazing rates of heterotrophic dinoflagellates in oceanic waters, measured with a dual-label radioisotope technique. Marine Biology, 87(3): 289-296.

Modigh M, Castaldo S, Saggiomo M, et al. 2003. Distribution of tintinnid species from 42°N to 43°S through the Indian Ocean. Hydrobiologia, 503(1): 251-262.

Pérez M T, Dolan J R, Rassoulzadegan F, et al. 1996. Predation on marine picoplankton populations examined with an 'add-in' approach. Journal of Plankton Research, 18(4): 635-641.

Pitta P, Giannakourou A, Christaki U. 2001. Planktonic ciliates in the oligotrophic Mediterranean Sea: longitudinal trends of standing stocks, distributions and analysis of food vacuole contents. Aquatic Microbial Ecology, 24(3): 297-311.

Rassoulzadegan F. 1978. Dimensions et taux d'ingestion des particules consommées par un tintinnide: *Favella ehrenbergii*(Clap.& Lachm.) Jörg., cilié pélagique marin. Annales de l'Institut Oceanographique, Paris(Nouvelle Serie), 54: 17-24.

Rassoulzadegan F. 1982. Dependence of grazing rate, gross growth efficiency and food size range on temperature in a pelagic oligotrichous ciliate *Lohmanniella spiralis* Leeg., fed on naturally occurring particulate matter. Annales de l'Institut Oceanographique, Paris(Nouvelle Serie), 58(Suppl.): 177-184.

Rassoulzadegan F, Etienne M. 1981. Grazing rate of the tintinnid *Stenosemella ventricosa*(Clap. & Lachm.) Jörg. on the spectrum of the naturally occurring particulate matter from a Mediterranean neritic area. Limnology and Oceanography, 26(2): 258-270.

Rassoulzadegan F, Lavalpeuto M, Sheldon R W. 1988. Partitioning of the food ration of marine ciliates between pico- and nanoplankton. Hydrobiologia, 159(1): 75-88.

Rivier A, Brownlee D C, Sheldon R W, et al. 1985. Growth of microzooplankton: a comparative study of bactivorous zooflagellates and ciliates. Marine Microbial Food Webs, 1: 51-60.

Rose J M, Fitzpatrick E, Wang A, et al. 2013. Low temperature constrains growth rates but not short-term ingestion rates of Antarctic ciliates. Polar Biology, 36(5): 645-659.

Sherr B F, Sherr E B, Fallon R D. 1987. Use of monodispersed, fluorescently labeled bacteria to estimate *in situ* protozoan bacterivory. Applied and Environmental Microbiology, 53(5): 958-965.

Sherr E, Sherr B, McDaniel J. 1991. Clearance rates of less <6μm fluorescently labeled algae(FLA) by estuarine protozoa-potential grazing impact of flagellates and ciliates. Marine Ecology Progress Series, 69(1-2): 81-92.

Sherr E B, Sherr B F, Fallon R D, et al. 1986. Small, aloricate ciliates as a major component of the marine heterotrophic nanoplankton. Limnology and Oceanography, 31(1): 177-183.

Shimeta J, Jumars P A, Lessard E J. 1995. Influences of turbulence on suspension feeding by planktonic protozoa: experiments in laminar shear fields. Limnology and Oceanography, 40(5): 845-859.

Spittler P. 1973. Feeding experiments with tintinnids. Oikos, 15: 128-132.

Stoecker D K. 1984. Particle production by planktonic ciliates. Limnology and Oceanography, 29(5): 930-940.

Stoecker D K, Guillard R R L. 1982. Effects of temperature and light on the feeding rate of *Favella* sp. (ciliated protozoa, suborder Tintinnina). Annales de l'Institut Oceanographique, Paris (Nouvelle Serie), 58(Suppl.): 309-318.

Stoecker D K, Guillard R R L, Kavee R M. 1981. Selective predation by *Favella ehrenbergii*(tintinnia) on and among dinoflagellates. The Biological Bulletin, 160(1): 136-145.

Strom S L. 2001. Light-aided digestion, grazing and growth in herbivorous protists. Aquatic Microbial Ecology, 23(3): 253-261.

Taniguchi A, Kawakami R. 1985. Feeding activity of a tintinnid ciliate *Favella taraikaensis* and its variability observed in laboratory cultures. Marine Microbial Food Webs, 1(1): 17-34.

Verity P G. 1985. Grazing, respiration, excretion, and growth rates of tintinnids. Limnology and Oceanography, 30(6): 1268-1282.

Verity P G, Villareal T A. 1986. The relative food value of diatoms, dinoflagellates, flagellates, and cyanobacteria for tintinnid ciliates. Archiv für Protistenkunde, 131(1-2): 71-84.

第十章 海洋桡足类对纤毛虫的摄食

微食物网概念提出后,微食物网的生产力如何被传递到高营养级就成为海洋浮游生态学的关键问题之一。桡足类(copepods)是中型浮游动物(体长>200μm)的优势类群,约占浮游动物生物量的80%(Verity and Paffenhofer,1996),其摄食的饵料粒径通常大于5μm。在微食物网中,异养细菌、蓝细菌和鞭毛虫等类群粒径微小,难以被桡足类直接摄食。浮游纤毛虫是粒级最大的微食物网生物,可以被桡足类等中型浮游动物摄食,是连接海洋浮游微食物网和经典食物链的关键环节。

桡足类的饵料通常包括浮游植物、纤毛虫等微型浮游生物,其中浮游植物被认为是桡足类的主要饵料(Calbet and Saiz,2005)。但浮游植物很难提供桡足类自身生长代谢所需的全部营养物质(Dam et al.,1994;Carlotti and Radach,1996;Peterson and Dam,1996),需要有其他来源如纤毛虫等补充。有实验表明,在浮游植物和纤毛虫共存的情况下,桡足类会优先摄食纤毛虫(Stoecker and Sanders,1985;Stoecker and Egloff,1987)。

第一节 桡足类摄食纤毛虫的早期证据

20世纪60年代起,就开始有研究发现一些桡足类摄食纤毛虫的间接或直接证据。间接的证据包括:①围隔培养与自然海区中的纤毛虫生物量与桡足类生物量常呈现负相关关系(Smetacek,1981;Sheldon et al.,1986);②有研究用沼泽植物碎屑培养桡足类 *Eurytemora affinis*,当碎屑饵料中不含细菌和原生动物时,*E. affinis* 的生存和产卵状态较差,当碎屑饵料中含有大量原生动物,主要是纤毛虫时,*E. affinis* 能够良好生存和产卵,据此推断桡足类能够摄食纤毛虫(Heinle et al.,1977)。

直接的证据是在一些桡足类的肠道(Mullin,1966;Harding,1974)和粪便(Turner and Anderson,1983;Turner,1984)中观察到了砂壳纤毛虫的壳。Berk等(1977)用粒径 8.88~17.30μm 的海洋纤毛虫 *Uronema nigricans* 投喂桡足类 *Eurytemora affinis*,当饵料丰度为 $4.17×10^3$~$7.44×10^3$ind./ml 时,可检测到 *E. affinis* 对 *U. nigricans* 的摄食速率为 717~5688prey/(grazer·h),清滤速率为 0.068~0.934ml/(ind.·h)。

无壳纤毛虫很脆弱,极易解体,所以很难判断摄食实验中无壳纤毛虫的丰度降低是由于被摄食还是解体造成的。直到 Merrell 和 Stoecker(1998)用活细胞示踪染料 5-氯甲基荧光素二乙酸酯（5-chloromethylfluorescein diacetate，CMFDA）对无壳纤毛虫进行染色并投喂桡足类 *Eurytemora affinis*，随后用显微镜观察到桡足类体内有荧光，才首次确认桡足类确实能够摄食无壳纤毛虫。

第二节 桡足类摄食纤毛虫的室内培养研究

一、桡足类摄食纤毛虫的摄食强度和功能反应

多数室内培养研究是将一种桡足类和一种纤毛虫进行共同培养，根据培养前后纤毛虫丰度的变化估算桡足类摄食的强度。表 10-1 展示了主要的桡足类与纤毛虫室内共培养实验研究，常见的桡足类摄食者有 6 种，其中研究最多的是汤氏纺锤水蚤（*Acartia tonsa*）；常见的纤毛虫饵料有 13 种，包括 7 种砂壳类和 6 种无壳类，使用最多的是具沟急游虫（*Strombidium sulcatum*）。摄食者对饵料的清滤速率最小是 0，因此该类研究给出的通常是最大清滤速率。室内培养发现桡足类对纤毛虫的最大清滤速率范围主要在 0.1~58.1ml/（ind.·h）（表 10-1）。假设每升海水中有一个桡足类，取桡足类对纤毛虫的清滤速率值为 10ml/（ind.·h），则每个桡足类每天可以清滤 240ml 海水。

表 10-1 室内培养实验获得的桡足类对纤毛虫的清滤速率

桡足类	纤毛虫	纤毛虫大小 （μm）	最大清滤速率 [ml/(ind.·h)]	参考文献
	Tintinnopsis tubulosa	48×148	3.96~12.03	Robertson（1983）
	Favella panamensis	82×265	4.5	
	Strombidium sulcatum		1.5~5.6	Broglio 等（2003）
	Strombidium sulcatum		1.5~12.0	Saiz 和 Kiørboe（1995）；Kiørboe 等（1996）
	Favella sp.	65×150	2.5~8.7	Stoecker 和 Sanders（1985）
Acartia tonsa	*Favella* sp.	65×150	0.3~10.4	
	Tintinnopsis sp.	32×65	1.2~2.8	
	Strombilidium sp.	50~60	1.9	Stoecker 和 Egloff（1987）
	Strombidium sp.	42×43	2.5~3.1	
	Balanion sp.	32×34	4.2~4.4	
	Urotricha sp.	9×12	2.7	
	Strobilidium spiralis	65×65	3.58	
	Strombidium reticulatum	43×30	1.96	Jonsson 和 Tiselius（1990）
	Mesodinium rubrum	43×38	0.63	
Acartia hudsonica	*Eutintinnus pectinis*	20×50	0.19~0.38	Turner 和 Anderson（1983）

续表

桡足类	纤毛虫	纤毛虫大小* (μm)	最大清滤速率 [ml/(ind.·h)]	参考文献
Acartia clausi	Helicostomella fusiformis	23×110	0.3~0.9	Ayukai（1987）
	Favella taraikaensis	77×210	2.0	
	Strombidium sulcatum	30	1.4~26.3	Wiadnyana 和 Rassoulzadegan（1989）
	Lohmanniella oviformis	18	8.2	Gismervik（2006）
	Strombidium sp.	40	10.9	
	Strombidium conicum	48	2.6	
	Strombidium vestitum	23	2.4	
	Strobilidium undinum		0.4~0.8	Gismervik 和 Andersen（1997）
Centropages typicus	Strombidium sulcatum	30	5.4~58.1	
	Strombidium sulcatum		4.16~41.6	Caparroy 等（1998）
Eurytemora affinis	Uronema nigricans	9~17	0.1	Berk 等（1977）
Pseudocalanus sp.	Lohmanniella oviformis	18	17.7	Gismervik（2006）
	Strombidium sp.	40	16.8	
	Strombidium conicum	48	8.8	

*表示宽×长，或长度

功能反应（functional response）指摄食者的摄食速率对饵料丰度变化的反应，可通过培养实验获得功能反应曲线，以定量分析摄食者与饵料间的相互作用关系。图 10-1 展示了典型的桡足类摄食纤毛虫的功能反应曲线。当培养体系中饵料纤毛虫的丰度很低时，浮游桡足类清滤速率很低，摄食速率为零，桡足类几乎停止摄食活动，此时的饵料丰度称为摄食阈值丰度；随饵料丰度升高，摄食速率和清滤速率都迅速增加；饵料浓度继续升高，清滤速率保持不变，而摄食速率持续增加。当饵料丰度升高到一定程度后，摄食速率不再增加，而清滤速率下降，此时的饵料丰度称为摄食饱和丰度（图 10-1）。

为数不多的室内培养研究表明桡足类摄食纤毛虫时，摄食饱和丰度远远高于自然水体中的纤毛虫丰度。例如，纤毛虫 *Strombidium sulcatum* 的丰度达到 $2×10^4$ind./L 后，桡足类 *Acartia tonsa* 的摄食速率才不再增加（Saiz and Kiørboe，1995），而在自然海区 *S. sulcatum* 的丰度通常只有 500~1000ind./L，比摄食实验获得的摄食饱和丰度低 2 个数量级。另一个研究发现，*S. sulcatum* 丰度在 0~$2.5×10^4$ind./L 时，*A. tonsa* 的清滤速率都没有降低的趋势（Kiørboe et al.，1996）。其他研究发现，在纤毛虫 *Tintinnopsis tubulosa* 的丰度达到 $2×10^3$ind./L（Robertson，1983）、*Favella* sp.的丰度达到 $3.4×10^3$ind./L（Stoecker and Sanders，1985）时，桡足类 *A. tonsa* 的摄食速率仍呈线性增加。在纤毛虫 *Eutintinnus pectinis* 丰度达

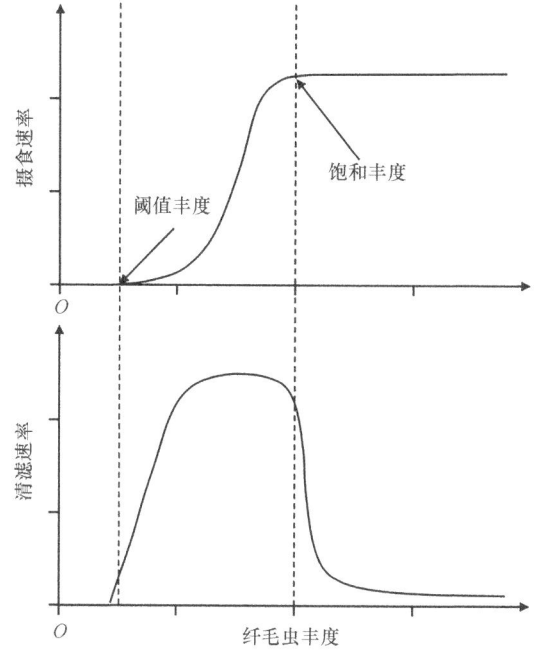

图 10-1 浮游桡足类摄食纤毛虫的功能反应曲线示意图

9×10^4ind./L 时，桡足类 Acartia hudsonica 的清滤速率一直在增加（Turner and Anderson，1983）。这些研究表明，在自然海区，桡足类对纤毛虫的摄食在大多数情况下都没有达到饱和。

对桡足类摄食纤毛虫的摄食阈值丰度的研究更少。在桡足类 Acartia tonsa 摄食纤毛虫 Strombidium reticulatum 的实验中，当 S. reticulatum 的丰度大于 900ind./L（生物量 1μg C/L）时，A. tonsa 保持较高的清滤速率；当饵料丰度低于 900ind./L 时，清滤速率开始逐渐降低；当 S. reticulatum 丰度低于 100ind./L（生物量约 0.1μg C/L）时，桡足类完全停止摄食纤毛虫（Jonsson and Tiselius，1990）。

二、桡足类的摄食方式与纤毛虫的应对策略

不同桡足类对纤毛虫的摄食方式不同。有的桡足类通过游泳来过滤海水中的纤毛虫，如桡足类 Centropages typicus 的游泳速度为 3.5mm/s，远高于其饵料无壳纤毛虫 Strombidium sulcatum 的游泳速度（0.2mm/s），因此 S. sulcatum 无法逃离 C. typicus 的摄食流场（Caparroy et al.，1998）。对于部分个体较小的桡足类如 Acartia tonsa，其摄食流场速度低，所以采取伏击摄食（ambush feeding）的策略。在伏击摄食时，A. tonsa 在水中自由沉降，同时使用大触角（first antenna）寻找周边是否存在饵料，当纤毛虫距离大触角 0.1～0.7mm 时即可被 A. tonsa 感知到，而无须真正触碰到大触角。一旦发现纤毛虫，A. tonsa 立即将身体扭转向饵料，使用

第二触角（second antenna）、胸肢（thoracopod）和第二小颚（second maxilla）等器官截获饵料，整个伏击摄食过程耗时不到 0.1s（Jonsson and Tiselius，1990）。

部分纤毛虫种类，如 *Mesodinium rubrum*，*Strobilidium velox* 和 *Halteria grandinella* 等有"弹跳"行为，即突然迅速游动。在未受扰动的自然条件下，*S. velox* 和 *H. grandinella* 平均每分钟分别弹跳 1.7～3.6 次和 8 次，分别占其游泳时间的 0.8%和 1.0%（Gilbert，1994）。这类纤毛虫能感受到桡足类摄食者带来的水流变化，在与桡足类摄食者接触前启动弹跳行为，逃避捕食。但弹跳逃避并不是对所有摄食者都有效，如伏击摄食纤毛虫的桡足类 *Acartia tonsa* 在自由沉降时的流场信号很弱，不足以触发纤毛虫的弹跳行为（Kiørboe and Visser，1999）。而且由于弹跳对周边水流造成很大的扰动，反而可能增加纤毛虫被摄食者发现的可能性。

由于桡足类和纤毛虫行为模式的不同，不同种类桡足类和纤毛虫之间的摄食关系也有差异。例如，桡足类 *Tortanus setacaudatus* 能有效摄食砂壳纤毛虫 *Favella panamensis*（体长 265μm），但是却不摄食砂壳纤毛虫 *Tintinnopsis tubulosa*（体长 148μm）。这种差异不仅是由于大颗粒饵料会被优先摄食，而且纤毛虫的行为模式也会对桡足类摄食产生影响。桡足类 *T. setacaudatus* 是伏击摄食者，纤毛虫 *T. tubulosa* 比 *F. panamensis* 的游泳速度慢，因此 *T. tubulosa* 对水流的扰动小，与桡足类的相遇概率也小（Robertson，1983）。

桡足类和纤毛虫本身都是可以运动的，因此二者的接触概率还与水体的扰动情况有关。小尺度扰动可以增加桡足类对纤毛虫的摄食强度，在较低的扰动强度下（能量耗散率 ε 在 0.001～$0.01 cm^2/s^3$），桡足类 *Acartia tonsa* 对纤毛虫 *Strombidium sulcatum* 的清滤速率可达到静水中的 4 倍，当扰动强度增大时（$\varepsilon=1$～$10 cm^2/s^3$）清滤速率下降，但仍显著高于静水中的清滤速率（Saiz and Kiørboe，1995）。桡足类 *Centropages typicus* 在低扰动（$\varepsilon=0.029$～$0.3 cm^2/s^3$）时，对纤毛虫 *S. sulcatum* 的清滤速率可达静水中的 4 倍，但当扰动强度升高至 $\varepsilon=4.4 cm^2/s^3$ 时，清滤速率下降到静水中的水平（Caparroy et al.，1998）。

桡足类是吞食整个砂壳纤毛虫，还是只摄食砂壳纤毛虫的肉体，这一问题并无定论。有研究未在培养中发现破损的壳，但在桡足类的粪便中观察到完整的壳，因此认为桡足类可吞食整个砂壳纤毛虫（Robertson，1983）。也有研究在添加桡足类的实验瓶中发现了砂壳纤毛虫的空壳，且空壳的比例高于没有加桡足类的对照瓶，此外，对照瓶中的空壳没有变形，而实验瓶中的空壳很多都被弄皱变形（Turner and Anderson，1983）。这一现象有两种可能的解释，一种解释是桡足类只摄食了砂壳纤毛虫的肉体，而壳被丢弃；另一种解释是当桡足类触碰到砂壳纤毛虫时，砂壳纤毛虫肉体从壳中逃逸。

三、浮游植物对桡足类摄食纤毛虫的影响

培养体系中存在高丰度浮游植物会对桡足类摄食纤毛虫产生一定影响。随培养体系中浮游植物的丰度升高，桡足类 *Acartia tonsa* 对浮游植物的清滤速率没有显著增加，但是对纤毛虫的清滤速率降低（Stoecker and Sanders，1985；Stoecker and Egloff，1987），这可能是由于高丰度的浮游植物影响桡足类对纤毛虫的探测。

有研究发现与浮游植物相比，桡足类更倾向于摄食纤毛虫（Stoecker and Sanders，1985）。这种倾向性可能是多重因素共同导致的：首先，桡足类通常会优先摄食大粒级的饵料，而纤毛虫的粒级通常高于浮游植物；其次，桡足类对纤毛虫有机械感知（Wiadnyana and Rassoulzadegan，1989）和化学感知（Stoecker and Egloff，1987），与浮游植物相比桡足类发现周边纤毛虫的概率更高。

第三节 自然海区桡足类对纤毛虫的摄食

一、海上原位培养

室内摄食培养实验通常是针对个别桡足类和纤毛虫种类开展的研究，所获得的数据不足以推算自然海区复杂的桡足类和纤毛虫物种之间的摄食情况，因此需要在海上进行原位培养实验，以研究自然海区桡足类对纤毛虫群落的摄食。

海上原位培养实验首先需要收集自然海区的桡足类摄食者。桡足类是浮游动物的优势类群，可使用浮游生物网过滤海水收集，随后在解剖镜下挑取活泼、健康、趋光性强的个体进行驯化培养，饥饿 24h 后再挑选出活力较好的个体进行摄食实验。桡足类摄食纤毛虫的实验分为对照组和实验组，对照组只含有自然海水，实验组则是在自然海水中添加桡足类。将培养瓶放在模拟自然温度、扰动和光照等环境的条件下进行培养，一段时间后取样统计对照组和实验组中纤毛虫的丰度。与对照组相比较，实验组中纤毛虫丰度的减少量即被视为桡足类的摄食量。

大多数海上原位培养实验的培养体积都比较小，通常在 2~3L，个体小的桡足类使用的培养瓶小些。大多数培养实验会用筛绢过滤自然海水以去除其他摄食者，使用的筛绢孔径最小为 20μm，最大为 400μm。原位培养实验将活生物体从自然栖息地人为转移到更小的非自然环境中，需要考虑环境改变对生物体生理活动可能产生的影响。可通过控制桡足类动物所在容器的水温、盐度、扰动、光照和黑暗等条件来最小化这类影响。温度模拟主要有恒温室和流水控温培养，扰动模拟使用浮游生物轮（plankton wheel），光照模拟要考虑的问题主要是模拟自然的光照周期和避免强光照射。培养时间大多为 24h，最长为 78h。还有培养实验为了消除桡足类代谢营养盐造成的影响，在培养瓶中加富营养盐。

在烧杯或培养瓶等小容器中培养浮游生物有一种影响，称为"瓶子效应"（bottle effect）或牢笼效应（containment effect）。瓶子效应已被证明会对桡足类的摄食等产生影响，其原因可能是生物拥挤、与容器壁相互作用，而且湍流和扩散条件、营养盐与光照供应可能在封装之后立即发生变化。在桡足类摄食浮游植物的实验中，培养过程中桡足类会释放营养盐，促进培养瓶中浮游植物的生长，从而导致低估桡足类对浮游植物的摄食（Roman and Rublee，1980）。在桡足类摄食纤毛虫的培养实验中，瓶子效应也可能促进培养瓶中纤毛虫的生长，但没有研究对此进行评估，已有的研究也都没有考虑瓶子效应。

海上原位培养还有一个前提假设，即认为培养瓶中纤毛虫的减少全部由桡足类摄食造成。但是这一假设本身存在一定不确定性，其他因素也可能导致纤毛虫的减少，如遇到气泡或过度扰动（Gifford，1985，1993）。培养过程中无壳纤毛虫的死亡也是个问题，无壳纤毛虫本身对环境变动就很敏感，在一个培养实验中，未添加桡足类的对照瓶中无壳纤毛虫的死亡率即可高达80%（Tiselius，1989）。

二、自然海区桡足类对纤毛虫的清滤速率

尽管海上原位培养实验存在一定不确定性，还是有很多研究使用这一方法估算了自然海区桡足类对纤毛虫群体的摄食强度。这些研究的海区主要是离岸很近的港湾或海边实验室，在开阔水域的研究资料很少。详细的研究地点、培养条件和桡足类对纤毛虫的清滤速率可参见张武昌等（2014）。

已有的实验发现浮游桡足类对纤毛虫的清滤速率普遍低于40ml/(ind.·h)，清滤速率最高为 *Calanus hyperboreus* 的162.5ml/(ind.·h)（Levinsen et al.，2000），其次为 *Calanus helgolandicus* 的98ml/(ind.·h)（Fileman et al.，2010）和 *Neocalanus cristatus* 的70.8ml/(ind.·h)（Liu et al.，2005）。随着桡足类摄食者的体长（或体重）增加，桡足类对纤毛虫的清滤速率升高，但是单位体重的清滤速率却减小，即小个体的桡足类摄食更为活跃（Levinsen et al.，2000）。

部分培养实验会把纤毛虫分成不同的粒级组，以研究桡足类对纤毛虫的清滤速率与纤毛虫体长的关系。例如，Levinsen等（2000）、Liu等（2005）和Fileman等（2007）把纤毛虫分为>20μm和<20μm两个粒级组，Vincent和Hartmann（2001）将其分为40μm以上和以下粒级组，Koski等（2002）将其分为30μm以上和以下粒级组。这些分粒级研究都证实一个结论，即随着纤毛虫的粒级增大，桡足类对纤毛虫的清滤速率相应升高。

与室内培养实验的结果类似，多个海上原位培养实验都发现桡足类对纤毛虫的清滤速率大于对浮游植物的清滤速率，表明自然海区的桡足类也倾向于选择摄食纤毛虫（Gifford and Dagg，1988；Atkinson，1996；Castellani et al.，2005）。这

种选择性摄食的原因可能是纤毛虫的粒级比较大，悬浮摄食的桡足类对大颗粒的清滤速率较大，也可能是伏击摄食的桡足类的行为具有选择性。与浮游植物饵料相比，纤毛虫饵料对桡足类有很多优点，如纤毛虫 C：N 值比浮游植物低，蛋白质含量较高（Kiørboe et al., 1985；Stoecker and Capuzzo, 1990），纤毛虫还可以合成桡足类不能合成的不饱和脂肪酸（Sanders and Wickham, 1993；Breteler et al., 1999；Breteler et al., 2004），这些优点都有利于桡足类的生长。

三、自然海区桡足类对纤毛虫的摄食压力

根据自然海区桡足类对纤毛虫的清滤速率结果，再结合该海区桡足类的丰度，可估算桡足类群体对纤毛虫的摄食压力。研究表明，桡足类群体对纤毛虫现存量的摄食压力约为每天 5%，但在个别海区也可高达 45%～200%（表 10-2）。例如，在北大西洋切萨皮克湾的表层富营养水体中，桡足类（主要是幼体）的丰度高达 80ind./L，因此该海区桡足类对纤毛虫的摄食压力较高（Dolan，1991）。

表 10-2　自然海区的中型浮游动物（主要是桡足类）对纤毛虫的摄食压力

海区	对纤毛虫现存量的摄食压力（%）	对纤毛虫生产力的摄食压力（%）	参考文献
北大西洋加里西亚沿岸*	0.91～3.25	2.38～51.4	Batten 等（2001）
北大西洋切萨皮克湾	34～200		Dolan（1991）
北太平洋俄勒冈沿岸	25～45		Fessenden 和 Cowles（1994）
南太平洋副热带锋区	6.4	21.3	Zeldis 等（2002）
地中海阿尔沃兰海	0.3～3.5（平均 1.4）		Calbet 等（2002）
地中海西北部	0.4～6.9（平均 2.4）		Broglio 等（2004）
南大洋南乔治亚岛	5.7	57	Atkinson（1996）
南大洋罗斯海**	0.3～4.8		Lonsdale 等（2000）

*表示桡足类对微型浮游动物的摄食压力；**表示桡足类 Oithona spp.对纤毛虫的摄食压力

在全球海洋范围内，尽管浮游植物的生物量高出纤毛虫 1 个数量级，但纤毛虫却贡献了桡足类饵料生物量的 30%（Calbet and Saiz, 2005）。与室内培养的发现类似，在自然海区桡足类对纤毛虫和浮游植物摄食压力的相对重要性取决于水体的营养状态。在北大西洋凯尔特海，桡足类群体对原生动物（主要为纤毛虫和异养甲藻）的摄食压力为 12%～17%，但在该海区出现水华的采样站，桡足类对原生动物的摄食压力仅有 2%（Fileman et al., 2007）。在浮游植物浓度较低的海区（<50μgC/L），纤毛虫和浮游植物对桡足类饵料的贡献基本相等；随浮游植物浓度升高，桡足类对纤毛虫的摄食压力下降，在浮游植物浓度分别为 50～500μgC/L 和>500μgC/L 的海区，纤毛虫仅分别占桡足类饵料的 25%和 22%（Calbet and Saiz, 2005）。

桡足类对不同粒级纤毛虫的摄食压力有差异,桡足类通常会选择性摄食粒级较大的纤毛虫。在北海的卡特加特海峡(Kattegat),桡足类对大个体(等效球体直径>50μm)纤毛虫的摄食基本抵消了纤毛虫的生长,但对小个体纤毛虫的摄食压力低于纤毛虫的生长率,因此在该海区大粒级纤毛虫丰度保持不变,小粒级纤毛虫丰度增加(Nielsen and Kiørboe,1994)。

桡足类对原生动物的摄食压力还存在年际差异,在英吉利海峡,2002 年桡足类 *Calanus helgolandicus* 和 *Acartia clausi* 对原生动物的摄食压力为 3%~32%,平均为 17%;但在 2003 年同一站点的摄食压力则低至 0.1%~15%,平均仅为 3%,不足前一年摄食压力的两成(Fileman et al.,2010)。

有研究根据桡足类摄食能力数据,在一定程度上估计自然海区桡足类对纤毛虫丰度/生物量的下行控制能力。例如,Stoecker 和 Sanders(1985)估算认为,当桡足类 *Acartia* spp.的丰度为 3~4ind./L 时,就可以阻止砂壳纤毛虫 *Favella* sp.的丰度出现净增长。Calbet 和 Saiz(2005)总结了 23 个桡足类摄食相关研究的数据,提出若要消耗纤毛虫现存量的 50%,假设桡足类平均体重为 10μgC/ind.,则需要水体中桡足类的丰度达到 7ind./L。

参 考 文 献

张武昌,陈雪,李海波,等. 2014. 海洋浮游桡足类摄食纤毛虫的研究. 海洋与湖沼, 45(4): 764-775.

Atkinson A. 1996. Subantarctic copepods in an oceanic, low chlorophyll environment: ciliate predation, food selectivity and impact on prey populations. Marine Ecology Progress Series, 130: 85-96.

Ayukai T. 1987. Predation by *Acartia clausi*(Copepoda: Calanoida) on two species of tintinnids. Marine Microbial Food Webs, 2(1): 45-52.

Batten S D, Fileman E S, Halvorsen E. 2001. The contribution of microzooplankton to the diet of mesozooplankton in an upwelling filament off the north west coast of Spain. Progress in Oceanography, 51(2): 385-398.

Berk S G, Brownlee D C, Heinle D R, et al. 1977. Ciliates as a food source for marine planktonic copepods. Microbial Ecology, 4(1): 27-40.

Breteler W C M K, Koski M, Rampen S W. 2004. Role of essential lipids in copepod nutrition: no evidence for trophic upgrading of food quality by a marine ciliate. Marine Ecology Progress Series, 274: 199-208.

Breteler W C M K, Schogt N, Baas M, et al. 1999. Trophic upgrading of food quality by protozoans enhancing copepod growth: role of essential lipids. Marine Biology, 135(1): 191-198.

Broglio E, Jonasdottir S, Calbet A, et al. 2003. Effect of heterotrophic versus autotrophic food on feeding and reproduction of the calanoid copepod *Acartia tonsa*: relationship with prey fatty acid composition. Aquatic Microbial Ecology, 31(3): 267-278.

Broglio E, Saiz E, Calbet A, et al. 2004. Trophic impact and prey selection by crustacean zooplankton on the microbial communities of an oligotrophic coastal area(NW Mediterranean Sea). Aquatic

Microbial Ecology, 35(1): 65-78.

Calbet A, Broglio E, Saiz E, et al. 2002. Low grazing impact of mesozooplankton on the microbial communities of the Alboran Sea: a possible case of inhibitory effects by the toxic dinoflagellate *Gymnodinium catenatum*. Aquatic Microbial Ecology, 26(3): 235-246.

Calbet A, Saiz E. 2005. The ciliate-copepod link in marine ecosystems. Aquatic Microbial Ecology, 38(2): 157-167.

Caparroy P, Perez M T, Carlotti F. 1998. Feeding behaviour of *Centropages typicus* in calm and turbulent conditions. Marine Ecology Progress Series, 168: 109-118.

Carlotti F, Radach G. 1996. Seasonal dynamics of phytoplankton and *Calanus finmarchicus* in the North Sea as revealed by a coupled one-dimensional model. Limnology and Oceanography, 41(3): 522-539.

Castellani C, Irigoien X, Harris R P, et al. 2005. Feeding and egg production of *Oithona similis* in the North Atlantic. Marine Ecology Progress Series, 288: 173-182.

Dam H G, Peterson W T, Bellantoni D C. 1994. Seasonal feeding and fecundity of the calanoid copepod *Acartia tonsa* in Long Island Sound: is omnivory important to egg production? Hydrobiologia, 292(1): 191-199.

Dolan J R. 1991. Microphagous ciliates in mesohaline Chesapeake Bay waters: estimates of growth rates and consumption by copepods. Marine Biology, 111(2): 303-309.

Fessenden L, Cowles T J. 1994. Copepod predation on phagotrophic ciliates in Oregon coastal waters. Marine Ecology Progress Series, 107: 103-111.

Fileman E, Petropavlovsky A, Harris R. 2010. Grazing by the copepods *Calanus helgolandicus* and *Acartia clausi* on the protozooplankton community at station L4 in the Western English Channel. Journal of Plankton Research, 32(5): 709-724.

Fileman E, Smith T, Harris R. 2007. Grazing by *Calanus helgolandicus* and *Para-Pseudocalanus* spp. on phytoplankton and protozooplankton during the spring bloom in the Celtic Sea. Journal of Experimental Marine Biology and Ecology, 348(1): 70-84.

Gifford D J. 1985. Laboratory culture of marine planktonic oligotrichs(Ciliophora, Oligotrichida). Marine Ecology Progress Series, 23: 257-267.

Gifford D J. 1993. Consumption of protozoa by copepods feeding on natural microplankton assemblages//Kemp P F, Sherr B F, Sherr E B, et al. Handbook of Methods in Aquatic Microbial Ecology. Boca Raton: Lewis Publishing: 723-729.

Gifford D J, Dagg M J. 1988. Feeding of the estuarine copepod *Acartia tonsa* dana: carnivory vs. herbivory in natural microplankton assemblages. Bulletin of Marine Science, 43(3): 458-468.

Gilbert J J. 1994. Jumping behavior in the oligotrich ciliates *Strobilidium velox* and *Halteria grandinella*, and its significance as a defense against rotifer predators. Microbial Ecology, 27(2): 189-200.

Gismervik I. 2006. Top-down impact by copepods on ciliate numbers and persistence depends on copepod and ciliate species composition. Journal of Plankton Research, 28(5): 499-507.

Gismervik I, Andersen T. 1997. Prey switching by *Acartia clausi*: experimental evidence and implications of intraguild predation assessed by a model. Marine Ecology Progress Series, 157: 247-259.

Harding G C H. 1974. The food of deep-sea copepods. Journal of the Marine Biological Association of the United Kingdom, 54(1): 141-155.

Heinle D R, Harris R P, Ustach J F, et al. 1977. Detritus as food for estuarine copepods. Marine

Biology, 40(4): 341-353.

Jonsson P R, Tiselius P. 1990. Feeding behaviour, prey detection and capture efficiency of the copepod *Acartia tonsa* feeding on planktonic ciliates. Marine Ecology Progress Series, 60: 35-44.

Kiørboe T, Mohlenberg F, Hamburger K. 1985. Bioenergetics of the planktonic copepod *Acartia tonsa*: relation between feeding, egg production and respiration, and composition of specific dynamic action. Marine Ecology Progress Series, 26: 85-97.

Kiørboe T, Saiz E, Viitasalo M. 1996. Prey switching behaviour in the planktonic copepod *Acartia tonsa*. Marine Ecology Progress Series, 143: 65-75.

Kiørboe T, Visser A W. 1999. Predator and prey perception in copepods due to hydromechanical signals. Marine Ecology Progress Series, 179: 81-95.

Koski M, Schmidt K, Engström-Öst J, et al. 2002. Calanoid copepods feed and produce eggs in the presence of toxic cyanobacteria *Nodularia spumigena*. Limnology and Oceanography, 47(3): 878-885.

Levinsen H, Turner J T, Nielsen T G, et al. 2000. On the trophic coupling between protists and copepods in Arctic marine ecosystems. Marine Ecology Progress Series, 204: 65-77.

Liu H, Dagg M J, Strom S L. 2005. Grazing by the calanoid copepod *Neocalanus cristatus* on the microbial food web in the coastal Gulf of Alaska. Journal of Plankton Research, 27(7): 647-662.

Lonsdale D J, Caron D A, Dennett M R, et al. 2000. Predation by *Oithona* spp. on protozooplankton in the Ross Sea, Antartica. Deep Sea Research Part II: Topical Studies in Oceanography, 47(15): 3273-3283.

Merrell J R, Stoecker D K. 1998. Differential grazing on protozoan microplankton by developmental stages of the calanoid copepod *Eurytemora affinis* Poppe. Journal of Plankton Research, 20(2): 289-304.

Mullin M M. 1966. Selective feeding by calanoid copepods from the Indian Ocean//Barnes H. Some Contemporary Studies in Marine Science. London: Allen and Unwin: 545-554.

Nielsen T G, Kiørboe T. 1994. Regulation of zooplankton biomass and production in a temperate, coastal ecosystem. 2. Ciliates. Limnology and Oceanography, 39(3): 508-519.

Peterson W T, Dam H G. 1996. Pigment ingestion and egg production rates of the calanoid copepod *Temora longicornis*: implications for gut pigment loss and omnivorous feeding. Journal of Plankton Research, 18(5): 855-861.

Robertson J R. 1983. Predation by estuarine zooplankton on tintinnid ciliates. Estuarine, Coastal and Shelf Science, 16(1): 27-36.

Roman M R, Rublee P A. 1980. Containment effects in copepod grazing experiments: a plea to end the black box approach. Limnology and Oceanography, 25(6): 982-990.

Saiz E, Kiørboe T. 1995. Predatory and suspension feeding of the copepod *Acartia tonsa* in turbulent environments. Marine Ecology Progress Series, 122: 147-158.

Sanders R, Wickham S A. 1993. Planktonic protozoa and metazoa: predation, food quality and population control. Marine Microbial Food Webs, 7(2): 197-223.

Sheldon R W, Nival P, Rassoulzadegan F. 1986. An experimental investigation of a flagellate-ciliate-copepod food chain with some observations relevant to the linear biomass hypothesis. Limnology and Oceanography, 31(1): 184-188.

Smetacek V. 1981. The annual cycle of protozooplankton in the Kiel Bight. Marine Biology, 63(1): 1-11.

Stoecker D K, Capuzzo J M. 1990. Predation on protozoa: its importance to zooplankton. Journal of Plankton Research, 12(5): 891-908.

Stoecker D K, Egloff D A. 1987. Predation by *Acartia tonsa* Dana on planktonic ciliates and rotifers. Journal of Experimental Marine Biology and Ecology, 110(1): 53-68.

Stoecker D K, Sanders N K. 1985. Differential grazing by *Acartia tonsa* on a dinoflagellate and a tintinnid. Journal of Plankton Research, 7(1): 85-100.

Tiselius P. 1989. Contribution of aloricate ciliates to the diet of *Acartia clausi* and *Centropages hamatus* in coastal waters. Marine Ecology Progress Series, 56: 49-56.

Turner J T. 1984. Zooplankton feeding ecology: contents of fecal pellets of the copepods *Acartia tonsa* and *Labidocera aestiva* from continental shelf waters near the mouth of the mississippi river. Marine Ecology, 5(3): 265-282.

Turner J T, Anderson D M. 1983. Zooplankton grazing during dinoflagellate blooms in a cape cod embayment, with observations of predation upon tintinnids by copepods. Marine Ecology, 4(4): 359-374.

Verity P G, Paffenhofer G. 1996. On assessment of prey ingestion by copepods. Journal of Plankton Research, 18(10): 1767-1779.

Vincent D, Hartmann H J. 2001. Contribution of ciliated microprotozoans and dinoflagellates to the diet of three copepod species in the Bay of Biscay. Hydrobiologia, 443: 193-204.

Wiadnyana N W, Rassoulzadegan F. 1989. Selective feeding of *Acartia clausi* and *Centropages typicus* on microzooplankton. Marine Ecology Progress Series, 53: 37-45.

Zeldis J, James M R, Grieve J, et al. 2002. Omnivory by copepods in the New Zealand Subtropical Frontal Zone. Journal of Plankton Research, 24(1): 9-23.

第三部分

第十一章 微食物网生物与海洋颗粒

海洋中存在着大量的颗粒。颗粒在海水中营造了不同于纯海水的小生境，异养细菌、蓝细菌、真核藻类、鞭毛虫、纤毛虫等微食物网生物可以黏附在海洋颗粒上，或生活在颗粒内部。海洋中的微生境可以视为一个连续存在的谱，海水在谱的一端，颗粒内的空间在谱的另一端。颗粒提供了适宜生长的营养环境，因此颗粒黏附的微食物网生物丰度高于周围水体。

微食物网生物在颗粒形成过程中起到重要的作用，它们可以直接促进颗粒形成，可以彼此结合形成颗粒，也可经由微型浮游动物摄食和排便形成颗粒。微食物网生物还能够对海洋颗粒进行转化，或对颗粒及其黏附生物进行摄食，从而影响颗粒的大小和沉降。微食物网生物与颗粒的相互作用使得它们成为海洋碳循环的重要参与者和贡献者，本章将介绍微食物网生物在海洋颗粒形成和沉降中的作用。

第一节 何为海洋颗粒

海水中除了生物之外，还有一些没有生命的碎屑（detritus），包括有机碎屑和无机碎屑。在海洋浮游生态学研究中，颗粒（particle）通常指有机碎屑。

科学家最初认为海水中的颗粒来自动物的粪便和浮游生物的尸体碎片，这些生物粪便和生物碎片中的物质逐渐溶解到海水中，因此随水深的增加，颗粒物质的总量会降低（Riley，1963）。但在 20 世纪 50 年代，日本科学家在深海发现了大量雪花状漂浮颗粒物，并将其命名为海雪（marine snow）（Suzuki and Kato，1953）。在深海中存在海雪这一现象，与浮游生物粪便等随水深减少的现象不一致，使科学家意识到除了生物粪便和生物碎片外还有其他的颗粒存在。

到 20 世纪 90 年代，随着新技术和新染料的应用，科学家发现海洋中有大量不同粒径和染色特性的颗粒。例如，Isao 等（1990）使用颗粒计数器发现了粒级为 0.3~1.3μm 的亚微米颗粒（sub-micrometer-sized particle）；Alldredge 等（1993）发现了粒级为 3~>100μm 的透明胞外聚合颗粒物（transparent exopolymer particles，TEP）；Long 和 Azam（1996）发现了考马斯蓝染色颗粒（Coomassie blue-stained particle，CSP）；Mostajir 等（1995）发现了可以被荧光染料 DAPI（4′,6-diamidino-2-phenylindole，4′,6-二脒基-2 苯基吲哚）染色的颗粒（DAPI yellow particle，DYP）。由于不能重复染色，因此不能确定一个颗粒是否同时是 TEP、

CSP 和 DYP。这些新发现的颗粒被统称为新颗粒（new particle），过去了解的颗粒（如生物粪便、碎片、海雪等）则被称为传统颗粒（classical particle）。由于新颗粒大多为透明状，也被称为海洋中的"暗物质"（dark matter）（Azam，1998）。

海雪和新颗粒都是高分子聚合物（polymer）聚合而成的颗粒物，其表面有细菌繁殖，并黏附了无机颗粒和浮游生物，因此被统称为有机聚合颗粒（organic aggregate），或简称为聚合颗粒（aggregate）（Simon et al.，2002）。这些聚合颗粒与粪便、生物碎片颗粒有明显区别：粪便和生物碎片颗粒有明显的边界，细菌等不能进入，过滤到滤膜上后，能够保持其内的液体，从而维持原有形态；而聚合颗粒中的高分子聚合物形成不同体积、不同孔径、不同结构的框架，海水充斥其中，细菌等生物可以进入，一旦过滤到滤膜上，聚合颗粒中的海水流出，框架就会塌缩，无法保持原来的形态，因此聚合颗粒也可被视为一种胶状物（gel）（Verdugo et al.，2004）。

根据是否能够被滤膜截留，海洋中的有机碳可以分为颗粒有机碳（particulate organic carbon，POC）和溶解有机碳（dissolved organic carbon，DOC）。新颗粒和传统颗粒都是 POC。DOC 可以划分为游离组分（free DOC，粒径＜1nm）和胶质组分（assembled DOC，粒径 1～1000nm）（Benner et al.，1992）。海水中的 DOC 保持可逆的组装/分散平衡，生成自组装微凝胶（self-assembled microgel，SAG），这些 SAG 是海洋 DOC 和 POC 之间的桥梁（Verdugo et al.，2004；Verdugo，2012）。

按照粒径，聚合颗粒分为大型聚合颗粒（macroaggregate）、小型聚合颗粒（microaggregate）和亚微米颗粒（submicron particle）（Simon et al.，2002）。大型聚合颗粒粒径＞500μm，主要是海雪；小型聚合颗粒粒径为 1～500μm，包括 TEP、CSP 和 DYP；亚微米颗粒粒径＜1μm。颗粒粒级越小，丰度就越大，这些颗粒与海水中的溶解有机成分从小到大形成了连续谱（图 11-1）。

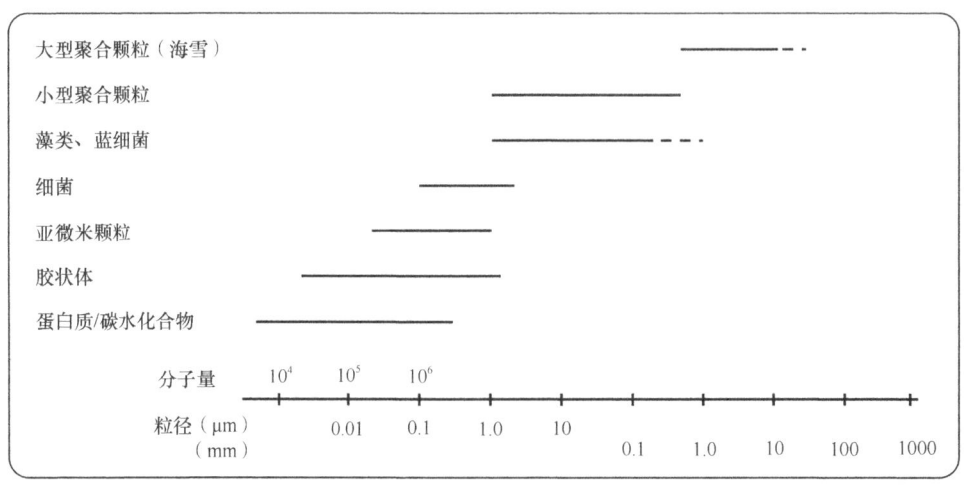

图 11-1　海洋主要颗粒及溶解有机成分的粒径谱 [改绘自 Simon 等（2002）]

第二节 颗粒上的微食物网生物

一、颗粒上的细菌

落射荧光显微镜计数细菌的方法发明以后,科学家观察到海水中的细菌有的营自由生活,称为自由生细菌(free bacteria),有的则黏附在颗粒上,称为黏附细菌(attached bacteria)。细菌能探测到化学物质的浓度梯度,并以每秒钟几百微米(相当于细菌体长的数百倍)的速度被吸引到颗粒上。海水中自由生细菌丰度通常在 10^6cell/ml,而颗粒黏附细菌的丰度可达 10^9cell/ml,相差 3 个数量级(Kiørboe and Jackson,2001)。

区分自由生细菌和黏附细菌的经典方法是用孔径 0.22μm 的滤膜过滤海水,用 DAPI 染色滤膜并在显微镜下观察,在滤膜上的细菌为自由生细菌,在颗粒上的细菌为黏附细菌。由于计数时只能观察颗粒在显微镜视野下的一面,与滤膜接触的另一面无法观察,所以只能假设颗粒两面的细菌数目相同以进行估算。另外需要注意的是,滤膜上的颗粒可能覆盖一部分自由生细菌,这将导致自由生细菌丰度的低估。

颗粒黏附细菌通常比自由生细菌大,活性也比自由生细菌高。颗粒黏附细菌在总细菌中的比例与颗粒物丰度、水流冲刷、颗粒物沉降和浮游动物摄食情况有关。总的来讲,淡水、河口和盐沼中颗粒黏附细菌所占的比例较大;在近海和大洋颗粒黏附细菌占比很低,90%以上的细菌都为自由生细菌。但颗粒黏附和自由生的名称只是为了区别细菌在被观察时的生活环境,在自然水体中颗粒黏附细菌和自由生细菌之间经常发生交换,颗粒上的细菌能够进入水体,水体中的细菌也能够附着到颗粒上。有些细菌种类可能进化出了复杂的生活方式,既有自由生活阶段,也有颗粒附着生活阶段(Grossart,2010)。

细菌具体黏附在哪些颗粒上,也是学者关心的一个重要问题。对亚微米颗粒的观察普遍未发现黏附细菌(Leppard,1992;Wells and Goldberg,1993;Schuster and Herndl,1995)。在小型聚合颗粒中,仅在 TEP 上发现有细菌黏附。TEP 的粒级较小,内部空间有限,因此细菌仅能够黏附在 TEP 的表面,无法生活在 TEP 内部。TEP 表面黏附细菌的丰度(N,cell/μm^2)与 TEP 的等效球直径(d,μm)的关系符合幂函数 $N=\alpha \times d^\beta$,其中 α 和 β 是幂函数常数,具体数值因研究而异(表 11-1)。不同海区 TEP 黏附细菌的丰度相差 10 倍,如等效球直径 5μm TEP 的黏附细菌丰度从 0.08cell/μm^2 到 0.7cell/μm^2 不等。TEP 黏附细菌占海水细菌总丰度的比例范围为 0.5%~25%,最大可达 89%(Passow,2002)。TEP 的粒级越小,其表面黏附细菌的丰度越高,这可能是由于:①TEP 越小,存在的时间就越长,因而有更多的时间让细菌黏附；②个体小的 TEP 密度较大，有较多的营养物质供养细菌。大型聚合颗粒（海雪）上黏附的细菌丰度通常为每个颗粒 $10^5 \sim 10^8$ cell，其丰度比自由生细菌高出 1~2 个数量级（Alldredge et al., 1986; Turley and Mackie, 1994; Silver et al., 1998），菌体大小也超过自由生细菌，这可能是由于颗粒提供了更适宜细菌生长的营养环境（Simon et al., 2002）。

表 11-1 TEP 表面黏附细菌的丰度

研究海区		α	β	颗粒黏附细菌丰度 (cell/μm²)		参考文献
				5μmTEP	100μmTEP	
北太平洋加利福尼亚沿岸		1.36	0.78	0.39	0.04	Passow 和 Alldredge (1994)
亚得里亚海北部	自然水体	0.28	0.70	0.09	0.01	Schuster 和 Herndl (1995)
	自然水体	0.30	0.82	0.08	0.007	
	培养实验	0.58	1.06	0.10	0.004	
北海卡特加特海峡	水华前	0.26	0.38	0.14	0.05	Mari 和 Kiørboe (1996)
	水华后	1.53	0.72	0.48	0.06	
多瑙河				0.70		Berger 等 (1996)

二、颗粒上的蓝细菌

已有研究在聚合颗粒上观察到大量的蓝细菌（Silver and Alldredge, 1981; Silver et al., 1986; Lochte and Turley, 1988; Thiel et al., 1989; Waite et al., 2000）。其中，在大西洋东北部 4500m 深海采集的聚合颗粒上，颗粒黏附聚球藻的丰度为 $8 \times 10^6 \sim 2 \times 10^7$ cell/ml，远高于表层海水（5×10^5 cell/ml）和 4500m 深度海水（3×10^2 cell/ml）中的自由生聚球藻丰度（Lochte and Turley, 1988; Thiel et al., 1989）。在西南太平洋 120m 水深处采集的聚合颗粒上，聚球藻丰度为 $8 \times 10^8 \sim 2 \times 10^9$ cell/ml，在 550m 水深聚合颗粒上聚球藻丰度为 7.5×10^6 cell/ml，均高出水体中自由生聚球藻的丰度 3 至 5 个数量级（Waite et al., 2000）。随着颗粒沉降，颗粒黏附蓝细菌的丰度降低，这可能是原生动物摄食导致的。

聚球藻个体微小，单个细胞的沉降速度极慢，因此只有形成颗粒才能大量沉降。聚球藻等微微型自养生物聚合成的颗粒空隙较小，其沉降速度快于由硅藻聚合的颗粒（Bach et al., 2016）。聚球藻等微微型自养浮游生物形成颗粒的途径主要有两个，第一个是被浮游动物摄食后生成粪便颗粒，第二个是被黏着整合到水体中已存在的聚合颗粒上。有研究使用激光共聚焦显微镜观察了西南太平洋沉积物中的聚合颗粒，发现聚球藻等微微型自养浮游生物主要位于聚合颗粒的内部，说明聚球藻主要是通过被摄食进入聚合颗粒的（Waite et al., 2000）。微型浮游动物是聚球藻的主要摄食者，但微型浮游动物的个体较小，形成的粪便颗粒也较小。

大中型浮游动物，如樽海鞘（salp）、尾海鞘（appendiculariae）、海樽（doliolid）和翼足类（pteropod）等，也能够摄食聚球藻等微微型自养浮游生物，并生成大型粪便颗粒（Fortier et al., 1994）。

三、颗粒上的原生动物

1975年即有研究报道海洋颗粒上存在原生动物（Paerl，1975）。随后，Caron等（1982）对颗粒上原生动物的类群和丰度展开了研究，在马尾藻海0~25m收集的颗粒上，颗粒黏附鞭毛虫的丰度为0.02~6200ind./ml，是周围海水自由生鞭毛虫的1~4倍，变形虫和纤毛虫的丰度为3~23ind./ml，也高于周围海水。

在加利福尼亚东边界流中，在沉积物捕获器采集到的颗粒上发现了纤毛虫，颗粒上纤毛虫的生物量远高于细菌。纤毛虫可以随颗粒沉降到2000m水深，在真光层以下，不同的水深有不同的颗粒黏附纤毛虫类群，且颗粒黏附纤毛虫的丰度并未随水深下降出现显著降低，表明深层颗粒黏附的纤毛虫并不是从表层沉降下来的，而是本地种（Silver et al., 1984）。学者认为原生动物会寻找并附着到正在沉降的颗粒上，在跟随颗粒下沉的同时，在颗粒上摄食和繁殖。当颗粒沉降出这些原生动物的生境时，它们会离开颗粒，上浮回到自己生境的上层。在上浮的过程中，原生动物可能还会寻找上升颗粒以搭便车（Smith et al., 1989；Toggweiler, 1989）。

变形虫的摄食主要依靠伪足的吞噬作用，所以它们必须附着在物体表面进行运动和摄食。在近岸海区，90%以上的变形虫是附着在颗粒上的（Rogerson et al., 2003）。在美国纽约哈德逊河口，在<200μm和>200μm的颗粒上变形虫的丰度范围分别为200~1452ind./L和213~1110ind./L，两组丰度间没有显著差异，但在>200μm的颗粒上变形虫的粒径更大，多样性更高（Anderson, 2015）。

有研究使用比斯开湾自然海水在实验室内制造海雪，并观察原生动物在海雪上的附着和繁殖过程。海雪形成后4天，异养鞭毛虫开始附着繁殖，再过1天纤毛虫才开始繁殖。鞭毛虫的主要类群是 *Bodo designis* 和 *Rhynchomonas nasuta*，纤毛虫的主要类群是 *Uronema marinum*、*Euplotes vannus* 和 *Aspidisca steini*（Artolozaga et al., 1997）。进一步的研究发现海雪造就了多样化的海洋小生境，可分为颗粒外水体、颗粒表面和颗粒内部三个小生境。有些原生动物仅存在于颗粒外水体，可以认为是真正自由游泳型，如变形虫 *Amoeba radiosa*、鞭毛虫 *Paraphysomonas* sp.、纤毛虫 *Cohnilembus punctatus*、*Uropedalium opisthosoma* 等；有些原生动物倾向于在颗粒表面生活，如鞭毛虫 *Cafeteria* sp.和 *Bicosoeca maris*，它们从不在颗粒内部和外部水体出现；有的原生动物类群虽然倾向于在颗粒表面生活，但是偶尔也会出现在颗粒内部（如鞭毛虫 *B. designis*、*R. nasuta* 和纤毛虫

E. vannus、*A. steini*）或颗粒外水体中（如鞭毛虫 *Oxyrrhis* sp.、*Pseudobodo tremulans*、*Salpingoeca* sp.和纤毛虫 *Uronema marinum*）（Artolozaga et al., 2000）。

第三节 微食物网生物促进海洋颗粒形成

一、细菌促进聚合颗粒形成

20 世纪 50~60 年代，科学家对聚合颗粒产生的原因展开了研究。有实验用滤膜过滤去除 POC，再对这些仅保留 DOC 的海水充气产生气泡，发现海水中会产生聚合物。这一实验表明，DOC 能够在界面（固体表面、气泡、海洋表面）通过物理、化学作用生成聚合颗粒（Barber, 1966）。

随着研究的深入，科学家发现海洋微食物网生物在 DOC 转变为聚合颗粒的过程中有重要作用。在实验模拟中，只有海水中存在活的细菌时，大型聚合颗粒才能生成；若用氯化汞固定杀死细菌，或者用过滤的方法去除细菌，海水中的 DOC 就不能转变为聚合颗粒（Biddanda, 1985）。细菌还能够促进小颗粒转化为更大的颗粒，如假交替单胞菌属（*Pseudoalteromonas*）细菌可以促进粒径 0.01cm 的多糖凝胶聚合形成粒径为 0.1~1cm 的大颗粒（Yamada et al., 2016）。

细菌可通过 3 种途径促进聚合颗粒形成：①DOC 高分子聚合物通过物理、化学作用自组织成 SAG，这些 SAG 处于聚合-解聚的动态平衡中，而细菌产生的双亲性（amphiphilic）胞外聚合物（exopolymer substance，EPS）在低浓度下即可诱导高分子聚合物的聚合作用，促进海水 DOC 生成小型聚合颗粒（Ding et al., 2008）；②细菌生长释放的 EPS 含有细胞膜的组分，黏度较高，有利于形成聚合颗粒（Stoderegger and Herndl, 1998, 1999），当加入细菌抑制剂后，海水中新形成的聚合颗粒浓度则很低（Sugimoto et al., 2007）；③细菌能够刺激藻类分泌胞外高分子聚合物形成 TEP，TEP 又通过凝集作用使 DOC、细菌、微藻、浮游植物等形成凝聚体，以海雪的形式沉降。例如，黏附在硅藻 *Thalassiosira weissflogii* 上的细菌能刺激硅藻分泌 TEP，进而形成硅藻团聚体，无菌培养的 *T. weissflogii* 则不发生聚集（Gärdes et al., 2011）。

二、微型浮游动物生产颗粒

微型浮游动物产生的粪便排出体外可直接生成海洋颗粒。在自然海水中能够观察到一些小型（粒径<150μm）粪便颗粒，称为"迷你粪球"（minipellet）。例如，赤道东太平洋水体中存在粒级为 3~50μm 的迷你粪球（Gowing and Silver, 1985），在南大洋罗斯海（Ross Sea）和威德尔海（Weddell Sea）观察到了粒级 20~

150μm 的迷你粪球（Nöthig and Bodungen，1989；Gowing et al.，2001），有些迷你粪球有自发荧光，有些则没有。南大洋海冰附着生活的裸甲藻（50μm×50μm）产生的迷你粪球可集聚在海冰中，这些粪便颗粒呈球形或椭球形，平均直径为 30μm，丰度可超过 10^5 个/L（Buck et al.，1990）。在美国华盛顿州的 Dabob 湾硅藻水华过程中，沉积物捕获器中可观察到甲藻 *Gyrodinium* sp.（107μm×68μm）的迷你粪球，颗粒平均尺寸为 83μm×69μm（Buck and Newton，1995）。

使用自然样品难以直接观察微型浮游动物排出迷你粪球的过程，但有研究通过分析现场的迷你粪球丰度与微型浮游动物丰度的相关性，推断这些迷你粪球是由微型浮游动物产生的（Beaumont et al.，2002；Buck et al.，2005）。在亚北极区 Baisfjorden 峡湾发现的迷你粪球的直径平均为 80μm，其直径相对于该海区存在的原生动物来说太大，可能是由桡足类幼体产生的（Pasternak et al.，2000）。

室内培养实验可观察到微型浮游动物排出粪便生成迷你粪球。砂壳纤毛虫 *Favella* sp.和无壳纤毛虫 *Balanion* sp.摄食甲藻产生的迷你粪球平均大小分别为 19μm×32μm 和 18μm×24μm，略大于其甲藻饵料 *Heterocapsa triquetra* 的粒径（16μm×22μm），迷你粪球内通常包含 1 个或多个饵料细胞残留（Stoecker，1984）。鞭毛虫 *Pteridomonas danica*[粒径（4.2±0.8）μm]摄食大肠杆菌[等效球直径（1.1±0.2）μm]产生的迷你粪球大小为 0.2~1μm（Pelegri et al.，1999）。

三、微型浮游动物促进颗粒融合

附着在颗粒上生活的微型浮游动物摄食产生的水流，可增加水体中颗粒间的接触机会，从而促使小粒径的颗粒融合形成更大的颗粒。

室内培养实验发现，颗粒上附着的鞭毛虫 *Paraphysomonas imperforata* 可以促使水体中粒径 0.5μm 的亚微米颗粒更快地转化为粒径 5~20μm 的微悬浮颗粒（micro-suspended particle）。该现象可能是鞭毛虫的摄食行为造成了朝向其所在附着颗粒的平流水流，夹杂在水流中的自由亚微米颗粒可能与鞭毛虫附着颗粒发生碰撞，随后黏附在鞭毛虫附着的颗粒上，聚集成更大的颗粒（Fukuda and Koike，2000）。类似地，附着在颗粒上生活的纤毛虫，如 *Uronema filificum*，也有较强的摄食水流（Fenchel and Blackburn，1999），同样会导致附着颗粒与周围水体中的较小颗粒发生聚集。这一现象在自然海水中也得到了证实，在富营养的近岸海水中加入真核生物抑制剂和细胞呼吸抑制剂后，亚微米颗粒向微悬浮颗粒转换的效率显著降低（Fukuda and Koike，2004）。上述结果表明，与物理作用导致的亚微米颗粒向微悬浮颗粒融合过程相比，生物作用的效率更高。在近岸高生产力的水体中存在更丰富的颗粒附着微型浮游动物和亚微米颗粒，生物过程是导致近岸水体微悬浮颗粒生成的关键机制。

第四节　微食物网生物对颗粒的转化

一、细菌改造颗粒

除了促进聚合颗粒形成，细菌还可以对颗粒进行改造，将颗粒的有机物质转化成溶解物质，将大颗粒分解成小颗粒。传统认识中，颗粒黏附细菌在将颗粒降解成溶解有机碳、减少有机质的沉降通量中起主要作用，但 Cho 和 Azam（1988）发现自由生细菌才是颗粒降解的主要作用者，它们消耗了沉降颗粒中的有机物，把大的颗粒转化成小颗粒。细菌只能利用溶解态的有机碳，因此颗粒的有机物首先要转化成溶解有机物才能被利用。细菌能够产生大量的胞外酶，将颗粒中的多糖和蛋白质等降解为溶解态小分子，从而将可沉降的固态颗粒转化为无法沉降的溶解有机物。在这一过程中，颗粒黏附细菌只使用了释放的溶解有机碳的一小部分，大部分的溶解有机碳释放到水体中被自由生细菌利用（Kiørboe and Jackson，2001）。

细菌还可以影响颗粒的沉降速度。聚合颗粒被细菌黏附后，细菌在颗粒表面和内部繁殖，消耗颗粒的有机物质使得颗粒变小。细菌还可以通过分泌黏性物质，填充颗粒内部的空隙，促进颗粒变大。随着细菌利用颗粒物质，颗粒的骨架逐渐被细菌细胞替代，使得颗粒多孔，从而降低了颗粒的沉降速度（Yamada et al.，2013）。

二、微型浮游动物摄食颗粒

研究微型浮游动物能否摄食碎屑颗粒，采取的主要方法是实验室内喂食实验，用硅藻、绿藻、桡足类和大型藻类等样品制造不同粒级的碎屑颗粒投喂微型浮游动物，通过微型浮游动物的生长情况，或者直接观察微型浮游动物体内是否含有碎屑颗粒，来判断是否存在摄食。

有研究用荧光染料 5-(4,6-二氯三嗪基)氨基荧光素[5-(4,6-dichlorotriazin-2-yl) aminofluorescein，DTAF]对人工制备的碎屑颗粒进行染色并投喂纤毛虫，经一段时间培养后用表面荧光显微镜观察到纤毛虫体内存在荧光颗粒，因此可以确定纤毛虫能够摄食碎屑颗粒（Posch and Arndt，1996）。

还有研究用人工制备的碎屑颗粒作为唯一碳源培养纤毛虫 *Tetrahymena pyriformis* 和鞭毛虫 *Chilomonas paramecium*，发现这两种微型浮游动物都能利用碎屑颗粒进行生长，而且生长速率与野外种群的报道相近，表明微型浮游动物可以摄食碎屑。以相同碳浓度的人工碎屑和纯培养细菌作为饵料分别培养鞭毛虫 *C.*

paramecium，碎屑支撑的鞭毛虫生长速率与细菌支撑的生长速率无显著差异，意味着碎屑颗粒可作为鞭毛虫在细菌之外的替代食物来源，并且可能对原生动物具有相当大的营养价值（Scherwass et al.，2005）。

三、微型浮游动物摄食颗粒黏附细菌

20 世纪 70~80 年代即有学者提出了微型浮游动物可摄食颗粒黏附细菌的假设（Fenchel and Jørgensen，1977；Fenchel，1986），此后，学者开展了多项实验以验证这一观点。

有研究制备了表面具黏性的海藻酸钙微球（Ca-alginate bead，粒径 3~50μm），再将这些微球放入细菌培养液中，这样细菌就能够在微球表面生长。将黏附了细菌的微球用荧光染料 DTAF 染色，投喂自然水样收集的和实验室纯培养的纤毛虫。经一段时间培养后用表面荧光显微镜观察，发现纤毛虫体内食物泡中存在发出荧光的细菌，这意味着纤毛虫摄食了微球表面的黏附细菌。不同纤毛虫种类对自由生细菌和颗粒黏附细菌的偏好性有差异，纤毛虫 *Euplotes* sp.对自由生细菌和颗粒黏附细菌的摄食速率基本一致，而纤毛虫 *Uronema* sp.和盾纤毛虫（scuticociliate）则倾向于摄食自由生细菌（Albright et al.，1987）。

另一个研究使用几丁质磨碎制成 20~40μm 的颗粒，与细菌 *Pseudomonas halodurans* 混合振荡培养 6 天后，有一些细菌黏附在几丁质颗粒上成为颗粒黏附细菌。向这些培养液中加入不同种的鞭毛虫进行培养，根据培养液中自由生细菌和颗粒黏附细菌的丰度变化可判断鞭毛虫的摄食情况。尽管培养液中自由生细菌的丰度远高于颗粒黏附细菌（约 5~90 倍），但鞭毛虫 *Rhynchomonas nasuta* 和 *Bodo* sp.对颗粒黏附细菌仍表现出明显的摄食偏好，对自由生细菌的摄食很有限，而鞭毛虫 *Monas* sp.和 *Cryptobia* sp.则主要摄食自由生细菌，对颗粒黏附细菌几乎没有摄食（Caron，1987）。

还有研究取自然海水加入荧光标记细菌（fluorescently labeled bacteria，FLB），随后放入滚动水槽（rolling tank）以较慢转速（2.5r/min）进行混合培养。自然海水中的细菌、颗粒、溶解有机物与 FLB 发生凝集，生成包含 FLB 的大型聚合颗粒。这些生物形成的聚合颗粒直径从 0.2mm 到 8.7mm 不等，平均 1.4mm。使用制备的聚合颗粒进行原生动物摄食实验，用表面荧光显微镜检查原生动物体内是否有 FLB，可判断原生动物是否摄食聚合颗粒黏附细菌。实验证明，原生动物能够摄食大型聚合颗粒上的黏附细菌，但纵然黏附细菌的丰度很高，原生动物对颗粒黏附细菌和自由生细菌的摄食速率却大体相当。这可能是由于原生动物需要付出额外努力才能将细菌从颗粒上分离下来，所以无法达到高摄食速率（Artolozaga et al.，2002）。

尽管如此，颗粒上原生动物的丰度仍比海水中高出 2～4 个数量级，表明原生动物在颗粒上生活有其他的获益，可能包括：①颗粒上的细菌丰度远高于海水中的自由生细菌，因此即便原生动物从颗粒上分离细菌需要花费额外能量，但是在海水中它们需要消耗更多能量来寻找细菌饵料，二者对比，在颗粒上摄食可能更为有利；②颗粒黏附细菌非常活跃，它们分解出超出自身使用量的 DOC，使颗粒周围的 DOC 更为富集，可促进颗粒周边细菌的丰度进一步增大，给在颗粒上生活的原生动物带来了额外福利（Artolozaga et al.，2002）。

四、桡足类"培养"颗粒黏附微生物

从真光层沉降到深海的颗粒每年将约 10 亿 t 的碳转移到深海中，有效降低了大气中的 CO_2 浓度。海洋暮色带（twilight zone，海面下 200～1000m 的一段水层）是沉降颗粒有机碳再矿化的重要区域，每年从海表输出的有机物有超过 90%在暮色带被呼吸变回 CO_2，其中大部分的呼吸作用是由细菌和浮游动物（主要是桡足类）进行的。有研究发现在北大西洋，细菌和浮游动物分别截获了约 50%的沉降有机颗粒，但细菌主导了暮色带的呼吸（79%），浮游动物的贡献仅占一小部分（21%）。导致这一差异的重要因素是桡足类对其捕捉到的大部分大粒级沉降颗粒进行了破碎分解，而不是直接吞食和消耗这些颗粒（Giering et al.，2014）。沉降颗粒中有些粒级过大，超过了桡足类最适的饵料粒径范围。还有一些颗粒的结构过于紧密，细菌和原生动物尚未侵入和在这些颗粒上定植，因而对桡足类的营养价值不高。通过破碎分解，桡足类将快速沉降的大颗粒转化为悬浮和缓慢下沉的小颗粒，促进细菌和微型浮游动物在这些小颗粒上定植，从而为桡足类提供了更适宜的饵料。此外，大量的细菌胞外酶还可以分解聚合颗粒中的难降解化合物，并使它们更容易被浮游动物消化。这一过程被称为桡足类"培养"微生物（microbial gardening）（Mayor et al.，2014）。

第五节 微食物网生物与海洋碳输出

一、微食物网生物在海洋 POC 中的比例

海洋 POC 的测量方法是用滤膜过滤海水，测定截留在滤膜上的有机碳。在过滤的过程中，海洋微食物网生物，包括异养细菌、蓝细菌、真核藻类、鞭毛虫和纤毛虫等也会被收集到滤膜上。因此，海洋 POC 不仅包括无生命的碎屑，还包括一部分微食物网生物。

微食物网生物个体微小，在测定 POC 过程中很难将其与碎屑分离开来，因此

估算微食物网生物在海洋 POC 中所占的比例难度较大。Pomeroy（1980）曾假设在 POC 中碎屑与生物的比例约为 10∶1。这一比例曾被后人多次引用，但真正测定过这个比例的研究极少。仅有的研究计数了马尾藻海 0～175m 水体内微食物网各个生物类群的丰度，再利用体积和生物量转换系数换算为微食物网生物量，估算结果表明，微食物网生物占 POC 的比例在 3～4 月为 55%，在 8 月为 24%（Caron et al.，1995；Roman et al.，1995）。

有些研究只估算了一部分微食物网生物的比例。例如，有研究估算大洋中细菌有机碳（bacterial organic carbon，BOC）占 POC 的比例平均为 43%，最高可达 70%（Cho and Azam，1988）。在北太平洋 ALOHA 观测站 0～80m 水体中，细菌（包括自养和异养）及细菌碎屑在总 POC 中的比例为 20%～30%（Kawasaki et al.，2011）。在北大西洋 BATS 观测站 0～65m 水体中，浮游植物和异养生物在总 POC 中的比例分别为 32% 和 15%（Gundersen et al.，2001）

二、微食物网生物的碳通量

微食物网生物碳通量的研究方法与上文 POC 中生物组分比例的研究类似，不同之处在于样品来自沉积物捕获器（sediment trap）。

真光层内微微型浮游生物（picoplankton）的丰度和生物量巨大，它们向深层的碳输出有直接和间接两种途径：直接途径是微微型浮游生物自身聚集，或被合并入其他大型颗粒，以碎屑颗粒形式直接沉降；间接途径是被浮游动物摄食，摄食者主要是微型浮游动物，桡足类等中型浮游动物难以直接捕获个体微小的微微型浮游生物。

传统观点中，微微型浮游生物个体微小，其自身形成的碎屑颗粒沉降很慢，而微型浮游动物摄食者产生的粪便颗粒也很小，沉降缓慢。因此碎屑颗粒和粪便颗粒在沉降过程中就会被细菌等分解掉，难以形成高效、大通量的沉降，微微型浮游生物对海洋上层碳输出的贡献相对较小，而大型、快速下沉的浮游植物（如硅藻）主导了来自上层海洋的碳通量。

部分研究获得的结果符合上述传统观点。在北太平洋环流区（26°N，155°W），有研究估算微食物网的碳通量约占 POC 通量的 6%～8%。在微食物网生物中，寡毛类纤毛虫占的比例（45%～79%）最大，细菌的比例为 2.7%～28.6%（Taylor，1989）。在东北太平洋真光层底部，小粒径的生命体（包括浮游植物和粒径>10μm 的原生动物等，但不包括细菌和粒径<10μm 的原生动物）对 POC 通量的贡献约为 36%，其中原生动物占的比例通常大于 10%。生命体对 POC 通量的贡献在真光层以下逐渐减小，在海洋中层（mesopelagic zone）为 22%，在海洋深层（bathypelagic zone）为 11%（Silver and Gowing，1991）。

近年来的一系列研究结果却发现，微微型浮游植物（picophytoplankton）不仅主导了大洋海区的初级生产，还是海洋上层碳输出的主要贡献者（Richardson and Jackson，2007）。例如，在赤道太平洋和阿拉伯海，研究发现不同粒级（pico级、nano级、micro级）浮游植物对碳输出的相对贡献与其对总净初级生产量（net primary production，NPP）的贡献成正比。其中，在赤道太平洋，微微型自养浮游生物的初级生产量占总NPP的70%以上，贡献了87%的POC通量，以及中型浮游动物碳输出的76%（Richardson et al.，2004）。在东北季风期间阿拉伯海微微型自养浮游生物的初级生产量占总NPP的86%，并贡献了97%的POC通量和75%的中型浮游动物碳输出（Richardson et al.，2006）。微微型浮游植物能够数百个细胞聚集成较大的碎屑颗粒（Albertano et al.，1997；Olli and Heiskanen，1999），增加了垂直沉降速度和由此产生的碳输出。此外，微微型浮游植物的聚集增加了有效尺寸，使桡足类等中型浮游动物可以摄食这些聚合颗粒，这也构成了微微型浮游植物碳输出的重要途径（Wilson and Steinberg，2010；Lomas and Moran，2011）。

参 考 文 献

Albertano P, Somma D D, Capucci E. 1997. Cyanobacterial picoplankton from the Central Baltic Sea: cell size classification by image-analyzed fluorescence microscopy. Journal of Plankton Research, 19(10): 1405-1416.

Albright L, Sherr E, Sherr B, et al. 1987. Grazing of ciliated protozoa on free and particle-attached bacteria. Marine Ecology Progress Series, 38(2): 125-129.

Alldredge A L, Cole J J, Caron D A. 1986. Production of heterotrophic bacteria inhabiting macroscopic organic aggregates (marine snow) from surface waters. Limnology and Oceanography, 31(1): 68-78.

Alldredge A L, Passow U, Logan B E. 1993. The abundance and significance of a class of large, transparent organic particles in the ocean. Deep Sea Research Part I: Oceanographic Research Papers, 40(6): 1131-1140.

Anderson O R. 2015. Particle-associated planktonic naked amoebae in the Hudson Estuary: size-fraction related densities, cell sizes and estimated carbon content. Acta Protozoologica, 50(1): 15-22.

Artolozaga I, Ayo B, Latatu A, et al. 2000. Spatial distribution of protists in the presence of macroaggregates in a marine system. FEMS Microbiology Ecology, 33(3): 191-196.

Artolozaga I, Santamaría E, López A, et al. 1997. Succession of bacterivorous protists on laboratory-made marine snow. Journal of Plankton Research, 19(10): 1429-1440.

Artolozaga I, Valcárcel M, Ayo B, et al. 2002. Grazing rates of bacterivorous protists inhabiting diverse marine planktonic microenvironments. Limnology and Oceanography, 47(1): 142-150.

Azam F. 1998. Microbial control of oceanic carbon flux: the plot thickens. Science, 280(5364): 694-696.

Bach L T, Boxhammer T, Larsen A, et al. 2016. Influence of plankton community structure on the sinking velocity of marine aggregates. Global Biogeochemical Cycles, 30(8): 1145-1165.

Barber R T. 1966. Interaction of bubbles and bacteria in the formation of organic aggregates in sea-water. Nature, 211: 257-258.

Beaumont K L, Nash G V, Davidson A T. 2002. Ultrastructure, morphology and flux of microzooplankton faecal pellets in an east Antarctic fjord. Marine Ecology Progress Series, 245: 133-148.

Benner R, Pakulski J D, McCarthy M, et al. 1992. Bulk chemical characteristics of dissolved organic matter in the ocean. Science, 255(5051): 1561-1564.

Berger B, Hoch B, Kavka G, et al. 1996. Bacterial colonization of suspended solids in the River Danube. Aquatic Microbial Ecology, 10(1): 37-44.

Biddanda B A. 1985. Microbial synthesis of macroparticulate matter. Marine Ecology Progress Series, 20: 241-251.

Buck K R, Bolt P A, Garrison D L. 1990. Phagotrophy and fecal pellet production by an athecate dinoflagellate in Antarctic sea ice. Marine Ecology Progress Series, 60: 75-84.

Buck K R, Newton J. 1995. Fecal pellet flux in Dabob Bay during a diatom bloom: contribution of microzooplankton. Limnology and Oceanography, 40(2): 306-315.

Buck K, Marin III R, Chavez F. 2005. Heterotrophic dinoflagellate fecal pellet production: grazing of large, chain-forming diatoms during upwelling events in Monterey Bay, California. Aquatic Microbial Ecology, 40(3): 293-298.

Caron D A. 1987. Grazing of attached bacteria by heterotrophic microflagellates. Microbial Ecology, 13(3): 203-218.

Caron D A, Dam H, Kremer P, et al. 1995. The contribution of microorganisms to particulate carbon and nitrogen in surface waters of the Sargasso Sea near Bermuda. Deep Sea Research Part I: Oceanographic Research Papers, 42(6): 943-972.

Caron D A, Davis P G, Madin L P, et al. 1982. Heterotrophic bacteria and bacterivorous protozoa in oceanic macroaggregates. Science, 218(4574): 795-797.

Cho B C, Azam F. 1988. Major role of bacteria in biogeochemical fluxes in the ocean's interior. Nature, 332(6163): 441-443.

Ding Y-X, Chin W-C, Rodriguez A, et al. 2008. Amphiphilic exopolymers from *Sagittula stellata* induce DOM self-assembly and formation of marine microgels. Marine Chemistry, 112(1-2): 11-19.

Fenchel T. 1986. The ecology of heterotrophic microflagellates//Marshall K C. Advances in Microbial Ecology. Boston: Springer: 57-97.

Fenchel T, Blackburn N. 1999. Motile chemosensory behaviour of phagotrophic protists: mechanisms for and efficiency in congregating at food patches. Protist, 150(3): 325-336.

Fenchel T M, Jørgensen B B. 1977. Detritus food chains of aquatic ecosystems: the role of bacteria//Alexander M. Advances in Microbial Ecology. Boston: Springer US: 1-58.

Fortier L, Le Fèvre J, Legendre L. 1994. Export of biogenic carbon to fish and to the deep ocean: the role of large planktonic microphages. Journal of Plankton Research, 16(7): 809-839.

Fukuda H, Koike I. 2000. Feeding currents of particle-attached nanoflagellates-A novel mechanism for aggregation of submicron particles. Marine Ecology Progress Series, 202: 101-112.

Fukuda H, Koike I. 2004. Microbial stimulation of the aggregation process between submicron-sized particles and suspended particles in coastal waters. Aquatic Microbial Ecology, 37(1): 63-73.

Gärdes A, Iversen M H, Grossart H P, et al. 2011. Diatom-associated bacteria are required for aggregation of *Thalassiosira weissflogii*. The ISME Journal, 5: 436-445.

Giering S L C, Sanders R, Lampitt R S, et al. 2014. Reconciliation of the carbon budget in the ocean's twilight zone. Nature, 507(7493): 480-483.

Gowing M M, Garrison D L, Kunze H B, et al. 2001. Biological components of Ross Sea short-term particle fluxes in the austral summer of 1995-1996. Deep Sea Research Part I: Oceanographic Research Papers, 48(12): 2645-2671.

Gowing M M, Silver M W. 1985. Minipellets: a new and abundant size class of marine fecal pellets. Journal of Marine Research, 43(2): 395-418.

Grossart H P. 2010. Ecological consequences of bacterioplankton lifestyles: changes in concepts are needed. Environmental Microbiology Reports, 2(6): 706-714.

Gundersen K, Orcutt K M, Purdie D A, et al. 2001. Particulate organic carbon mass distribution at the Bermuda Atlantic Time-series Study(BATS) site. Deep Sea Research Part II: Topical Studies in Oceanography, 48(8): 1697-1718.

Isao K, Hara S, Terauchi K, et al. 1990. Role of sub-micrometre particles in the ocean. Nature, 345(6272): 242-244.

Kawasaki N, Sohrin R, Ogawa H, et al. 2011. Bacterial carbon content and the living and detrital bacterial contributions to suspended particulate organic carbon in the North Pacific Ocean. Aquatic Microbial Ecology, 62(2): 165-176.

Kiørboe T, Jackson G A. 2001. Marine snow, organic solute plumes, and optimal chemosensory behavior of bacteria. Limnology and Oceanography, 46(6): 1309-1318.

Leppard G G. 1992. Size, morphology and composition of particulates in aquatic ecosystems: solving speciation problems by correlative electron microscopy. Analyst, 117(3): 595-603.

Lochte K, Turley C. 1988. Bacteria and cyanobacteria associated with phytodetritus in the deep sea. Nature, 333(6168): 67-69.

Lomas M, Moran S. 2011. Evidence for aggregation and export of cyanobacteria and nano-eukaryotes from the Sargasso Sea euphotic zone. Biogeosciences, 8(1): 203-216.

Long R A, Azam F. 1996. Abundant protein-containing particles in the sea. Aquatic Microbial Ecology, 10: 213-221.

Mari X, Kiørboe T. 1996. Abundance, size distribution and bacterial colonization of transparent exopolymeric particles(TEP) during spring in the Kattegat. Journal of Plankton Research, 18(6): 969-986.

Mayor D J, Sanders R, Giering S L C, et al. 2014. Microbial gardening in the ocean's twilight zone: detritivorous metazoans benefit from fragmenting, rather than ingesting, sinking detritus. BioEssays: News and Reviews in Molecular, Cellular and Developmental Biology, 36(12): 1132-1137.

Mostajir B, Dolan J R, Rassoulzadegan F. 1995. Seasonal variations of pico-and nano-detrital particles(DAPI Yellow Particles, DYP) in the Ligurian Sea(NW Mediterranean). Aquatic Microbial Ecology, 9(3): 267-277.

Nöthig E-M, Bodungen B V. 1989. Occurrence and vertical flux of faecal pellets of probably protozoan origin in the southeastern Weddell Sea(Antarctica). Marine Ecology Progress Series, 56: 281-289.

Olli K, Heiskanen A-S. 1999. Seasonal stages of phytoplankton community structure and sinking loss in the Gulf of Riga. Journal of Marine Systems, 23(1): 165-184.

Paerl H W. 1975. Microbial attachment to particles in marine and freshwater ecosystems. Microbial Ecology, 2(1): 73-83.

Passow U. 2002. Transparent exopolymer particles(TEP) in aquatic environments. Progress in

Oceanography, 55(3): 287-333.

Passow U, Alldredge A. 1994. Distribution, size and bacterial colonization of transparent exopolymer particles(TEP) in the ocean. Marine Ecology Progress Series, 113: 185-198.

Pasternak A, Arashkevich E, Riser C W, et al. 2000. Seasonal variation in zooplankton and suspended faecal pellets in the subarctic Norwegian Baisfjorden, in 1996. Sarsia, 85(5-6): 439-452.

Pelegri S, Christaki U, Dolan J, et al. 1999. Particulate and dissolved organic carbon production by the heterotrophic nanoflagellate *Pteridomonas danica* Patterson and Fenchel. Microbial Ecology, 37(4): 276-284.

Pomeroy L R. 1980. Detritus and its role as a food source//Barnes R, Mann K. Fundamentals of Aquatic Ecosystems. London: Blackwell Scientific Publications: 84-102.

Posch T, Arndt H. 1996. Uptake of sub-micrometre-and micrometre-sized detrital particles by bacterivorous and omnivorous ciliates. Aquatic Microbial Ecology, 10(1): 45-53.

Richardson T L, Jackson G A. 2007. Small phytoplankton and carbon export from the surface ocean. Science, 315(5813): 838-840.

Richardson T L, Jackson G A, Ducklow H W, et al. 2004. Carbon fluxes through food webs of the eastern equatorial Pacific: an inverse approach. Deep Sea Research Part I: Oceanographic Research Papers, 51(9): 1245-1274.

Richardson T L, Jackson G A, Ducklow H W, et al. 2006. Spatial and seasonal patterns of carbon cycling through planktonic food webs of the Arabian Sea determined by inverse analysis. Deep Sea Research Part II: Topical Studies in Oceanography, 53(5): 555-575.

Riley G A. 1963. Organic aggregates in seawater and the dynamics of their formation and utilization. Limnology and Oceanography, 8(4): 372-381.

Rogerson A, Anderson O R, Vogel C. 2003. Are planktonic naked amoebae predominately floc associated or free in the water column? Journal of Plankton Research, 25(11): 1359-1365.

Roman M R, Caron D A, Kremer P, et al. 1995. Spatial and temporal changes in the partitioning of organic carbon in the plankton community of the Sargasso Sea off Bermuda. Deep Sea Research Part I: Oceanographic Research Papers, 42(6): 973-992.

Scherwass A, Fischer Y, Arndt H. 2005. Detritus as a potential food source for protozoans: utilization of fine particulate plant detritus by a heterotrophic flagellate, *Chilomonas paramecium*, and a ciliate, *Tetrahymena pyriformis*. Aquatic Ecology, 39(4): 439-445.

Schuster S, Herndl G J. 1995. Formation and significance of transparent exopolymeric particles in the northern Adriatic Sea. Marine Ecology Progress Series, 124: 227-236.

Silver M W, Alldredge A L. 1981. Bathypelagic marine snow: deep sea algal and detrital community. Journal of Marine Research, 39: 501-530.

Silver M W, Coale S L, Pilskaln C H, et al. 1998. Giant aggregates: Importance as microbial centers and agents of material flux in the mesopelagic zone. Limnology and Oceanography, 43(3): 498-507.

Silver M W, Gowing M M. 1991. The "particle" flux: origins and biological components. Progress in Oceanography, 26(1): 75-113.

Silver M W, Gowing M M, Brownlee D C, et al. 1984. Ciliated protozoa associated with oceanic sinking detritus. Nature, 309(5965): 246-248.

Silver M W, Gowing M M, Davoll P J. 1986. The association of photosynthetic picoplankton and ultraplankton with pelagic detritus through the water column(0-2000m). Canadian Bulletin of Fisheries and Aquatic Sciences, 214: 311-341.

Simon M, Grossart H P, Schweitzer B, et al. 2002. Microbial ecology of organic aggregates in aquatic ecosystems. Aquatic Microbial Ecology, 28(2): 175-211.

Smith K, Williams P, Druffel E. 1989. Upward fluxes of particulate organic matter in the deep North Pacific. Nature, 337(6209): 724-726.

Stoderegger K E, Herndl G J. 1998. Production and release of bacterial capsular material and its subsequent utilization by marine bacterioplankton. Limnology and Oceanography, 43(5): 877-884.

Stoderegger K E, Herndl G J. 1999. Production of exopolymer particles by marine bacterioplankton under contrasting turbulence conditions. Marine Ecology Progress Series, 189: 9-16.

Stoecker D K. 1984. Particle production by planktonic ciliates. Limnology and Oceanography, 29(5): 930-940.

Sugimoto K, Fukuda H, Baki M A, et al. 2007. Bacterial contributions to formation of transparent exopolymer particles(TEP) and seasonal trends in coastal waters of Sagami Bay, Japan. Aquatic Microbial Ecology, 46: 31-41.

Suzuki N, Kato K. 1953. Studies on suspended materials marine snow in the sea: Part I. Sources of marine snow. Bulletin of the Faculty of Fisheries, Hokkaido University, 4(2): 132-137.

Taylor G T. 1989. Variability in the vertical flux of microorganisms and biogenic material in the epipelagic zone of a North Pacific central gyre station. Deep Sea Research Part I: Oceanographic Research Papers, 36(9): 1287-1308.

Thiel H, Pfannkuche O, Schriever G, et al. 1989. Phytodetritus on the deep-sea floor in a central oceanic region of the Northeast Atlantic. Biological Oceanography, 6(2): 203-239.

Toggweiler J. 1989. Are rising and falling particles microbial elevators? Nature, 337(6209): 691-692.

Turley C, Mackie P. 1994. Biogeochemical significance of attached and free-living bacteria and the flux of particles in the NE Atlantic Ocean. Marine Ecology Progress Series, 115: 191-203.

Verdugo P. 2012. Marine microgels. Annual Review of Marine Science, 4: 375-400.

Verdugo P, Alldredge A L, Azam F, et al. 2004. The oceanic gel phase: a bridge in the DOM-POM continuum. Marine Chemistry, 92(1): 67-85.

Waite A M, Safi K A, Hall J A, et al. 2000. Mass sedimentation of picoplankton embedded in organic aggregates. Limnology and Oceanography, 45(1): 87-97.

Wells M L, Goldberg E D. 1993. Colloid aggregation in seawater. Marine Chemistry, 41(4): 353-358.

Wilson S, Steinberg D. 2010. Autotrophic picoplankton in mesozooplankton guts: evidence of aggregate feeding in the mesopelagic zone and export of small phytoplankton. Marine Ecology Progress Series, 412: 11-27.

Yamada Y, Fukuda H, Inoue K, et al. 2013. Effects of attached bacteria on organic aggregate settling velocity in seawater. Aquatic Microbial Ecology, 70(3): 261-272.

Yamada Y, Fukuda H, Tada Y, et al. 2016. Bacterial enhancement of gel particle coagulation in seawater. Aquatic Microbial Ecology, 77(1): 11-22.

第十二章　微食物网生物对营养盐的再生作用

海洋中的氮、磷等营养盐随生物地球化学循环在生物体和环境间周转。异养生物（含混合营养生物）吸收利用水体有机营养物质或摄食其他生物进行生长（同化），同时会将溶解态营养盐释放回海水（异化），这一营养盐的释放过程称为营养盐再生（nutrient regeneration）。

在海洋微食物网的各生物类群中，除聚球藻、原绿球藻和微微型自养真核生物外，其他类群如病毒、细菌、鞭毛虫、纤毛虫等都是异养生物（部分微型浮游动物为混合营养，也有异养过程）。海洋微食物网的营养盐再生过程主要由这些异养（含混合营养，本章下同）生物类群来完成。其中，微型浮游动物对营养盐的再生主要通过排泄（excretion）来完成，而细菌对营养盐的再生又被称为再矿化（remineralization）。

营养盐再生效率（regeneration efficiency，R）可以用来衡量生物对营养盐再生的贡献情况。R 为在一定时间段内摄食者排出的营养盐（E）和摄入的营养盐（I）的比值（$R=E/I\times100\%$），或者是培养体系在一定培养时间段内排出的所有的营养盐和培养开始时颗粒态营养盐的比值（Goldman et al.，1985）。

第一节　异养细菌对营养盐的再生作用

在微食物网概念提出之前，早已有研究发现海洋异养细菌能够吸收无机磷营养盐（Johannes，1964b）。随着微食物网概念的提出，人们逐渐意识到单细胞异养生物（细菌和原生动物）是大部分营养盐再生的贡献者（Caron，1994）。

海洋中没有被摄食消耗的有机物质被称为碎屑（detritus）（Fenchel and Jørgensen，1977）。海洋异养细菌依靠分解海水中的碎屑获取营养物质进行生长和代谢，这些碎屑被称为细菌的底物（substrate）。碎屑的化学组成变化很大，如大型海藻和海草的残体是一类高分子量、结构复杂的惰性颗粒碎屑，而海洋浮游生物释放的单糖、氨基酸等则是极易被氧化的溶解性碎屑。在大洋中，后者是异养细菌的主要底物。传统观点认为细菌是分解者，利用这些溶解和颗粒碎屑物质并向海水释放氮、磷营养盐。但是 20 世纪 80 年代起，多项研究表明，这些异养细菌还可以直接吸收利用海水中的无机氮、磷营养盐，与浮游植物是竞争关系（Wheeler and Kirchman，1986；Kirchman，1994）。

异养细菌是吸收还是释放营养盐取决于细菌和底物的元素比例（Caron，

1994）。若底物的C∶N值远高于细菌的C∶N值，意味着底物中氮营养盐缺乏，此时细菌主要是吸收海水中的氮，以补充底物中不足的氮营养盐；反之若底物的C∶N值远低于细菌，细菌分解底物时会向海水中释放氮营养盐。例如，有研究用不同C∶N（1.5∶1到10∶1）的底物培养细菌以研究氮元素的再生效率，所用细菌菌株的C∶N值为5∶1。当底物的C∶N值为10∶1时，细菌吸收海水中的氮营养盐，氮的再生效率为0%；当底物的C∶N值为1.5∶1时，细菌释放氮营养盐，氮的再生效率为86%（Goldman et al.，1987b）。

虽然，至今尚未严格确定浮游异养细菌生长所需底物的化学组成，但是生态学家通常认为海洋中的碳水化合物是细菌生长和代谢的主要底物，其C∶N值可能大于10∶1。与底物相比，异养细菌个体微小，单位体重的核酸含量很高，即细菌的氮元素含量较高。在生长速率较高的阶段，细菌的C∶N值可低至4∶1。因此，海洋中快速生长的细菌主要是吸收海水中的氮营养盐，不太可能对氮再生有显著贡献。

在海洋里，细菌底物中的溶解有机碳主要来自浮游植物。由于底物C∶N值是细菌是否吸收无机营养盐的决定因素，有假说认为在寡营养海区，由于缺乏氮营养盐，浮游植物生成的溶解有机碳C∶N值较高，因此细菌会从海水中吸收更多无机营养盐，又加剧了水体的寡营养化；在近岸富营养海区，氮营养盐丰富，浮游植物产生的底物C∶N值较低，细菌能从底物中获得足够的氮营养盐，对海水中无机氮营养盐的吸收会较少（Kirchman，1994）。类似地，在层化较好的海区，真光层中营养盐较缺乏，浮游植物产生高C∶N值的底物，细菌会吸收海水中的无机营养盐；而在真光层以下营养盐丰富，浮游植物产生低C∶N值的底物，此时异养细菌有可能是营养盐的再生者（Caron，1994）。此后的研究在以上假说的基础上进行了细化，在近岸河口等富营养海区也发现异养细菌吸收营养盐，但都是有机形态的营养盐，而非无机营养盐（白洁等，2005；王秋璐等，2010）。

总的来说，目前海洋生态学家普遍认为异养细菌在绝大多数海区都吸收营养盐，海洋浮游原生动物才是营养盐再生的主要贡献者。

第二节　病毒对营养盐的再生作用

海洋浮游病毒会导致浮游细菌和浮游植物死亡，部分情况下病毒介导的细菌和浮游植物死亡率甚至高于浮游动物摄食导致的死亡率。病毒裂解宿主，导致宿主细胞中的有机和无机物质释放到周围海水中。宿主释放的营养盐主要是有机物的形态，这一点不同于浮游动物排泄的无机营养盐，因此，病毒介导的营养盐再生过程又称营养盐再活化（remobilization）（Wilhelm and Suttle，2000）。

直接测定病毒对营养盐的再活化速率是非常困难的。有研究使用放射性标记法研究病毒侵染宿主后的营养元素释放情况，发现病毒侵染单胞藻 *Aureococcus anophagefferens* 后，氮、磷和铁都以颗粒态形式释放，培养体系中没有新增的溶解态氮、磷、铁营养盐，且 *A. anophagefferens* 被病毒裂解后培养体系内细菌丰度显著升高，表明病毒介导释放的营养盐可能迅速被培养体系中的细菌利用（Gobler et al., 1997）。

因为直接测定比较困难，间接估算病毒对营养元素的再活化是另一种研究策略。间接法主要根据病毒的周转率和裂解量、细菌的死亡率和生长率、细菌的元素含量等参数，估算病毒裂解细菌释放到水体中的营养盐含量，进而评估营养盐的再活化速率。利用这种方法，Wilhelm 和 Suttle（2000）估计在墨西哥湾病毒对氮营养盐的再活化速率为 0.02~1.0μg/（L·d），在佐治亚海峡病毒对氮、磷和铁营养盐的再活化速率分别为 0.25~1.98μg/（L·d）、0.02~0.18μg/（L·d）和 0.05~0.17μg/（L·d）。

除了病毒裂解宿主释放营养盐外，病毒自身的衰亡也能够向水体释放营养元素。与其宿主相比，单位重量病毒颗粒的磷元素含量更高。因此，与氮元素相比，病毒衰亡对海洋磷循环的重要性要更大（Jover et al., 2014）。

第三节　微型浮游动物对营养盐的再生作用

在能够实际测定微型浮游动物营养盐的排泄速率之前，有学者先用理论推算了不同粒级海洋浮游动物对磷营养盐再生的贡献（Johannes，1964a）。这一理论用等体重代谢时间（body-weight equivalent excretion time，BEET），即动物排出等同于自身体内所含营养盐总量的营养盐所需的时间，来衡量不同粒级生物代谢速率的快慢。根据当时已有的资料，可确立几种浮游动物排泄磷的 BEET 与其体重的关系，包括小型甲壳类、丰年虾、纤毛虫和鞭毛虫等。根据这些数据，Johannes（1964a）发现浮游动物的粒径越小，其单位体重的代谢速率越高，BEET 越短，排泄营养盐的相对速率越高。其中，体积为 1μm³、干重为 2.5×10⁻⁷μg 的鞭毛虫排泄磷的 BEET 为 2min，这意味着体积为 1μm³ 的鞭毛虫仅需 2min 即可排出等同于自身总含磷量的磷营养盐。考虑到微型浮游动物的营养盐代谢速率高于大、中型浮游动物，且微型浮游动物在浮游动物总生物量中所占比例也高于大、中型浮游动物，可认为海洋微型浮游动物对营养盐再生作用的重要性远大于大、中型浮游动物。

在理论推算之外，也有大量针对微型浮游动物营养盐再生作用的实测研究，主要集中在两个方面：①针对某一微型浮游动物种类，研究其对营养盐的排泄速率（excretion rate）；②针对某一海区的所有微型浮游动物，研究其对营养盐的再生速率（regeneration rate）。前者主要在实验室内进行，使用的是已有的纯培养微型浮游动物和饵料生物；后者则通常在自然海区进行原位培养，或采集自然海区微型浮游动物和饵料生物样品带回实验室培养。

一、微型浮游动物的营养盐排泄

室内测定微型浮游动物营养盐排泄速率的研究不多，主要困难有两个：微型浮游动物的室内培养较难，测定微型浮游动物代谢速率的技术难度较高。

微型浮游动物的个体微小，尽管其单位体重的代谢速率很高，但单个个体的代谢速率却很低。要克服由于个体代谢速率低而难以检测的困难，就要加长培养时间。为使浮游动物在较长培养期内保持良好的生理代谢状态，需要以浮游植物、细菌等作为饵料，这就带来了一个额外问题，即饵料在培养过程中会吸收利用浮游动物释放的营养盐。消除投喂饵料对营养盐的影响主要有 3 个方法：①在持续黑暗条件下进行培养，以避免浮游植物饵料在培养过程中吸收营养盐；②添加细菌抑制剂阻止细菌生长，以避免细菌吸收营养盐；③设立对照组，估计饵料对营养盐的吸收速率。

在微型浮游动物的排泄物中，无机铵盐（NH_4^+）和无机磷酸盐（PO_4^{3-}）是营养盐的主要存在形式，此外还有溶解有机氮和溶解有机磷，如尿素、氨基酸、多肽、嘌呤、嘧啶等，这些溶解有机营养盐也很重要，但是其绝对量比无机营养盐要小，因此，本部分仅介绍 NH_4^+ 和 PO_4^{3-} 等无机营养盐的排泄速率。

（一）鞭毛虫的营养盐排泄

现有的研究测定了部分鞭毛虫物种的氮、磷排泄速率和/或再生效率，包括 *Bodo designis*、*Jakoba libera*、*Monas* sp.、*Ochromonas* sp.、*Oxyrrhis* sp.、*Paraphysomonas imperforata*、*Poterioochromonas malhamensis*、*Pseudobodo* spp.、*Spumella* sp. 和 *Stephanoeca diplocostata* 等（张武昌等，2016）。这些鞭毛虫物种的单位体重（干重）排氮速率范围在 2.8~140μg NH_4^+-N/（mg·h），单个细胞最大排氮速率为 $7.0×10^{-9}$~$13.8×10^{-6}$μg NH_4^+-N/（cell·h），无机铵盐的再生效率为 0%~100%。最大排磷速率为 $3.8×10^{-9}$~$6.6×10^{-7}$μg PO_4^{3-}-P/（cell·h），无机磷酸盐再生效率为 0%~100%。各鞭毛虫物种的具体排泄速率和再生效率数据可参见张武昌等（2016）。

鞭毛虫的营养盐排泄速率和再生效率因培养时期不同而有差异。在饵料（硅藻和细菌）不受营养盐限制的培养体系中，鞭毛虫 *Paraphysomonas imperforata* 的铵含量在指数生长期最高，为 11~19pg NH_4^+-N/cell，进入到稳定生长期后明显降低，为 6~16pg NH_4^+-N/cell。*P. imperforata* 的 NH_4^+ 再生效率则在指数生长期最低，为 15%~30%，在稳定生长期则为 28%~100%。值得注意的是，当饵料生物受到氮限制时，指数生长期间鞭毛虫的氮再生效率进一步降低至 8%（Goldman et al.，1985）。

饵料生物的营养元素组成对鞭毛虫的营养盐再生也有影响。例如，浮游植物的氮、磷含量比可影响鞭毛虫 *Paraphysomonas imperforata* 的氮、磷再生。饵料的氮缺乏越严重，指数生长期鞭毛虫对氮再生的滞后越长，甚至到稳定生长期才有氮再生；在饵料磷缺乏但氮不受限的情况下，NH_4^+的排出在 *P. imperforata* 的指数生长期即开始，直到稳定生长期后还有 NH_4^+排出。当饵料中缺乏氮时，鞭毛虫 *P. imperforata* 在整个生长周期中都有磷的排泄，且排泄速度大体相同；当饵料中缺乏磷时，在鞭毛虫的任何生长时期都没有磷排泄（Goldman et al.，1987a）。

不同生长策略鞭毛虫的营养盐再生效率不同。例如，*Paraphysomonas imperforata* 和 *Bodo designis* 是高生长率鞭毛虫（生长率分别为 0.2/h 和 0.14/h），*Jakoba libera* 和 *Stephanoeca diplocostata* 为低生长率鞭毛虫（生长率分别为 0.029/h 和 0.026/h）。在相同的饵料、营养盐浓度等培养条件下，上述 4 种鞭毛虫对氮的再生效率没有显著差异（46%～69%），但是对磷的再生效率却有显著差异，高生长率鞭毛虫为 63%～85%，而低生长率鞭毛虫仅有 33%～35%（Eccleston-Parry and Leadbeater，1995）。

大多数研究采取的培养温度都较为温和，主要处于 15～25℃，但也有部分研究选取了低温和高温进行培养。有研究在 3℃、18℃、23.5℃和 30℃条件下培养鞭毛虫 *Monas* sp.，发现 *Monas* sp.在 18℃和 23.5℃时排铵速率较低，在 3℃和 30℃时排铵速率明显提高，这种变化可能是鞭毛虫在低温和高温下处于生理不适状态造成的（Sherr et al.，1983）。在 0～10℃，从南大洋分离的 *Paraphysomonas imperforata* 的生长速率随温度增加而显著提高，但是温度的变化对氮和磷的再生没有显著影响（Rose et al.，2008）。

（二）纤毛虫的营养盐排泄

目前有研究报道了 5 个属浮游纤毛虫的营养盐排泄速率，包括 *Codonella*、*Colpidium*、*Stokesia*、*Strombidium* 和 *Tintinnopsis*（张武昌等，2016）。这些研究表明，纤毛虫单位体重的排氮速率为 0.25～178μg NH_4^+-N/（mg·h），最大个体排氮速率为 $1.59×10^{-7}$～$1.2×10^{-4}$μg NH_4^+-N/（cell·h）；纤毛虫单位体重排磷速率为 13～363μg PO_4^{3-}-P/（mg·h），最大个体排磷速率为 0～$1.3×10^{-5}$μg PO_4^{3-}-P/（cell·h）。各属种纤毛虫的具体排泄速率数据可参见张武昌等（2016）。

纤毛虫的营养盐排泄速率和再生效率也随培养时期不同而有差异。纤毛虫的排铵速率一般在指数生长期较高，例如，无壳纤毛虫 *Strombidium sulcatum* 的排铵速率在指数生长期分别为 $2.72×10^{-6}$μg NH_4^+-N/（cell·h）（活细菌饵料）和 $8.72×10^{-6}$μg NH_4^+-N/（cell·h）（热杀死细菌饵料），进入稳定生长期时迅速降低到不足 $1×10^{-6}$μg NH_4^+-N/（cell·h）。*S. sulcatum* 的氮再生效率在指数生长期前期一直较低（<30%），在指数生长期后期开始升高（>100%），并在稳定生长期保持较高水平（30%~100%）(Ferrier-Pages and Rassoulzadegan, 1994)。*S. sulcatum*的排磷速率在培养实验刚开始的停滞期最高，为 $13\times10^{-6}\mu g\ PO_4^{3-}$-P/（cell·h），之后随培养时间延长而逐渐降低，到稳定生长期后期降至 0（Allali et al., 1994）。

温度对纤毛虫的营养盐排泄速率也有一定影响。砂壳纤毛虫 *Tintinnopsis acuminata* 和 *T. vasculum* 的营养盐排泄速率都随温度升高而升高，*T. acuminata* 的排铵速率在 15℃时为 9pg NH_4^+-N/（cell·h），25℃时上升至 21pg NH_4^+-N/（cell·h）；*T. vasculum* 的排铵速率在5℃时为 32pg NH_4^+-N/（cell·h），15℃时上升为 82pg NH_4^+-N/（cell·h）(Verity, 1985)。

（三）营养盐排泄速率与干重的关系

根据多个研究测定的原生动物排泄速率资料，对氮和磷的最大排泄速率[Excr, μg/（cell·h）]与干重（DW, mg/cell）进行回归，可获得如下关系（Dolan, 1997）：

$$\lg \text{Excr } N = -1.388 + 0.622\lg DW (n=15, r=0.899) \quad (12\text{-}1)$$
$$\lg \text{Excr } P = -2.101 + 0.570\lg DW (n=12, r=0.847) \quad (12\text{-}2)$$

在后生动物营养盐再生研究中也有类似的"最大排泄速率-相对个体大小"回归关系（Hargrave and Geen, 1968; Mullin et al., 1975; Ikeda et al., 1982; Wen et al., 1994），如在大堡礁，大中型浮游动物的氮、磷的最大排泄速率[Excr, ng/（ind.·h）]与其个体相对大小(size, μg DW/ind.)符合关系 $\lg \text{Excr } N=0.518+0.525\lg \text{size}$ 和 $\lg \text{Excr } P= -0.174+0.429\lg \text{size}$（Ikeda et al., 1982）。

根据上述回归关系，可发现原生动物的单位体重最大排磷率高于后生动物，而原生动物的单位体重最大排氮率与后生动物较接近。表明，原生动物在海洋磷再生过程中的作用可能高于在氮再生过程中的作用。

二、自然海区微型浮游动物对营养盐的再生作用

（一）营养盐再生速率的测定方法

同位素稀释法（isotope dilution technique）是测定自然海区微型浮游动物群落营养盐再生速率的常用方法之一。自然海区的微型浮游动物吸收水体中天然存在的 ^{14}N 营养盐进行生命活动，与此同时也向水体排泄出 ^{14}N 营养盐。若在自然海水中人为添加 ^{15}N 同位素标记的氮营养盐进行培养，并假设：①在培养过程中微型浮游动物对 ^{14}N 和 ^{15}N 的吸收速率没有差别，即对 ^{14}N 和 ^{15}N 按照海水中的原子比例进行吸收；②吸收的 ^{15}N 没有被微型浮游动物排泄到海水中，那么经一段时

间培养后，海水中 ^{15}N 标记的氮营养盐所占比例通常会变小，这种方法因此得名同位素稀释法（Caperon et al.，1979）。测量海水氮营养盐中 ^{14}N 和 ^{15}N 的原子比例变化，即可估算氮营养盐的吸收速率和再生速率[μg N/（L·h）]（Glibert，1982；Laws，1984）。

类似地，在自然海水中添加 ^{33}P 标记的磷营养盐进行培养，测量 ^{32}P 和 ^{33}P 的原子比例变化即可估算微型浮游动物对磷的吸收速率和再生速率（Harrison，1983）。

除直接测定外，也可以根据其他参数间接估算微型浮游动物的营养盐再生速率。例如，有研究根据微型浮游动物的碳和氮代谢模型，用微型浮游动物的摄食速率、生长效率等来估计其营养盐生产率，以此来反映其对营养盐的再生能力（Caron and Goldman，1988；Landry，1993）。以氮营养盐为例，这一方法计算氮营养盐排泄速率[E_N, μg N/（L·d）]的公式为

$$E_N = (AE_N \times I_C \times N:C_{PP}) - (I_C \times GGE_C \times N:C_{MZ}) \qquad (12-3)$$

式中，AE_N 是微型浮游动物对氮的同化效率，I_C 是微型浮游动物的碳摄食速率，$N:C_{PP}$ 和 $N:C_{MZ}$ 分别是浮游植物饵料（PP）和微型浮游动物（MZ）体内的氮碳比，GGE_C 是微型浮游动物体碳的总生长效率（gross growth efficiency）。AE_N、$N:C_{PP}$、$N:C_{MZ}$ 和 GGE_C 等参数可通过查询文献获得。I_C 可通过微型浮游动物摄食的稀释培养实验（参见本书第七章）获得，I_C 等于微型浮游动物摄食率与培养期间培养瓶中饵料平均浓度的乘积。

此外，还有研究根据微型浮游动物生物量、呼吸率和 P：C 的值（质量比，约为 1.4%）来估计磷的再生速率（Sorokin，1985），或使用环境中磷浓度的变化率和吸收率来计算磷再生速率（Sorokin，2002；Sorokin and Dallocchio，2008）。

（二）自然海区微型浮游生物的氮再生

在自然海区开展营养盐再生研究时，通常难以将细菌、浮游植物和浮游动物严格区分开。因此多数研究使用一定孔径的滤膜和筛绢对微食物网生物进行筛选，研究特定粒级微型浮游生物的营养盐再生，所获得的结果主要反映该粒级内微型浮游动物的营养盐再生能力。

多数野外实验结果验证了 Johannes（1964a）的假说，即微型浮游动物是海洋中营养盐最主要的再生者。在北太平洋的南加利福尼亚湾，超过 90%的再生 NH_4^+ 来自粒级小于 183μm 的微型浮游生物（Harrison，1978）。在南大西洋的本格拉上升流系统（Benguela upwelling system），大于 95%的 NH_4^+ 再生由粒级小于 15μm 的微型浮游生物贡献（Probyn，1987）。在挪威奥斯陆峡湾（Oslofjord），粒级 45～200μm 的微型浮游动物是最主要的营养盐再生者（Paasche and Kristiansen，1982）。在夏威夷的卡内奥赫湾（Kaneohe Bay），NH_4^+ 主要的排泄者是粒级小于 35μm 的

微型浮游动物，它们排泄的 NH_4^+ 是大型浮游动物的 30~40 倍（Caperon et al., 1979）。在北大西洋的切萨皮克湾（Chesapeake Bay），粒级小于 10μm 的浮游生物对 NH_4^+ 的再生速率远远高于粒级超过 200μm 的浮游生物（Glibert et al., 1992）。

微型浮游动物对 NH_4^+ 的再生速率存在季节变化。在英吉利海峡，冬季微型浮游动物 NH_4^+ 再生速率最低，不足 1ng NH_4^+-N/（L·h），夏季速率最高，达 25ng NH_4^+-N/（L·h）（Le Corre et al., 1996）。NH_4^+ 再生速率的季节变化还与浮游生物的粒级有关，如在切萨皮克湾，10~200μm 粒级浮游生物的 NH_4^+ 再生速率从冬季到夏季呈现持续升高的趋势，而粒级小于 10μm 的浮游生物的 NH_4^+ 再生速率在冬季最低，到初春时 NH_4^+ 再生速率升高 10 倍，并持续到夏末（Glibert et al., 1992）。

在垂直方向上，微型浮游动物 NH_4^+ 再生速率通常在表层较高，如在本格拉上升流区（Probyn, 1987）和英吉利海峡（Le Corre et al., 1996）。这种垂直方向上的变化可能是由于光照的影响。在密西西比河的羽流区（the Mississippi River plume），有光照的条件下 NH_4^+ 的再生速率为 0.08~0.75μmol/（L·h），在避光的条件下只有 0.02~0.3μmol/（L·h）（Gardner et al., 1997），但是在同海区的另一项研究却未发现光照对 NH_4^+ 的再生速率有影响（Jochem et al. 2004）。

在多数调查海区，NH_4^+ 的再生和被吸收是平衡的。本格拉上升流区真光层中 NH_4^+ 的再生基本支撑了 NH_4^+ 的需求（Probyn, 1987）。在夏威夷的卡内奥赫湾（Caperon et al., 1979）和北大西洋的伊比利亚上升流区（Iberian upwelling region）（Varela et al., 2003），NH_4^+ 每日的排泄量和吸收量大致相等。

但在部分海区发现 NH_4^+ 再生和被吸收不平衡。英吉利海峡的水体终年混合良好，NH_4^+ 的再生大于吸收，在春天和夏天水体中有 NH_4^+ 的积累（Le Corre et al., 1996）。在富营养的北大西洋贝德福德海盆（Bedford Basin），NH_4^+ 的再生速率仅为被吸收速率的 36%（La Roche, 1983）。

粒级差异也会影响 NH_4^+ 再生和被吸收的平衡。在英吉利海峡，粒级为 1μm 的微型浮游生物 NH_4^+ 的再生速率与吸收速率基本相等，1~15μm 粒级生物的 NH_4^+ 吸收速率为再生速率的 1.8 倍，15~200μm 粒级生物的 NH_4^+ 吸收速率为再生速率的 2.4 倍（Le Corre et al., 1996）。

有研究在同位素稀释法培养中添加桡足类，发现 NH_4^+ 再生速率先随桡足类丰度增加而升高，当桡足类丰度为 20ind./L 时，NH_4^+ 再生速率达到最高值，之后随桡足类丰度继续增加 NH_4^+ 再生速率降低（Glibert et al., 1992），这说明，浮游动物各个营养级之间的摄食作用也会影响该海区 NH_4^+ 的再生。

尿素再生是自然海区氮营养盐再生研究关注的重点之一。微型浮游动物尿素的再生速率也可用 ^{15}N 同位素稀释法进行测定。在切萨皮克湾，微型浮游动物尿素的再生速率为 0.42~2.58μg N/（L·h），且尿素的再生速率始终大于被吸收速率（Lomas et al., 2002）。在英吉利海峡，微型浮游动物尿素的再生速率为 0.6~

20.6nmol N/（L·h），在冬季最低，夏季最高。nano 级和 micro 级浮游动物是尿素最主要的再生者，分别贡献了 51%和 36%的再生尿素，pico 级生物的尿素再生速率很低，只在 4 月和 10 月才能检测到。各季节、水层的尿素再生速率和吸收速率比值都接近 1，说明微型浮游动物再生的尿素几乎完全被浮游植物吸收。在周年内，再生尿素占总再生氮（NH_4^+和尿素）的 33%（L'Helguen et al., 2005）。

（三）自然海区微型浮游动物的磷再生

氮营养盐通常被认为是在生物时间尺度上限制海洋初级生产的因素，而磷营养盐则是在地质时间尺度上限制海洋初级生产的因素（Tyrrell, 1999）。因此与氮营养盐相比，海洋浮游生物对磷营养盐的再生和吸收受到的关注相对较少。

目前，在自然海区进行的微型浮游动物磷再生研究不多。已有的研究发现，海洋磷再生速率和吸收速率的变化范围都很大，分别为 0.02～2100μg P/（L·d）和 0.01～3000μg P/（L·d），最大值和最小值可相差 6 个数量级（表 12-1）。海区的富营养化程度越高，磷再生速率越快。磷再生速率和吸收速率的比值范围为 0.1～1.4，总体来说再生和被吸收处于平衡状态（Sorokin and Dallocchio, 2008）。

表 12-1 自然海区微型浮游动物的磷再生速率

海区	营养状况	磷再生速率 [μg P/（L·d）]	磷吸收速率 [μg P/（L·d）]	再生速率/吸收速率	参考文献
北太平洋中部	寡营养	0.02～0.07	0.01～0.18	0.8～1.3	Perry 和 Eppley（1981）
热带东太平洋	寡营养	0.03～0.4	0.04～0.4	0.9～1.4	Sorokin（1985）
北大西洋贝德福德海盆	中营养	2.1～4.1	1.8～7.1	0.9～1.0	Harrison（1983）
南太平洋纳斯卡海岭	中营养	0.8～1.1	0.5～3.9	0.3～0.6	Sorokin（1985）
鄂霍次克海中央港口	中营养	0.6～1.6	0.7～3.4	0.4～1.4	Sorokin（2002）
南太平洋秘鲁上升流区	富营养	0.9～6.8	1.2～9.1	0.4～0.9	Harrison（1983）
南太平洋秘鲁上升流区	富营养	0.7～3.1	6.2～11.8	0.1～0.5	Sorokin（1985）
鄂霍次克海 Kashevarov 浅滩	富营养	0.8～2.9	3.7～5.4	0.3～0.5	Sorokin（2002）
亚得里亚海 Comacchio 潟湖	超富营养	400～1500	600～1800		Sorokin 等（1996）
威尼斯潟湖	超富营养	800～2100	1000～3000		Sorokin and Dallocchio（2008）

参 考 文 献

白洁, 李岿然, 张昊飞, 等. 2005. 胶州湾异养浮游细菌对磷的吸收作用及影响因素研究. 中国海洋大学学报(自然科学版), 35(5): 835-838.

王秋璐, 周燕遐, 王江涛, 等. 2010. 海洋异养细菌对无机氮吸收的研究. 海洋通报, 29(2): 231-234.

张武昌, 陈雪, 李海波, 等. 2016. 海洋浮游微食物网对氮、磷营养盐的再生研究综述. 海洋通报, 35(3): 241-251.

Allali K, Dolan J R, Rassoulzadegan F. 1994. Culture characteristics and orthophosphate excretion of a marine oligotrich ciliate, *Strombidium sulcatum*, fed heat-killed bacteria. Marine Ecology Progress Series, 105: 159-165.

Caperon J, Schell D, Hirota J, et al. 1979. Ammonium excretion rates in Kaneohe Bay, Hawaii, measured by a ^{15}N isotope dilution technique. Marine Biology, 54(1): 33-40.

Caron D A. 1994. Inorganic nutrients, bacteria, and the microbial loop. Microbial Ecology, 28(2): 295-298.

Caron D A, Goldman J C. 1988. Dynamics of protistan carbon and nutrient cycling. The Journal of Protozoology, 35(2): 247-249.

Dolan J R. 1997. Phosphorus and ammonia excretion by planktonic protists. Marine Geology, 139: 109-122.

Eccleston-Parry J D, Leadbeater B S C. 1995. Regeneration of phosphorus and nitrogen by four species of heterotrophic nanoflagellates feeding on three nutritional states of a single bacterial strain. Applied and Environmental Microbiology, 61(3): 1033-1038.

Fenchel T M, Jørgensen B B. 1977. Detritus food chains of aquatic ecosystems: the role of bacteria//Alexander M. Advances in Microbial Ecology. Boston: Springer US: 1-58.

Ferrier-Pages C, Rassoulzadegan F. 1994. N remineralization in planktonic protozoa. Limnology and Oceanography, 39(2): 411-419.

Gardner W S, Cavaletto J F, Cotner J B, et al. 1997. Effects of natural light on nitrogen cycling rates in the Mississippi River plume. Limnology and Oceanography, 42(2): 273-281.

Glibert P M. 1982. Regional studies of daily, seasonal and size fraction variability in ammonium remineralization. Marine Biology, 70(2): 209-222.

Glibert P M, Miller C A, Garside C, et al. 1992. NH_4^+ regeneration and grazing: interdependent processes in size-fractionated $^{15}NH_4^+$ experiments. Marine Ecology Progress Series, 82: 65-74.

Gobler C J, Hutchins D A, Fisher N S, et al. 1997. Release and bioavailability of C, N, P, Se, and Fe following viral lysis of a marine chrysophyte. Limnology and Oceanography, 42(7): 1492-1504.

Goldman J C, Caron D A, Anderson O K, et al. 1985. Nutrient cycling in a microflagellate food chain: I. Nitrogen dynamics. Marine Ecology Progress Series, 24: 231-242.

Goldman J C, Caron D A, Dennett M R. 1987a. Nutrient cycling in a microflagellate food chain : IV. Phytoplankton-microflagellate interactions. Marine Ecology Progress Series, 38: 75-87.

Goldman J C, Caron D A, Dennett M R. 1987b. Regulation of gross growth efficiency and ammonium regeneration in bacteria by substrate C：N ratio. Limnology and Oceanography, 32(6): 1239-1252.

Hargrave B T, Geen G H. 1968. Phosphorus excretion by zooplankton. Limnology and Oceanography, 13(2): 332-342.

Harrison W G. 1978. Experimental measurements of nitrogen remineralization in coastal waters. Limnology and Oceanography, 23(4): 684-694.

Harrison W G. 1983. Uptake and recycling of soluble reactive phosphorus by marine microplankton. Marine Ecology Progress Series, 10: 127-135.

Ikeda T, Fay E H, Hutchinson S A, et al. 1982. Ammonia and inorganic phosphate excretion by zooplankton from inshore waters of the Great Barrier Reef, Queensland.I. Relationship between excretion rates and body size. Marine and Freshwater Research, 33(1): 55-70.

Jochem F J, McCarthy M J, Gardner W S. 2004. Microbial ammonium cycling in the Mississippi River plume during the drought spring of 2000. Journal of Plankton Research, 26(11): 1265-1275.

Johannes R E. 1964a. Phosphorus excretion and body size in marine animals: microzooplankton and nutrient regeneration. Science, 146(3646): 923-924.

Johannes R E. 1964b. Uptake and release of dissolved organic phosphorus by representatives of a coastal marine ecosystem. Limnology and Oceanography, 9(2): 224-234.

Jover L F, Effler T C, Buchan A, et al. 2014. The elemental composition of virus particles: implications for marine biogeochemical cycles. Nature Reviews Microbiology, 12(7): 519-528.

Kirchman D L. 1994. The uptake of inorganic nutrients by heterotrophic bacteria. Microbial Ecology, 28(2): 255-271.

L'Helguen S, Slawyk G, Le Corre P. 2005. Seasonal patterns of urea regeneration by size-fractionated microheterotrophs in well-mixed temperate coastal waters. Journal of Plankton Research, 27(3): 263-270.

La Roche J. 1983. Ammonium regeneration: its contribution to phytoplankton nitrogen requirements in a eutrophic environment. Marine Biology, 75(2): 231-240.

Landry M R. 1993. Predicting excretion rates of microzooplankton from carbon metabolism and elemental ratios. Limnology and Oceanography, 38(2): 468-472.

Laws E. 1984. Isotope dilution models and the mystery of the vanishing ^{15}N. Limnology and Oceanography, 29(2): 379-386.

Le Corre P, Wafar M, L'Helguen S, et al. 1996. Ammonium assimilation and regeneration by size-fractionated plankton in permanently well-mixed temperate waters. Journal of Plankton Research, 18(3): 355-370.

Lomas M W, Trice T M, Glibert P M, et al. 2002. Temporal and spatial dynamics of urea uptake and regeneration rates and concentrations in Chesapeake Bay. Estuaries, 25(3): 469-482.

Mullin M M, Perry M J, Renger E H, et al. 1975. Nutrient regeneration by oceanic zooplankton: a comparison of methods. Marine Science Communications, 1: 1-13.

Paasche E, Kristiansen S. 1982. Ammonium regeneration by microzooplankton in the Oslofjord. Marine Biology, 69(1): 55-63.

Perry M J, Eppley R W. 1981. Phosphate uptake by phytoplankton in the central North Pacific Ocean. Deep Sea Research Part A. Oceanographic Research Papers, 28(1): 39-49.

Probyn T A. 1987. Ammonium regeneration by microplankton in an upwelling environment. Marine Ecology Progress Series, 37: 53-64.

Rose J M, Vora N M, Caron D A. 2008. Effect of temperature and prey type on nutrient regeneration by an Antarctic bacterivorous protist. Microbial Ecology, 56(1): 101-111.

Sherr B F, Sherr E B, Berman T. 1983. Grazing, growth, and ammonium excretion rates of a heterotrophic microflagellate fed with four species of bacteria. Applied and Environmental Microbiology, 45(4): 1196-1201.

Sorokin Y I. 1985. Phosphorus metabolism in planktonic communities of the eastern tropical Pacific Ocean. Marine Ecology Progress Series, 27: 87-97.

Sorokin Y I, Dallocchio F. 2008. Dynamics of phosphorus in the Venice lagoon during a picocyanobacteria bloom. Journal of Plankton Research, 30(9): 1019-1026.

Sorokin Y I, Dallocchio F, Gelli F, et al. 1996. Phosphorus metabolism in anthropogenically transformed lagoon ecosystems: the Comacchio lagoons(Ferrara, Italy). Journal of Sea Research,

35(4): 243-250.

Sorokin Y I. 2002. Dynamics of inorganic phosphorus in pelagic communities of the Sea of Okhotsk. Journal of Plankton Research, 24(12): 1253-1263.

Tyrrell T. 1999. The relative influences of nitrogen and phosphorus on oceanic primary production. Nature, 400(6744): 525-531.

Varela M M, Barquero S, Bode A, et al. 2003. Microplanktonic regeneration of ammonium and dissolved organic nitrogen in the upwelling area of the NW of Spain: relationships with dissolved organic carbon production and phytoplankton size-structure. Journal of Plankton Research, 25(7): 719-736.

Verity P G. 1985. Grazing, respiration, excretion, and growth rates of tintinnids. Limnology and Oceanography, 30(6): 1268-1282.

Wen Y H, Vézina A, Peters R H. 1994. Phosphorus fluxes in limnetic cladocerans: coupling of allometry and compartmental analysis. Canadian Journal of Fisheries and Aquatic Sciences, 51(5): 1055-1064.

Wheeler P A, Kirchman D L. 1986. Utilization of inorganic and organic nitrogen by bacteria in marine systems. Limnology and Oceanography, 31(5): 998-1009.

Wilhelm S W, Suttle C A. 2000. Viruses as regulators of nutrient cycles in aquatic environments//Microbial Biosystems: new Frontiers. Proceedings of the 8th International Symposium on Microbial Ecology. Atlantic Canada Society of Microbial Ecology: 551-556.

第十三章　微食物网和经典食物链的相对重要性

微食物网概念的提出改变了人们对海洋浮游生物群落间营养关系的认识。以往人们认为海洋中的物质和能量是沿着经典食物链直线传递的，现在发现除经典食物链外，海洋生态系统中还有很多物质和能量的可选途径，这些途径相互交织形成海洋食物网。

由于微食物网和经典食物链在海洋中同时存在，它们在某特定海区的相对重要性自然引起关注。尽管学者普遍认为微食物网和经典食物链在不同生境中的重要性不同，但没有一个统一的标准区分谁更重要。整体来说可从两个方面来考虑微食物网的重要性：第一，微食物网生物个体比经典食物链小，因此在有利于小个体生物生存的环境，如寡营养海区，微食物网的重要性大一些；第二，微食物网的异养细菌所占的比例较大时，微食物网的重要性较大。

第一节　自养微食物网生物的相对重要性

在自然海区，寡营养通常与水体层化相联系，层化的水体往往是寡营养的，寡营养的水体往往是层化的。寡营养和层化海区的初级生产力较低，可通过比较小粒级自养浮游生物占总浮游植物生物量的比例，来评价寡营养和层化海区微食物网的相对重要性。在总浮游植物生产力和生物量较低的寡营养水体中，微微型浮游植物（≤2μm）的生物量占比更高，重要性更大（Legendre and Rassoulzadegan, 1996; Bell and Kalff, 2001）。部分水体中微微型浮游植物甚至可以贡献接近100%的总浮游植物生物量，表明此时微食物网的重要性远超经典食物链（图13-1）。

寡营养和水体层化都有利于小粒径自养生物的原因可从如下几个角度阐述。

一、水体层化和扰动与细胞粒级的关系

根据斯托克斯定律（Stokes' law），液体中球形物体的下沉速度（v）与物体半径的平方（r^2）成正比，与物体与液体的密度差（$\rho-\rho'$）成正比，与重力加速度（g, 982cm/s^2）成正比，与液体的黏度（η，海水约为 10^{-2}cm^2/s）成反比，即：

$$v = 0.222gr^2(\rho - \rho')/\eta \qquad (13\text{-}1)$$

依照式(13-1)，假设海水中生物与海水的密度差值为0.05g/cm^3，则直径10μm的浮游生物沉降速度理论值约为每天0.2m，直径1μm的生物每天仅沉降0.02m，

表明大粒径生物的沉降速度相对较快，而小粒径生物在海水中的沉降速度微不足道（Kiørboe，1993）。

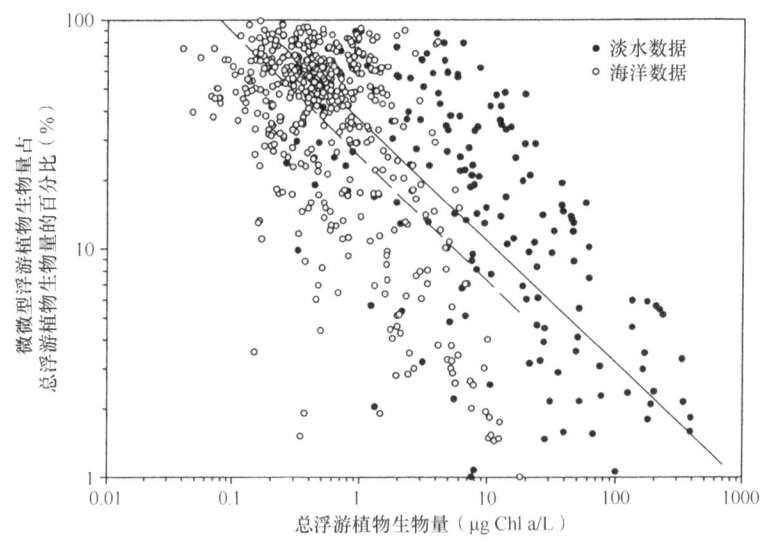

图 13-1 微微型浮游植物生物量占比与总浮游植物生物量的关系[改绘自 Bell 和 Kalff（2001）]
实线为淡水数据的回归线，虚线为海洋数据的回归线

一个光合自养生物在静水中从表层沉降到真光层底部的时间称为 t_s，其繁殖时间称为 t_r。光合自养生物需要光照才可以生存繁殖，因此对于静水中的 $t_r > t_s$ 光合自养生物，个体在繁殖之前就沉降出真光层，将导致该种群减少直至消失。若水体是扰动的，湍流可将生物向各个方向输送，尽管部分个体会沉降出真光层，但该生物仍有一些个体有机会繁殖，如果繁殖量大于因沉降造成的损失，则该物种仍会在真光层存在。对于 $t_r < t_s$ 的光合自养生物，无论静水还是扰动的海水，该生物都可在真光层存在，但此时扰动却成了该物种存在的负面因素，将导致部分个体被输送出真光层。

以真光层深度为 50m 的水体来举例，一个直径几微米的光合自养生物下沉速度约为 0.1m/d，即 t_s 为 500d，若其 t_r 为 5d，则该类生物不需要扰动就能在真光层水体生存，而扰动对其有负面影响。对于一个直径为 200μm 的光合自养生物，沉降速度约为 20m/d，即 t_s 约为 2.5d，若 t_r 也为 5d，这类生物就需要扰动才能维持真光层内的种群。

因此，水体是否有扰动对光合自养生物能否在真光层存活有选择性，静水有利于 $t_r < t_s$ 的生物类群，扰动水体有利于 $t_r > t_s$ 的生物类群。个体微小的微食物网自养生物繁殖快、沉降慢，属于 $t_r < t_s$ 生物，更适宜在海洋层化水体中生存；大个体浮游藻类繁殖慢、沉降快，则更适宜在海洋扰动水体中生存。

二、营养盐吸收速率与细胞粒级的关系

人们通常认为浮游生物的营养吸收速率取决于细胞的表面积，小粒径细胞具有更大的比表面积（specific surface area），即表面积与体积的比值，因此养分吸收效率高于大粒径细胞。若单个球形细胞（半径 r）对体外营养物质的饱和吸收速率（saturated nutrient uptake rate, U）与细胞的表面积（A）成正比，即：

$$U \propto A = 4\pi r^2 \tag{13-2}$$

那么，球形细胞的比吸收速率（specific uptake rate, U/V，其中 V 为细胞体积），即单位体积吸收体外营养物质的速率可表示为（Kiørboe，1993）

$$U/V \propto \frac{4\pi r^2}{4/3\pi r^3} = 3/r \tag{13-3}$$

所以球形细胞的比吸收速率与半径成反比，细胞越大，单位体积细胞吸收营养物质的饱和速率越小，然而这一关系只在无机营养物质浓度较高的环境下成立。

细胞周围海水的营养盐浓度（C'）高的情况下，营养盐在水体中的扩散系数（D）高，细胞周围的营养盐能够很快补充，不会形成营养盐限制。但当营养盐浓度低时，营养盐向细胞表面扩散的速率将成为细胞养分供应的限制因素。如果细胞吸收营养盐的速度大于营养盐在水体中的扩散速度，将在细胞表面形成一个营养匮乏的薄层，称为边界层（boundary layer）。此时细胞的比吸收速率主要受营养盐扩散速率制约，可表示为（Kiørboe，1993）

$$U/V = 3DC'/r^2 \tag{13-4}$$

这意味着在营养盐浓度较低时，球形细胞的比吸收速率与细胞半径的平方成反比，小个体的生物在寡营养海区具有生存优势。

三、运动和扰动对营养盐吸收的影响

当细胞在海水中运动，如沉降或游动时，细胞边界层内营养匮乏的海水将比细胞静止不动的情况下更快地被替换。这种向细胞表面的营养物质平流运输可使边界层内营养盐浓度梯度增大，营养物质吸收率增加。

舍伍德数（Sherwood number，Sh）是流体力学中的无量纲数，指平流传质与扩散传质系数之比，可表示细胞沉降或游动造成的平流传质导致的养分吸收率相对增加。不同粒径球形细胞在运动时的舍伍德数见表13-1。显然，当细胞运动时，小粒径细胞的营养盐吸收增加不大，大粒径细胞的营养盐吸收增加则更明显。但对于微食物网生物来说，粒径增大可更显著地增加营养扩散吸收，如粒径为 1μm 和 100μm 的生物相比，后者的表面积是前者的 10 000 倍，即营养的扩散吸收是前者的 10 000 倍，但是运动导致的营养吸收增加只有约 6 倍（表13-1）。这意味

着对于在低营养盐环境中的大个体细胞，通过运动增加营养盐吸收的效果有限，不如迁移到高营养盐区域吸收营养盐，所以寡营养海区的甲藻通常的策略是通过昼夜垂直迁移到下层的营养盐浓度高的水体。

表 13-1　不同粒径球形细胞在运动和扰动条件下的舍伍德数（Kiørboe，1993）

细胞粒径（μm）	沉降*	游动#	扰动†
1	1.002	1.23	1.001
10	1.113	5.26	1.013
100	6.78	6.86	1.13
1000	297		2.30

* 假设细胞密度为 0.05g/cm³，细胞沉降速度 $v(cm/s)=1091×r^2$，r 表示细胞半径
\# 假设细胞游动速度 $v(cm/s)=9.3×10^{-2}ESD^{0.24}$，ESD 表示等效球形直径（equivalent spherical diameter）
† 假设流体剪切速率为 0.1/s，营养盐扩散系数为 $10^{-5}cm^2/s^2$

与运动类似，湍流也可以增加营养物质向细胞表面的平流运输，因此海水扰动也能增加细胞对营养物质的吸收率。层流场中球形细胞的舍伍德数可使用流体剪切速率和营养盐的扩散系数计算获得（表 13-1），理论上只有粒径大于 100μm 的细胞能从扰动中获益。大多数海洋自养浮游生物粒径小于 100μm，所以扰动水体并不能帮助自养浮游生物吸收营养盐，静水和扰动水体都是利于小粒径的自养生物。

四、细胞粒径对光子吸收的影响

光合自养浮游生物收集和利用光能进行光合作用的能力受到光子通量密度[photon flux density，I，μmol photon/（m²·s）]、光子效率（photon efficiency，ψ，mol C/mol photon）、叶绿素 a 特异性光能吸收截面（chlorophyll a-specific absorption cross-section，a_{chl}，m²/mg Chla）和叶绿素 a 的相对浓度（θ，mg Chla/mg C）等参数的影响，可以用碳比光合速率[carbon-specific gross photosynthetic rate，$P_c(I)$，/s]表示（Geider et al.，1986）。

$$P_c(I) = Z \times \psi \times a_{Chl} \times \theta \times I \tag{13-5}$$

式中，Z 是无量纲常数，等于碳的原子量（0.012 mg C/μmol C）。

在对硅藻的研究中发现 ψ 和 θ 与细胞粒径无关，而 a_{chl} 与细胞粒径有关，细胞粒径越大，a_{chl} 值越小（Geider et al.，1986）。对于相同数量的色素体（如叶绿体），当它们被包装在离散的颗粒（如细胞）中，而不是以悬浮液状态均匀分散时，这些色素体吸收光能的总效率将不可避免地下降，这种现象被称为包装效应（package effect）（Kirk，1986），也称自我遮蔽效应（self-shading effect）。包装效应相当于降低了细胞的 a_{chl} 值，在给定光照强度的条件下，细胞越大，每毫克光合色素所收集的光能就越少，因此，大粒径细胞的单位体积光合作用效率低于小

粒径细胞。与大粒径生物相比，微微型自养浮游生物的直径（600~2000nm）与光合有效辐射波长（400~700nm）很接近，其包装效应最小，可以很好地吸收光子（Raven，1998）。

小粒径细胞的包装效应较小，意味着与大粒径细胞相比具有以下特点。

（1）对于在单位体积内给定的色素含量，小粒径细胞的色素特异性光能吸收截面更大，单位体积光合作用效率高于大粒径细胞。

（2）在给定的辐射强度下，小粒径细胞偿还色素体合成能量成本的时间更短。

（3）仅需较低的入射光子通量密度进行光合作用，即可补偿呼吸作用造成的能量损失。

（4）较低的入射光子通量密度即可提供足够的能量使光合作用下游反应饱和，因此光合电势差导致光抑制损失的风险较大，增加了细胞修复光抑制性损伤的资源成本。

（5）使用细胞内 UV-B 吸收色素（UV-B absorbing pigment）以屏蔽破坏性 UV-B 辐射的能力较弱（Raven，1998）。

因此在其他条件相同的情况下，与大粒径细胞相比，小粒径细胞在低光子通量密度下有更高的比生长率，但也增加了光合有效辐射和 UV-B 损坏光合装置的风险。这些影响综合决定了微微型光合浮游生物在自然海区中的分布模式。

五、高温对小粒级生物优势度的放大

根据范托夫定律（van't Hoff's law），温度制约着生物体内物理化学过程的反应速度，因此水温直接影响着所有水生生物的生长和代谢活动。范托夫定律可以用温度系数 Q_{10} 来近似描述，即温度每升高 10℃，反应速率增加的系数。在最适生存条件下，海洋浮游生物的 Q_{10} 值通常在 2 和 3 之间，如单细胞藻类生长速率的 Q_{10} 值为 1.88~2.08，硅藻生长速率的 Q_{10} 值约为 2.3，浮游动物呼吸速率的 Q_{10} 值约为 2。

寡营养和层化水体在热带高温海区最为典型，高温可放大生物代谢率的异速增长差异（具体见本书第十四章），因此进一步增强了小粒径自养生物的竞争优势。以稳态条件下的最大日生产力来比较，在 0℃ 条件下，粒径 1μm 浮游植物细胞的倍增速率（doubling rate）为 2.10/d，粒径 100μm 细胞的倍增速率仅为 0.14/d，倍增时间分别为 11.4h 和 166.1h；在 30℃ 时，倍增速率则分别增加到 4.77/d（粒径 1μm 细胞）和 0.88/d（粒径 100μm 细胞），倍增时间分别为 5.0h 和 27.2h（Lenz，1992）。若以各粒级浮游植物生物量增长百分比来代表日生产力变化，可发现温度每增加 10℃，微微型浮游植物比微型和小型浮游植物的异速优势增加 2 倍。在 0℃ 时，微微型浮游植物的生物量每天可以增加 279%，而小型浮游植物只能增加 16%，两者相差263%，在30℃时则相差2104%（表13-2）。所以在水温较高的海区，小粒径生物具有更强的竞争优势，寡营养海区的浮游微食物网相对重要性增加还得益于温度的协同作用。

表13-2 各粒级浮游植物日生产力*随温度升高增长的百分比（%）（Lenz，1992）

	浮游植物	温度			
		0℃	10℃	20℃	30℃
粒级	微微型浮游植物（1~2μm）	279	558	1116	2232
	微型浮游植物（2~20μm）	112	224	448	896
	小型浮游植物（20~100μm）	16	32	64	128
差值	微微型与微型浮游植物	167	334	668	1336
	微型与小型浮游植物	96	192	384	768
	微微型与小型浮游植物	263	526	1052	2104

*日生产力以各粒级浮游植物生物量增长百分比表示

六、自养微食物网生物已经达到最小

综上，个体微小的自养微食物网生物在层化、寡营养和高温条件下比经典食物链生物更具竞争优势，那么这些自养生物是否已经达到粒径最小的极限呢？

染色体和细胞膜是细胞的最基本组成部分，它们在细胞物质中所占比例随细胞粒径变小而升高。当细胞直径低于2μm时，染色体和细胞膜占比升高的速度骤然变大。若仅考虑细胞生存所需的最小基因组和细胞膜厚度，真核自养生物的理论最小直径约为0.3μm，革兰氏阴性原核自养生物（如蓝细菌）为0.1μm（Raven，1998）。在自然环境中，已知个体最小的浮游真核自养生物 Ostreococcus tauri 的直径约为0.8μm，已知最小的浮游原核自养生物原绿球藻的直径约为0.6μm，均已接近其理论最小值，因此海洋浮游微食物网的生物向小个体方面的选择已经到达极限（Raven，1998），微食物网生物已经小到不能再小了。

第二节 异养微食物网生物的相对重要性

另一个衡量微食物网重要性的指标是海区的异养情况。除自养微食物网生物外，海洋中还存在着丰度更高的异养细菌。细菌的能量和物质来源是溶解有机碳（DOC），DOC的主要来源则是浮游植物。浮游植物向水体释放DOC的速率（leakage rate，E，mol/s）与细胞的表面积/体积相关，可表示为

$$E=S \times P \times A/V \tag{13-6}$$

式中，P为穿透率（permeability，cm/s），S为细胞内物质存量（mol），A为细胞表面积（cm^2），V为细胞体积（cm^3）（Bjørrisen，1988）。寡营养和层化海区的浮

游植物粒径小，而小的浮游植物细胞会释放更多的 DOC，可导致异养细菌生物量提高，该海区微食物网的重要性增加。

海区的异养情况可通过对照异养与自养的比例来衡量，包括异养与自养生物量的比例、细菌生产力消耗初级生产力的比例、呼吸作用与光合作用大小的比较。

一、异养与自养生物量的比例

不同海区异养细菌生物量与浮游植物生物量的比值（BactC/PhytoC）有较大差异，可以作为反映微食物网重要性的一个指标。当比值高时，微食物网在生态系统中更为重要。

有研究比较了不同海区细菌生物量和浮游植物生物量的关系，发现浮游植物的生物量为 1000μgC/L 时，BactC/PhytoC 约为 0.03，当浮游植物生物量为 10~20μgC/L 时，BactC/PhytoC 约为 1，当浮游植物生物量下降到 1μgC/L 时，BactC/PhytoC 提高到 10，即随着浮游植物生物量下降，BactC/PhytoC 值升高，微食物网重要性增大（Simon et al.，1992）。

生物量金字塔也可用来表示不同情形下细菌生物量与浮游植物生物量的关系。Gasol 等（1997）使用归一化方法，将浮游植物的生物量作为 1，总结了异养细菌、异养原生生物和浮游动物的相对生物量，发现在寡营养大洋异养细菌的相对生物量与浮游植物基本相等，在近海则降至浮游植物生物量的 62%（图 13-2）。Uye 等（1999）发现寡营养条件时细菌生物量相当于浮游植物生物量的 63%，在富营养条件下则只相当于浮游植物生物量的 10%。

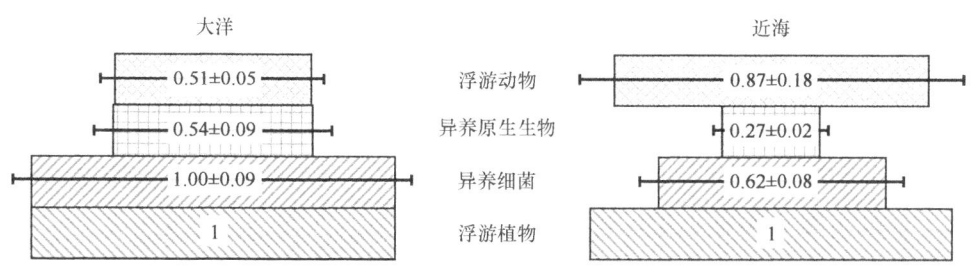

图 13-2　大洋和近海浮游植物、异养细菌、异养原生生物和浮游动物的相对生物量金字塔[改绘自 Gasol 等（1997）]

图中数字为各生物类群的相对生物量值

营养盐浓度的高低是影响细菌和浮游植物相对生物量的重要因素。通过对寡营养海水进行氮浓度梯度培养实验，培养两周后可观察到在营养盐浓度低的培养组，细菌生物量大于浮游植物生物量，随着营养盐浓度的增加，细菌生物量与浮游植物生物量比值总体降低（图 13-3）（Duarte et al.，2000）。

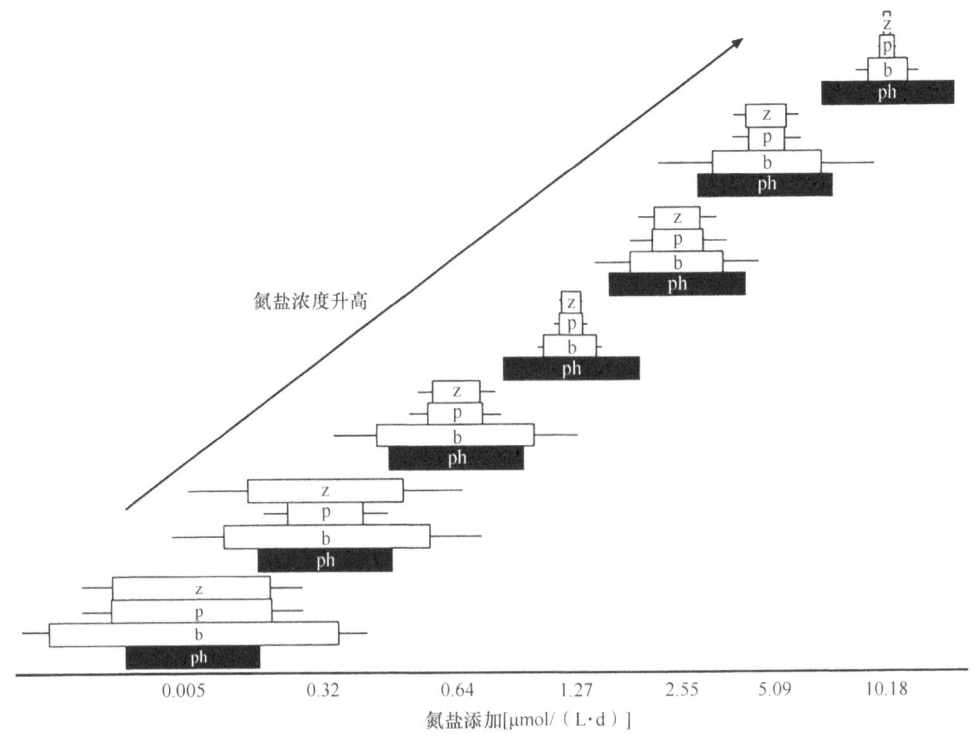

图 13-3 不同氮盐浓度条件下浮游植物（ph）、异养细菌（b）、异养原生生物（p）和浮游动物（z）的相对生物量金字塔[改绘自 Duarte 等（2000）]

温度也会影响细菌和浮游植物的相对生物量大小。当水温高于 20℃时，BactC/PhytoC 大于 1，当水温低于 20℃时，BactC/PhytoC 随温度的降低而降低（Li et al.，2004）。这意味着在极地海区浮游细菌的重要性降低，极地浮游生态系统为自养系统，而热带的浮游生态系统为异养系统。温度对 BactC/PhytoC 值的影响还反映在季节变化上，夏季加那利群岛海域异养生物和浮游植物生物量比值比其他季节高 3 倍（Baltar et al.，2009）。

二、细菌生产力消耗初级生产力的比例

细菌生产力消耗初级生产力的比例也可以用来衡量微食物网的重要性，细菌生产力消耗的初级生产力比例越大，则该海区微食物网的重要性越高。

通常在初级生产力低的寡营养海区，细菌生产力消耗初级生产力的比例较高，微食物网重要性大；当营养盐浓度增加和初级生产力升高时，细菌生产力消耗初级生产力的比例降低，经典食物链的重要性超过微食物网。在泛北冰洋的寡营养区域，细菌生产力消耗了初级生产力的 2/3，而在营养盐浓度较高的锋面区域，只有 1/20 的初级生产力被细菌生产力消耗（Andersen，1988）。在地中海（Conan et al.，1999）和东海（Shiah et al.，2001）的研究也同样显示，随着初级生产力的增

加，细菌生产力消耗初级生产力的比例减少。

在大西洋，细菌生产力消耗初级生产力的比例与温度呈正相关关系。在大西洋寒带和温带海区，细菌生产力与初级生产力的比率平均为 2%～10%，而在热带海区则升高至约 40%，个别采样位点甚至可达 100%，这表明随温度升高，大西洋从净自养状态转变为有一定比例的异养状态，微食物网的重要性越来越高（Hoppe et al.，2002）。

三、群落呼吸消耗初级生产力的比例

浮游细菌通过生产和呼吸作用消耗有机碳，因此也可通过比较群落呼吸作用消耗初级生产力的比例来衡量微食物网的重要性。

在自然海区，随着初级生产力的升高，细菌的呼吸作用通常也增强。但是呼吸作用增强的速度慢于初级生产力升高的速度，最终造成的效果是随初级生产力的升高，呼吸作用与初级生产力的比值变小，微食物网重要性下降。

在部分寡营养大洋，其初级生产力过低，细菌呼吸作用消耗的有机碳将可能高于初级生产力，使该生态系统处于异养状态（del Giorgio et al.，1997；Duarte and Agustí，1998）。对于寡营养亚热带环流是否处于异养状态是最近十多年来海洋生态学研究的热点问题之一（Ducklow and Doney，2013）。Williams 等（2013）认为，这些环流区处于净自养状态，即若考虑到较大的海区范围和较长的研究时间，环流区的总初级生产力将超过群落呼吸。Duarte 等（2013）持相反观点，认为这些环流区处于净异养的状态，群落呼吸超过总初级生产力。因此，对寡营养亚热带环流区的营养状态仍需进一步研究。

四、原生动物和桡足类饵料组成比例

微食物网与经典食物链的连接是通过原生动物和桡足类完成的。原生动物可摄食微食物网的浮游细菌，以及经典食物链的浮游植物，桡足类则可摄食微食物网的原生动物和经典食物链的浮游植物，因此，可通过比较原生动物和桡足类饵料组成比例来衡量微食物网的重要性（de Laender et al.，2010）。细菌在原生动物饵料中的比例越高、原生动物在桡足类饵料中的比例越高，意味着微食物网重要性越大。

在巴伦支海南部，春季是经典食物链为主导，浮游植物生物量大于浮游细菌，原生动物的饵料中细菌和浮游植物的量大体相当，而桡足类对原生动物没有明显的选择性。夏季该海区微食物网占主导，浮游植物生物量小于浮游细菌，原生动物对细菌的摄食量是对浮游植物摄食量的 4 倍，桡足类的饵料中 80%～90% 是原生动物，表明夏季微食物网比春季重要性高（de Laender et al.，2010）。

第三节 自然海区浮游食物网的结构

一、浮游食物网结构的划分

如前所述，水体的营养状况影响着浮游食物网的结构。随着水体营养盐浓度增加，浮游植物的粒径增大，异养生物量在浮游生态系统中的比例变小，系统的呼吸速率与自养生产速率的比值减小，系统偏向自养；而低营养盐浓度的环境中浮游植物粒径小，异养生物量比例大，系统的呼吸速率与自养生产速率的比值高，系统偏向异养。这两种极端的浮游食物网结构分别被称为经典食物链为主的食物网（"classical type" of food web）（Lenz, 1992）和微食物环为主的食物网（"microbial loop" type of food web）（Kiørboe et al., 1990）。

在大多数情况下，自然海区浮游食物网的结构处于这两个极端之间的过渡状态。在任意水体的浮游食物网中，微食物网和经典食物链共存，根据两者的重要性不同可对水体食物网进行分类。

根据细菌生产力占初级生产力的比值，可将白令海北部和楚科奇海的浮游食物网分为：①微食物环型，初级生产者主要为微微型和微型自养生物，约64%的初级生产力被细菌生产所消耗；②大型浮游动物/底栖生物型，初级生产者主要为硅藻，约5%的初级生产力被细菌生产所消耗；③微食物环/大型浮游动物/底栖生物型，初级生产者由微微型和微型自养生物及硅藻混合组成，约28%的初级生产力被细菌生产所消耗。其中，①型和②型是两个极端，③型则是介于其间的中间类型（Andersen, 1988）。

根据浮游食物网中能流的变化可将浮游食物网分为4个类型：经典食物链主导、多重摄食者主导、微食物网主导及微食物环主导的浮游食物网（图13-4）（Legendre and Rassoulzadegan, 1995）。这一分类方法主要依据中型浮游动物摄食浮游植物和微型浮游动物的比例，以及微型浮游动物摄食细菌和浮游植物的比例。4种浮游食物网结构组成连续的谱，其中以经典食物链（图13-4a）和微食物环（图13-4d）为主导的系统是两个极端情况。

在最简单的经典食物链主导的食物网中，较大的浮游植物（>5μm组分）利用硝酸盐，中型浮游动物摄食浮游植物，再由鱼类摄食中型浮游动物从而完成能流传递（图13-4a），细菌在这一过程中的贡献几乎可以忽略。在多重摄食者主导的系统中，中型浮游动物摄食微型浮游动物、微型浮游动物摄食细菌的比例高于经典食物链主导的系统（图13-4b）。在微食物网主导的浮游食物网中，中型浮游动物摄食浮游植物的比例下降，微型浮游动物摄食细菌的比例进一步提高（图13-4c）。在微食物环主导的浮游食物网中，中型浮游动物的作用几乎可以忽略，能流主要在浮游植物、异养细菌和微型浮游动物之间循环（图13-4d）（Legendre and

Rassoulzadegan，1995）。

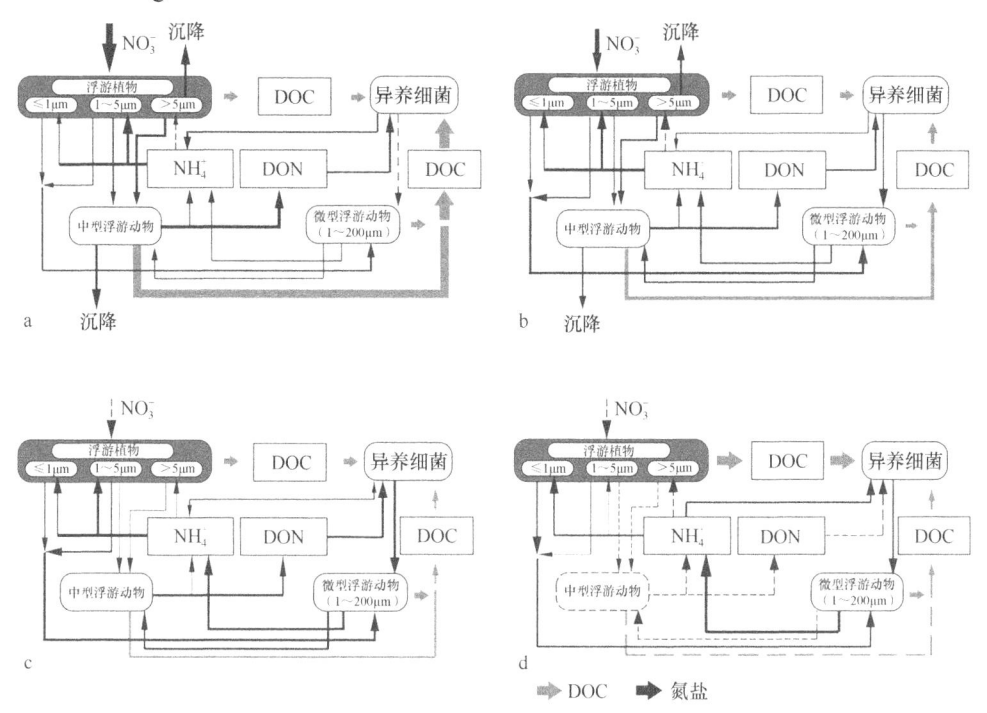

图 13-4　海洋浮游食物网的营养途径示意图[改绘自 Legendre 和 Rassoulzadegan（1995）]
a. 经典食物链主导的浮游食物网；b. 多重摄食者主导的浮游食物网；c. 微食物网主导的浮游食物网；d. 微食物环主导的浮游食物网。浅色箭头表示 DOC 通量，深色箭头表示氮盐通量，箭头的粗细与通量大小成正比，虚线表示极微弱或无效的通量

经典食物链和微食物网分别主导的两个系统是短暂的，如春季水华期和上升流海区系统都是经典食物链为主导，细菌生物量显著超过浮游植物生物量的寡营养水体是微食物网为主导，这些系统本质上是不稳定的，不能长期存在。而多重摄食者主导的北太平洋"高营养盐低叶绿素"（high nutrient low chlorophyll，HNLC）系统，以及微食物网主导的南极洲冰川消退边缘水体等系统具有较高的稳定性，可以长时间存在。

Legendre 和 Rassoulzadegan（1995）的食物网结构划分需要测定太多的参数，在实际的研究中很难实现，所以他们简化提出了 6 组相互关联的比值：浮游植物对铵盐和硝酸盐吸收率的比值、小粒级和大粒级浮游植物初级生产力的比值、中型浮游动物摄食微型浮游动物和浮游植物的比值、细菌底物 DOC 和 DON 的浓度比、细菌对铵盐的吸收和释放比、水体铵盐和硝酸盐浓度比（图 13-5）。上述比值在微食物网占主导的海区最大，在经典食物链占主导的海区最小。有研究使用这 6 组比值，评估了加拿大圣劳伦斯湾不同季节的浮游食物网状态，发现微食物网在夏季占主导，经典食物链在 12 月至次年 5 月占主导，其他时段则为多重摄食者主导的浮游食物网占主导（Mousseau et al., 2001）。除这6组比值外，也有研究选取了其他参数研究浮游食物网的结构（Hlaili et al., 2014）。

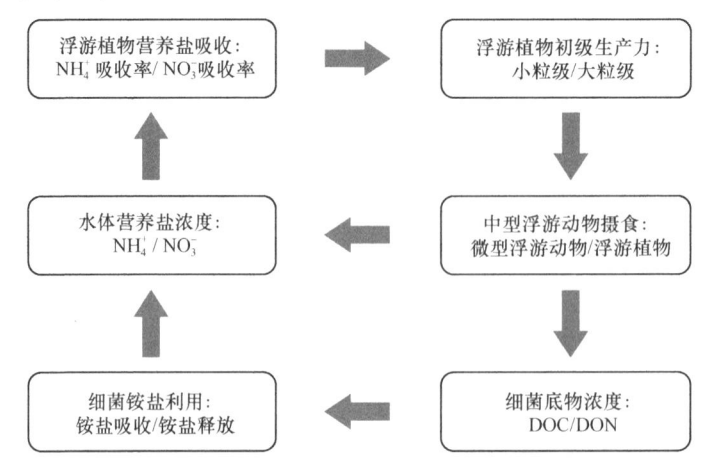

图 13-5　浮游食物网结构划分所需的6组比值[改绘自 Legendre 和 Rassoulzadegan（1995）]

总之，自然海区的浮游食物网处于经典食物链占主导和微食物网占主导之间的过渡状态。但对于这些过渡状态的划分还没有通用、完善的参数依据，仅有一些概念模型的认识。

二、浮游食物网生物的演化历程

现代自然海区中微食物网和经典食物链同时存在，但是在地质演化过程中，微食物网生物的出现要远远早于经典食物链生物。

化能自养原核微生物是生命起源之初的主要生命形态。随后不产氧的光合自养原核生物，如光合氢氧化微生物、光合硫氧化微生物、光合铁氧化微生物等，在太古宙（Archean Eon）早期出现并成为主要的初级生产者（图13-6）。在太古宙和元古宙（Proterozoic Eon）之交产氧的光合蓝细菌出现之前，海洋中的有机质主要由非产氧光合自养原核生物生产。这一阶段的海洋初级生产力是地质历史时期中最低的，比元古宙低1~2个数量级，比显生宙（Phanerozoic Eon）低3~4个数量级。

元古宙早期（约25亿年前），产氧的光合蓝细菌出现（图13-6），海洋和大气中出现了第一次大氧化事件（the Great Oxidation Event），光合蓝细菌逐渐占据主导地位，元古宙也因此被称为蓝细菌时代。在中元古代（Mesoproterozoic Era）晚期（约13亿年前），单细胞真核生物出现。早期的真核生物没有运动器官，形态类似变形虫，主要营底栖生活，并随着鞭毛的出现逐渐演化成类似鞭毛虫的生物（Mitchell, 2007）。约10.5亿年前，多细胞真核生物出现。在新元古代（Neoproterozoic

第十三章 微食物网和经典食物链的相对重要性

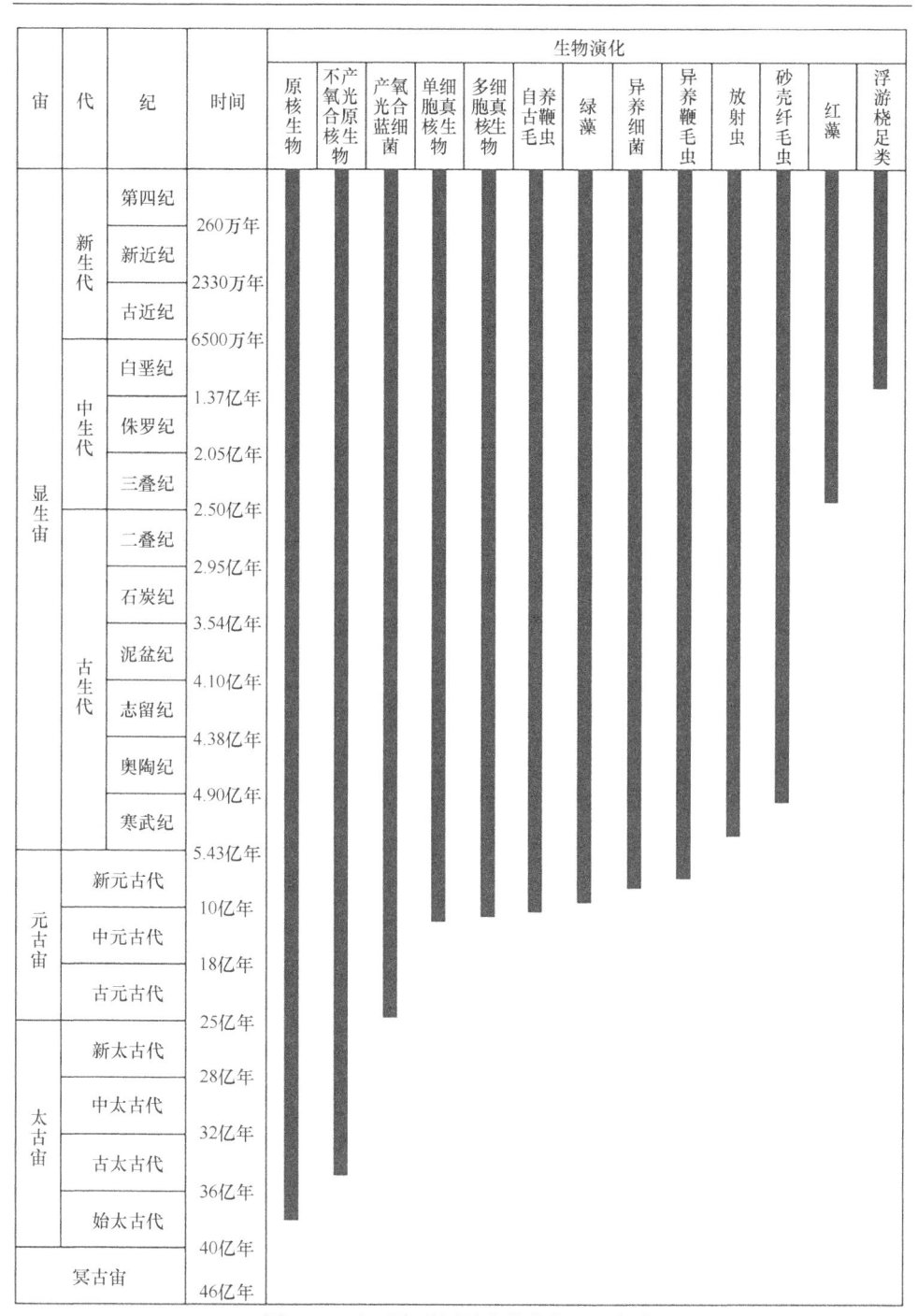

宙	代	纪	时间	生物演化											
				原核生物	不产氧光合原核生物	产氧光合蓝细菌	单细胞真核生物	多细胞真核生物	自养鞭毛虫	绿藻	异养细菌	异养鞭毛虫	放射虫	砂壳纤毛虫	红藻	浮游桡足类
显生宙	新生代	第四纪	260万年													
		新近纪	2330万年													
		古近纪	6500万年													
	中生代	白垩纪	1.37亿年													
		侏罗纪	2.05亿年													
		三叠纪	2.50亿年													
	古生代	二叠纪	2.95亿年													
		石炭纪	3.54亿年													
		泥盆纪	4.10亿年													
		志留纪	4.38亿年													
		奥陶纪	4.90亿年													
		寒武纪	5.43亿年													
元古宙	新元古代		10亿年													
	中元古代		18亿年													
	古元古代		25亿年													
太古宙	新太古代		28亿年													
	中太古代		32亿年													
	古太古代		36亿年													
	始太古代		40亿年													
冥古宙			46亿年													

图 13-6 主要海洋浮游生物的起源年代表

Era, 约 10 亿~5 亿年前)真核绿藻出现并成为主要的初级生产者。由于氧气的出现,海洋中出现了异养细菌降解溶解有机碳,但是其进化的进程落后于自养生物，加之氧浓度不高，溶解有机碳的生产速度大大高于异养细菌的降解速度，到新元古代晚期形成了巨大的溶解有机碳库，其浓度可能是现代海洋的数百倍甚至上千倍（Rothman et al.，2003；Fakhraee et al.，2021）。随着水体中饵料颗粒的增加，一部分古鞭毛虫会进行吞噬活动成为混合营养鞭毛虫，并最终成为完全异养的鞭毛虫（图 13-6）。

在显生宙寒武纪（Cambrian Period）的早期出现了原生动物放射虫，寒武纪后期出现了砂壳纤毛虫（Rigby and Milsom，2000）。由于没有化石记录，无壳纤毛虫的出现年代不详，按照砂壳纤毛虫由无壳纤毛虫分泌外壳进化而来的假设，无壳纤毛虫的起源年代应早于砂壳纤毛虫，因此，在寒武纪完整的微食物网中已经出现。在中生代（Mesozoic Era）的早期，地球的海陆格局处于盘古大陆时期，真核红藻占据主导地位，并逐渐分支出甲藻、颗石藻和硅藻。浮游桡足类在白垩纪（Cretaceous Period）才开始出现，此后逐渐发展成目前海洋中微食物网和经典食物链的格局（图 13-6）。

从微食物网和经典食物链各生物类群的起源过程来看，微食物网的演化历史更古老，而经典食物链则是比较年轻的。在现代海洋中，以经典食物链占主导的富营养系统只在有额外营养盐输入的海区出现，而且这个系统是不稳定的，大颗粒的沉降等自然过程会将系统拉回到寡营养的状态。因此，微食物网可以看作是海洋食物网的正常状态，是经典食物链的背景，而经典食物链是短时状态，是原始状态在受到扰动时的变形，需要被不断"重启"（Margalef，1974；Landry，1977）。

第四节　微食物网和经典食物链的能量传递效率

海洋为人类提供了大量的鱼、虾、蟹等优质蛋白质，是人类重要的食物来源和保障。只要海洋生态系统中的物质和能量流入大于流出，海洋食物产出就是有保障的。

20 世纪 80 年代初，随着浮游细菌精确计数问题的解决和微食物网概念的提出，海洋生态学者开始重新考虑海洋生态系统的物质和能量流动，尤其是细菌对上层营养级的贡献。若细菌的能量大部分都耗散而不是传递到上层营养级，被称为"汇"（sink），如果细菌的能量可以传递给上层营养级，被称为"链"（link）。这就是著名的 sink 和 link 之争。

从营养级传递的角度来看，细菌不能被桡足类直接摄食，而需要通过微型浮游动物才能传递至经典食物链。科学界通常认为营养级之间的传递效率只有 10%，因此，仅有 1% 的细菌能量被传递至大中型浮游动物，这意味着细菌对经典食物链的产出几乎没有贡献。有研究用 ^{14}C 标记示踪的方法，发现只有极小部分细菌的碳生物量进入大型桡足类，绝大多数的 ^{14}C 以 CO_2 形式释放到空气中，或以 DOC

形式存储在培养体系中（Ducklow et al.，1986）。另有研究发现，细菌最主要的捕食者为 1~3μm 的鞭毛虫，在＜12μm 的微食物网培养体系内即存在 4 个营养级，分别为细菌、1~5μm 的鞭毛虫、5~8μm 的鞭毛虫、8~10μm 的鞭毛虫及＞10μm 的纤毛虫和鞭毛虫，所以，细菌的碳向经典食物链的传递效率应该很低（Wikner and Hagström，1988）。Pomeroy 和 Wiebe（1988）认为，微食物网内的能量传递损失已足够大，即使仅有两个营养级的微食物网也是能量的汇。

也有学者提出不能仅从细菌向桡足类传递的能量比例小就断定细菌的作用是汇，从营养级传递看，浮游植物向桡足类传递的能量比例也很小。因此，要综合比较细菌和浮游植物向桡足类传递能量的比例，才能正确理解细菌的作用（Sherr et al.，1988）。这一观点得到许多学者的响应，但不同的研究得出了不同的能量传递效率结果。有研究发现细菌对中型浮游动物贡献较大，如在寡营养淡水湖泊中，细菌和浮游植物对桡足类饵料的贡献大体相等（Wylie and Currie，1991）。巴伦支海春季水华时，经典食物链主导，原生动物的饵料中细菌和浮游植物各占一半；夏季浮游植物少，微食物网为主导，浮游原生动物的饵料中 80%是细菌，而桡足类的饵料中 80%~90%是原生动物（de Laender et al.，2010）。Koshikawa 等（1996）认为，以细菌和藻类为基础的能量传递路径都是效率很低的，两个路径能量传递效率低的原因不同，细菌能量传递效率低是因为营养层级太多，藻类能量传递效率低是因为有的藻类适口性太差。

部分研究发现，细菌的能量传递效率确实低于浮游植物。Berglund 等（2007）使用围隔实验进行研究，将能量传递效率定义为中型浮游动物生产力与基础生产力（浮游植物+细菌）的比值。细菌生产力占 91%时，能量传递效率为 2%，浮游植物生产力占 74%时，能量传递效率为 22%，因此细菌的能量传递效率低于浮游植物。

Williams（1984）问到：为什么海洋浮游生态系统要维持这个耗费巨大，但对食物网的产出却几乎没有贡献的庞大的微生物群落？答案或许是，尽管浮游植物的生产力很大一部分都会在微食物网中矿化，导致海洋食物网能量耗散，对高营养层次生产力的贡献很小，但是从营养盐循环的角度来看，如果海水中只有经典食物链，所有的营养盐都会沉降到深层海洋，导致表层营养盐耗尽，不再有生物存活。微食物网的浮游细菌可同化浮游植物生产的有机废物，形成营养丰富的细菌个体，被原生动物摄食后释放出无机营养盐，再被浮游植物使用。对营养盐浓度低的生态系统而言，保持营养盐循环从而维持生物和生态系统的存续才是第一要务（Williams，1984；Hagström et al.，1988）。正因为如此，微食物网可被视为海洋浮游生态系统的飞轮（fly wheel），它消耗着能量，看似一种浪费，但是却保持了系统的稳定存续，使得真光层能维持一定（但较低）的营养盐浓度和生物量，等待着新营养盐的注入和经典食物链的下次重启（Strom，2000）。

参 考 文 献

Andersen P. 1988. The quantitative importance of the "microbial loop" in the marine pelagic: a case study from the North Bering/Chukchi seas. Archiv für Hydrobiologie Beihefte: Ergebnisse der Limnologie, 31: 243-251.

Baltar F, Arístegui J, Montero M F, et al. 2009. Mesoscale variability modulates seasonal changes in the trophic structure of nano-and picoplankton communities across the NW Africa-Canary Islands transition zone. Progress in Oceanography, 83(1-4): 180-188.

Bell T, Kalff J. 2001. The contribution of picophytoplankton in marine and freshwater systems of different trophic status and depth. Limnology and Oceanography, 46(5): 1243-1248.

Berglund J, Müren U, Båmstedt U, et al. 2007. Efficiency of a phytoplankton-based and a bacterial-based food web in a pelagic marine system. Limnology and Oceanography, 52(1): 121-131.

Bjørrisen P K. 1988. Phytoplankton exudation of organic matter: why do healthy cells do it? Limnology and Oceanography, 33(1): 151-154.

Conan P, Turley C M, Stutt E D, et al. 1999. Relationship between phytoplankton efficiency and the proportion of bacterial production to primary production in the Mediterranean Sea. Aquatic Microbial Ecology, 17(2): 131-144.

de Laender F, Oevelen D, Soetaert K, et al. 2010. Carbon transfer in herbivore- and microbial loop-dominated pelagic food webs in the southern Barents Sea during spring and summer. Marine Ecology Progress Series, 398: 93-107.

del Giorgio P A, Cole J J, Cimbleris A C P. 1997. Respiration rates in bacteria exceed phytoplankton production in unproductive aquatic systems. Nature, 385(6612): 148-151.

Duarte C M, Agustí S. 1998. The CO_2 balance of unproductive aquatic ecosystems. Science, 281(5374): 234-236.

Duarte C M, Agustí S, Gasol J M, et al. 2000. Effect of nutrient supply on the biomass structure of planktonic communities: an experimental test on a Mediterranean coastal community. Marine Ecology Progress Series, 206: 87-95.

Duarte C M, Regaudie-de-Gioux A, Arrieta J M, et al. 2013. The oligotrophic ocean is heterotrophic. Annual Review of Marine Science, 5: 551-569.

Ducklow H W, Doney S C. 2013. What is the metabolic state of the oligotrophic ocean? A debate. Annual Review of Marine Science, 5: 525-533.

Ducklow H W, Purdie D A, Williams P J L, et al. 1986. Bacterioplankton: a sink for carbon in a coastal marine plankton community. Science, 232(4752): 865-867.

Fakhraee M, Tarhan L G, Planavsky N J, et al. 2021. A largely invariant marine dissolved organic carbon reservoir across Earth's history. Proceedings of the National Academy of Sciences of the United States of America, 118(40): e2103511118.

Gasol J M, del Giorgio P A, Duarte C M. 1997. Biomass distribution in marine planktonic communities. Limnology and Oceanography, 42(6): 1353-1363.

Geider R J, Platt T, Raven J A. 1986. Size dependence of growth and photosynthesis in diatoms. Marine Ecology Progress Series, 30: 93-104.

Hagström Å, Azam F, Andersson A, et al. 1988. Microbial loop in an oligotrophic pelagic marine ecosystem: possible roles of cyanobacteria and nanoflagellates in the organic fluxes. Marine

Ecology Progress Series, 49: 171-178.
Hlaili A S, Niquil N, Legendre L. 2014. Planktonic food webs revisited: reanalysis of results from the linear inverse approach. Progress in Oceanography, 120: 216-229.
Hoppe H G, Gocke K, Koppe R, et al. 2002. Bacterial growth and primary production along a north-south transect of the Atlantic Ocean. Nature, 416(6877): 168-171.
Kiørboe T. 1993. Turbulence, phytoplankton cell size, and the structure of pelagic food webs. Advances in Marine Biology, 29: 1-72.
Kiørboe T, Kaas H, Kruse B, et al. 1990. The structure of the pelagic food web in relation to water column structure in the Skagerrak. Marine Ecology Progress Series, 59: 19-32.
Kirk J T O. 1986. Optical properties of picoplankton suspensions. Canadian Bulletin of Fisheries and Aquatic Sciences, 214: 501-520.
Koshikawa H, Harada S, Watanabe M, et al. 1996. Relative contribution of bacterial and photosynthetic production to metazooplankton as carbon sources. Journal of Plankton Research, 18(12): 2269-2281.
Landry M R. 1977. A review of important concepts in the trophic organization of pelagic ecosystems. Helgoländer Wissenschaftliche Meeresuntersuchungen, 30(1): 8-17.
Legendre L, Rassoulzadegan F. 1995. Plankton and nutrient dynamics in marine waters. Ophelia, 41(1): 153-172.
Legendre L, Rassoulzadegan F. 1996. Food-web mediated export of biogenic carbon in oceans: hydrodynamic control. Marine Ecology Progress Series, 145: 179-193.
Lenz J. 1992. Microbial loop, microbial food web and classical food chain: their significance in pelagic marine ecosystems. Archiv fur Hydrobiologie Beihefte: Ergebnisse der Limnologie, 37: 265-278.
Li W K, Head E J, Harrison G W. 2004. Macroecological limits of heterotrophic bacterial abundance in the ocean. Deep Sea Research Part I: Oceanographic Research Papers, 51(11): 1529-1540.
Margalef R. 1974. Ecologia. Barcelona: Ediciones Omega: 951.
Mitchell D R. 2007. The evolution of eukaryotic cilia and flagella as motile and sensory organelles//Jékely G. Eukaryotic Membranes and Cytoskeleton: Origins and Evolution. New York: Springer New York: 130-140.
Mousseau L, Klein B, Legendre L, et al. 2001. Assessing the trophic pathways that dominate planktonic food webs: an approach based on simple ecological ratios. Journal of Plankton Research, 23(8): 765-777.
Pomeroy L R, Wiebe W J. 1988. Energetics of microbial food webs. Hydrobiologia, 159(1): 7-18.
Raven J A. 1998. The twelfth Tansley Lecture. Small is beautiful: the picophytoplankton. Functional Ecology, 12(4): 503-513.
Rigby S, Milsom C V. 2000. Origins, evolution, and diversification of zooplankton. Annual Review of Ecology and Systematics, 31(1): 293-313.
Rothman D H, Hayes J M, Summons R E. 2003. Dynamics of the Neoproterozoic carbon cycle. Proceedings of the National Academy of Sciences of the United States of America, 100(14): 8124-8129.
Sherr B F, Sherr E B, Hopkinson C S. 1988. Trophic interactions within pelagic microbial communities: indications of feedback regulation of carbon flow. Hydrobiologia, 159(1): 19-26.
Shiah F K, Chen T Y, Gong G C, et al. 2001. Differential coupling of bacterial and primary production in mesotrophic and oligotrophic systems of the East China Sea. Aquatic Microbial

Ecology, 23(3): 273-282.

Simon M, Cho B C, Azam F. 1992. Significance of bacterial biomass in lakes and the ocean: comparison to phytoplankton biomass and biogeochemical implications. Marine Ecology Progress Series, 86: 103-110.

Strom S L. 2000. Bacterivory: interactions between bacteria and their grazers//Kirchman D L. Microbial Ecology of the Oceans. New York: Wiley-Liss: 351-386.

Uye S, Iwamoto N, Ueda T, et al. 1999. Geographical variations in the trophic structure of the plankton community along a eutrophic–mesotrophic–oligotrophic transect. Fisheries Oceanography, 8(3): 227-237.

Wikner J, Hagström Å. 1988. Evidence for a tightly coupled nanoplanktonic predator-prey link regulating the bacterivores in the marine environment. Marine Ecology Progress Series, 50(1): 137-145.

Williams P J L. 1984. Bacterial production in the marine food chain: the emperor's new suit of clothes//Fasham M J R. Flows of Energy and Materials in Marine Ecosystems: Theory and Practice. Boston: Springer US: 271-299.

Williams P J L B, Quay P D, Westberry T K, et al. 2013. The oligotrophic ocean is autotrophic. Annual Review of Marine Science, 5: 535-549.

Wylie J L, Currie D J. 1991. The relative importance of bacteria and algae as food sources for crustacean zooplankton. Limnology and Oceanography, 36(4): 708-728.

第十四章　全球变暖对微食物网的影响

地球上的环境一直在变化。人类活动排放的大量二氧化碳，导致全球气温升高，海洋也随之发生变化，主要表现为变暖、酸化和溶解氧浓度降低。这些变化对海洋微食物网生物也会产生影响，其中海洋变暖的影响最受关注。

海洋变暖对海洋浮游微食物网的影响表现在很多方面。随着全球变暖，微食物网生物正在变小。本章将主要介绍全球变暖对微食物网生物的代谢及生物地理分布的影响。

第一节　生态学代谢理论

代谢是生物对物质和能量的处理过程，生物的个体代谢速率与温度和个体大小有关：生物的代谢速度随着温度的升高而增加，随生物个体大小（即生物量）的增加而异速增长，这一关系被称为生态学代谢理论（metabolic theory of ecology，MTE）（Brown et al., 2004），其具体关系为，个体的代谢速率（$I_{ind.}$）与个体生物量（M）的 3/4 指数幂成正比（异速增长）（Kleiber, 1932; Brody et al., 1934; Kleiber, 1947），与温度（T）的指数幂成正比（Boltzmann, 1966; Gillooly et al., 2001）。

$$I_{ind.} = I_0 \times M^{3/4} \times e^{-E/(kT)} \quad (14\text{-}1)$$

式中，I_0 是不依赖于个体大小的常数，E 是活化能，k 是玻耳兹曼常数（8.62×10^{-5} eV/K），T 是绝对温度。

为消除个体大小的影响，可以使用个体代谢速率除以个体生物量，得到单位体重代谢率（$I_{mass\text{-}specific}$）与个体生物量和温度的关系（Brown et al., 2004）。

$$I_{mass\text{-}specific} = I_0 \times M^{-1/4} \times e^{-E/(kT)} \quad (14\text{-}2)$$

式（14-2）表明，单位体重代谢率随个体生物量的增加而减小，生物的个体越大，其能量利用效率越高。

根据 MTE 可预测自养过程与异养过程对温度的反应存在差异，主要是自养过程与异养过程活化能不同所导致的。多项研究发现异养过程的活化能高于自养过程。Allen 等（2005）的全球碳循环模型估算认为生物异养过程的活化能约为 0.65eV，而自养过程的活化能约为 0.32eV。有研究针对海洋浮游生物群落的代谢进行了研究，认为海洋异养过程的活化能为 0.56eV，自养过程为 0.33eV（López-Urrutia et al., 2006）。根据文献数据，López-Urrutia（2008）估算海洋异养生物的活化能为0.64～0.84eV，自养生物的活化能为0.39eV。还有研究根据大量实测数据，总结海洋异养过程活化能为(0.66±0.05)eV，自养过程为(0.32±0.04)eV（Regaudie-de-Gioux and Duarte, 2012）。

活化能不同导致异养代谢的低温抑制阈值高于自养代谢。这意味着与自养生物相比，异养生物的单个细胞代谢速率对温度的依赖性更强，低温对异养代谢的抑制更为明显，而当温度升高时，异养代谢速率的提高快于自养代谢。

MTE的本质是认为环境能量控制着生物的代谢速率和个体（即生物量）大小，进而控制着生物数量的多少。尽管MTE的一些细节还存在争议，但它已成为海洋微食物网生态学研究的主线之一。使用MTE可以对海洋浮游微食物网生物的代谢研究进行理论指导，解释个体、种群和群落，以及生态系统等不同水平上的现象和问题。自20世纪70年代起，研究主要关注低温是否会对微食物网的异养过程造成抑制。MTE为研究海洋浮游微食物网的物质循环提供了可检验的预测，进入21世纪之后，随着全球变暖相关研究兴起，海洋微食物网生态学研究更多聚焦于检验MTE对微食物网生物代谢的相关预测。

第二节 低温对微食物网生物代谢的影响

地球两极水温≤4℃的海区被称为冷海（cold ocean）。在冷海，春季水华可在水温–1.8～2℃的海区发生（Pomeroy and Wiebe, 1988），而且这些海区的渔业产量也很大。低温对自养和异养过程的影响研究最初源自对这些现象的思考。

一、低温对细菌生长的抑制

20世纪70年代前后，有学者提出低温对海洋异养细菌生长有抑制作用，–1～2℃的低温使有机物质的分解过程被抑制，从而阻碍了寒冷水域异养细菌的生长（Fedosov and Ermachenko, 1969；Sorokin, 1971）。随后的研究发现，海洋异养细菌在低温时可以存活，但在6℃以下时，细菌的异养代谢水平随温度降低而直线下降。低温对海洋浮游植物同样有影响，光合作用在0℃以下受到抑制。在高纬度海区，春季海水温度升高到–1.8～2℃，此时低温对浮游植物的影响已较小，海区发生春季水华，但此时细菌的生长仍受低温抑制。例如，在波罗的海春季浮游植物水华过后，细菌的生物量并没有随之升高（Legrand et al., 2015），而纽芬兰海域的水华在–1℃发生，这时的细菌呼吸速率极低，几乎无法测量（Pomeroy and Deibel, 1986）。

针对低温水华和高渔业产量耦合的现象，Pomeroy等发表了一系列论文，提出4℃以下细菌代谢和生长被抑制，其程度大于浮游植物受抑制的程度，导致初级生产力进入呼吸代谢的比例较小，被浮游动物和底栖生物摄食的比例较大，为鱼类提供了充足的饵料，因而支持了较高的冷海鱼类生物量（Pomeroy and Deibel，1986；Pomeroy and Wiebe，1988；Pomeroy et al.，1990；Pomeroy et al.，1991；Pomeroy and Wiebe，1993），这一观点被称为"冷海范式"（cold ocean paradigm）。北半球的低温水华发生在陆架区，此处渔业产量大，可能是因为"冷海范式"造成大多数浮游植物生产力未能被微食物网消耗，而是被浮游动物和陆架区的底栖生物摄食，进而为鱼类提供了充足的饵料；但在南半球，表层低温水华水体之下是深海，渔业产量就较低（Pomeroy and Wiebe，1988）。

随着"冷海范式"的提出，对其的质疑也接踵而来。有研究发现极地海区异养细菌丰度、代谢率和生长率与低纬度海区的结果较为相近（Fuhrman and Azam，1980；Hodson et al.，1981；Hanson et al.，1983）。"冷海范式"提出者Pomeroy等对此现象给出了两种可能的解释：①该现象是由培养条件不严格造成的结果误差，细菌代谢在0℃左右变化很大，即使只有1~2℃的差异其代谢率都会有较大的变化，实际培养操作中，温度难免会有小的变化，所以很难精确测量特定温度下的代谢率；②底物浓度会对代谢率产生影响，在低温情况下如果底物浓度较高，也可以维持异养细菌一定的生长率（Pomeroy and Wiebe，1988）。

但是随着更多数据显示极地海区异养细菌丰度和生长率与低纬度大洋区域的相近，"冷海范式"最终被否定（Azam et al.，1991；Rivkin et al.，1996）。目前学者认为温度和底物浓度是冷海异养细菌生长率低的原因（Pomeroy and Wiebe，2001）。但在永冷海（permanently cold sea，指周年温度≤4℃的海区，包括极地海区和大部分深海），绝大多数异养细菌都生活在远低于其最佳生长温度的环境中，此时究竟是温度还是底物浓度限制了细菌的生长率仍有争论。Kirchman等（2009）比较了南北极海域与温带、热带海洋的异养细菌生物量和生长率，指出尽管部分研究发现极地海区的细菌生物量和生长率与温带和热带海洋相近，但是极地的细菌生物量和生长率的平均值都明显低于温带和热带海洋，低温是造成这一差异的部分原因，但更重要的因素是极地海区的底物浓度较低。在深海，溶解有机碳（dissolved organic carbon，DOC）的浓度较低，可能是细菌生长缓慢的主要原因（Arrieta et al.，2015；Jiao et al.，2015），低温对深海细菌生长的抑制作用则研究很少。

二、低温对微型浮游动物生长的抑制

低温同样会抑制微型浮游动物的生长率，包括植食性（herbivore）和细菌食性（bacterivore）微型浮游动物。

有研究整合了多项培养实验的结果，总结了微型浮游动物、浮游植物的最大生长率与水温的关系，发现水温＞15℃时植食性微型浮游动物的最大生长率大于或等于浮游植物的最大生长率。随着温度降低，植食性微型浮游动物最大生长率降低，且降低速度快于浮游植物。在各个培养温度条件下，植食性微型浮游动物的最大生长率始终低于细菌食性微型浮游动物。在低温的高纬度海区，植食性微型浮游动物的最大生长率不足浮游植物最大生长率的一半，这是高纬度海区春季低温时发生水华的一个原因（Rose and Caron，2007）。

上述观点得到了一些质疑，因为该研究使用的微型浮游动物和浮游动物培养实验并非来自高纬度海区。有野外观测表明，极地微型浮游动物可能已经适应了低温的极端环境（Levinsen et al.，1999）。在北冰洋巴伦支海（Barents Sea）进行的纤毛虫和异养鞭毛虫原位培养发现，在 0℃左右微型浮游动物的最大生长率与浮游植物的最大生长率接近（Franzè and Lavrentyev，2014），因此，极地海区的微型浮游动物可能与浮游植物一样，对低温环境有了适应。

三、低温对微型浮游动物摄食的抑制

微型浮游动物对浮游植物的摄食率通常使用原位稀释培养法来估算，已有的研究结果表明，极地海区微型浮游动物的摄食率较低。在北冰洋海区，仅有的几项原位稀释培养研究均显示微型浮游动物对浮游植物的摄食率很低（Verity et al.，2002；Sherr et al.，2009；Calbet et al.，2011），如有研究进行了 40 次稀释培养，其中只有 20 次可以观察到有明显摄食，摄食率最大仅为 0.16/d（Sherr et al.，2009）。在南大洋海区进行的稀释培养研究则相对较多（Landry et al.，2001；Safi et al.，2007；Pearce et al.，2010，2011；Garzio et al.，2013；Yang et al.，2016），绝大多数在 5℃以下进行的稀释培养都显示摄食率接近 0，意味着此时微型浮游动物几乎没有摄食。但也有例外，如在南极半岛东部海区，微型浮游动物对浮游植物摄食率较高（0.3~2.4/d）（Pearce et al.，2010），在南大洋阿蒙森海（Amundsen Sea）水华期间微型浮游动物摄食率较高，消耗了大部分的浮游植物生产（Yang et al.，2016）。

除原位稀释培养实验外，还可以通过比较非极地海区微型浮游动物在不同温度梯度下的摄食率，推测低温对微型浮游动物摄食的抑制作用。例如，有研究发现微型浮游动物的摄食率随水温降低而下降，水温为 5~8℃时，摄食率为 0.2~0.5/d，水温<0℃时，摄食率为 0.1~0.3/d（Verity et al.，2002）。类似的低温抑制效应在许多研究中也有出现（Caron et al.，2000；Calbet and Landry，2004；Landry and Calbet，2004；Garzio et al.，2013）。

除低温抑制作用外，其他因素也可能影响极地海区微型浮游动物的摄食率，包括浮游植物的群落结构和微型浮游动物的生理状态。在南极半岛西部海区，微型浮游动物会选择性摄食粒径较小的（pico 级和 nano 级）浮游植物，而避免摄食大型硅藻,还倾向于优先摄食生长较快的浮游植物类群(Garzio et al., 2013)。南大洋纤毛虫的单位体积摄食率受其生理状态的影响,在饥饿状态下,纤毛虫可以达到很高的摄食率,但是在非饥饿状态下,摄食率很低(Rose et al., 2013)。还有研究认为,水华浮游植物会产生对其他浮游植物生长有抑制作用的物质,在制备稀释培养所需的过滤海水时,浮游植物细胞可能因为过滤而破碎,将这些抑制物质释放到海水中,从而使得稀释培养得出的摄食率偏低(Stoecker et al., 2015)。

第三节 全球变暖对微食物网生物代谢的影响

全球变暖相关研究预测 21 世纪内海洋表层温度会升高 2~6℃,其中北极是全球变暖最快的地区,北极地区变暖的速度是全球平均速度的 3 倍(Vaquer-Sunyer et al., 2010)。因此北冰洋是全球变暖对微食物网生物影响研究的首选海区。

根据 MTE,温度升高将对自养和异养过程有不同的影响。据此可预测变暖可对微食物网生物代谢造成一定影响,包括导致水华与细菌、微型浮游动物的时滞变小,影响细菌的生长效率和微型浮游动物的摄食效率,使得生态系统变得更为异养。针对这些预测,各国学者开展了大量围隔实验进行检验。

一、变暖导致水华与细菌生产的时滞变小

与自养生物的代谢速率相比,单细胞异养生物的代谢速率对温度的依赖性更强。春季水华可以在很低的水温条件下发生,而此时细菌受低温的抑制更显著,因此在浮游植物春季水华中,浮游植物自养生产和细菌异养呼吸过程在时间上是不耦合的,异养呼吸滞后于自养生产(Pomeroy and Deibel, 1986; Lignell, 1993; Bird and Karl, 1999)。根据 MTE,异养生物对升温的反应要快于自养生物,那么全球变暖导致的海水温度升高是否能减少浮游植物生长和细菌生产之间的时滞?

德国的 AQUASHIFT 计划在波罗的海基尔湾(Kiel Fjord)进行了非常系统的围隔实验,以期研究全球变暖对海洋生态系统的影响(Sommer et al., 2012)。在 AQUASHIFT 计划资助下,有学者研究了温度对春季水华期间浮游植物与细菌耦合关系的影响。该围隔培养实验有 4 个温度梯度,包括对照组(ΔT+0℃)和 3 个升温组(ΔT+2℃、ΔT+4℃和ΔT+6℃),实验共进行了 77d,对照组的温度为当天基尔湾的 10 年(1993~2002 年)平均温度,在 2 月实验开始时为 2℃。实验执行期间,每周至少测定 3 次初级生产力和细菌生产力。围隔实验结果表明,随温度升高,初级生产力峰值出现的时间没有显著改变,但细菌生产力峰值出现的时间显著提前;对照组细菌生产力和初级生产力峰值的时滞为 16d,而 6℃升温组的时滞仅为 4d,温度每升高 1℃,时滞平均减少 2d(Hoppe et al., 2008)。这一实验的结果支持了 MTE 的预测,即环境温度升高可使得细菌生产与浮游植物春季水

华的耦合更加紧密。

二、变暖导致水华与微型浮游动物生产的时滞变小

根据 MTE，浮游植物与微型浮游动物的最大生长率随温度的变化而不同，因此 Rose 和 Caron（2007）提出了如下假设：当温度升高时，微型浮游动物的最大生长率升高较快，将缩短浮游植物水华和微型浮游动物生产之间的时滞。

为验证上述假设，在 AQUASHIFT 计划资助下，有学者同样于基尔湾进行了围隔实验。实验持续 49d，分为对照组（ΔT+0℃）和升温组（ΔT+6℃）两个温度梯度，其中对照组的温度也设定为基尔湾的当日 10 年平均温度。实验发现，升温组微型浮游动物的生长率（0.25/d）高于对照组（0.12/d），且升温组微型浮游动物的生物量峰值比对照组大大提前。温度升高 6℃，微型浮游动物的丰度高峰期提前了 7～21d，即每升高 1℃，生物量峰值提前 1.1～3.5d（Aberle et al.，2012；Aberle et al.，2015），意味着温度升高可缩短水华与微型浮游动物生产的时滞。但另一个在挪威峡湾进行的围隔实验中，并未发现升温使水华与微型浮游动物生产时滞缩短的现象（Calbet et al.，2014）。

三、变暖对细菌生长效率的影响

细菌摄入的碳一部分用于生长，一部分通过呼吸释放到环境中。细菌生长效率（bacteria growth efficiency，BGE）指细菌生长利用的物质与摄入物质的比值，其中，摄入物质等于生长利用的碳与呼吸释放的碳之和，因此细菌生长效率可以表示为 BGE=$P/(P+R)$，其中 R 是呼吸率，P 是生长率。当温度升高时，细菌生长率和呼吸率都随之升高，升温对细菌生长效率的影响取决于温度对呼吸率和生长率的影响是否相同：当二者受到的影响相同时，细菌生长效率保持不变；若生长率增加的程度小于呼吸率的增加程度，细菌生长效率会降低，反之细菌生长效率将升高。

海洋细菌的生长效率是否受温度的影响有两种不同的观点。第一种观点认为细菌生长效率主要是受底物质量的影响，温度的影响很小（del Giorgio and Cole，1998）。有研究在地中海沿岸的 Blanes 湾进行海水升温培养实验，实验组海水比对照组升高 2.7℃，但细菌的生长效率并没有显著变化（Vázquez-Domínguez et al.，2007）。另有研究通过升温培养实验，发现细菌生长率对温度升高的反应与底物的质量有很大关系，底物不能满足细菌的需要时，细菌生长率不会随温度升高而升高，底物供应是细菌生长率的重要限制因子（Huete-Stauffer et al.，2015）。

第二种观点认为细菌生长效率与温度呈负相关关系，升温将导致细菌生长效

率降低。Rivkin 和 Legendre（2001）总结了极地、温带和热带海洋浮游细菌的生长效率与温度（T）的关系，发现温度与生长效率之间存在显著的负相关性，温度解释了 54%的细菌生长效率变化，并提出二者之间的关系可以表示为 BGE=0.374–0.0104×T。在实际观测研究中，通过比较切萨皮克湾同一站位细菌生长效率的季节变化，可发现温度越高细菌的生长效率越低（Apple et al., 2006）。这意味着与高纬度海区相比，低纬度海区细菌摄入的碳可能有更大的比例被呼吸消耗，而不是用于生长。

值得关注的是，现有的升温培养实验存在着一定的局限性。假设 21 世纪末海水温度升高 2℃，而细菌以约 1/d 的速度繁殖生长，从现在至 21 世纪末的过程中将约有 30 万代细菌。换句话说，每一代细菌平均仅经历 0.000 007℃的温度变化。如此微小的温度变化对生物几乎没有影响，细菌在这种缓慢的升温过程中很可能会适应新的环境。但升温培养实验则是将水温立即升高 2℃，急剧的温度变化可能会对细菌产生完全不同的影响，因此对升温实验的结果应谨慎考虑（Sarmento et al., 2010）。

四、变暖对微型浮游动物摄食的影响

温度升高促进生物代谢，因此升温可促进微型浮游动物摄食率提高。在水温较低时，升温能增加微型浮游动物对细菌的摄食率，在–1～5℃，微型浮游动物的摄食强度会随温度升高而增加（Vaqué et al., 2009）。但当环境温度超过一定数值后，微型浮游动物的摄食率就不再有明显的增加。例如，有研究在美国马萨诸塞州近岸使用荧光标记细菌法，检测了温度对不同季节微型浮游动物对细菌摄食率的影响，将冬季海水加温至 20℃后，可使微型浮游动物对细菌的摄食率升高至与夏季相当的水平，但是对于环境温度超过 13～15℃的水样，升温的影响就不明显（Marrasé et al., 1992）。另一个研究在地中海西北部寡营养的 Blanes 湾进行了 12 个月的升温培养实验，对照组为原位水温（12.5～25℃），实验组则是将对照组升温 2.7℃，发现升温增加了微型浮游动物对细菌的摄食率，且在寒冷的月份升温的影响较大（Vázquez-Domínguez et al., 2012）。在北冰洋巴伦支海（水温–1.19℃）进行的 7 个温度梯度（温度范围 1～10℃）升温实验发现，微型浮游动物对细菌的摄食率在温度升高 5.5℃时达到峰值，继续升温时摄食率不再提高（Lara et al., 2013）。

微型浮游动物对浮游植物的摄食率也因升温而增加。在南大洋罗斯海进行的升温培养实验发现，培养温度从 12℃升高至 16℃后，微型浮游动物对浮游植物的摄食率从 0.49/d 升高到 0.84/d（Rose et al., 2009）。还有研究发现，在叶绿素浓度较高的富营养环境中，微型浮游动物对浮游植物的摄食率（m）和浮游植物生长

率（μ）之间的比值 $m:\mu$ 将随温度升高而增加，当温度升高到一定程度后，$m:\mu$ 随温度升高反而减小；在叶绿素浓度低的寡营养环境中，$m:\mu$ 随温度升高持续减小（Chen et al.，2012）。

五、变暖使得海洋生态系统更加异养

MTE 预测海洋初级生产力将随着温度升高而增加，且呼吸作用的增加速率高于初级生产力的增加速率。这意味着有机物质在自养生产和异养消耗之间的平衡发生改变，使更多的物质和能量进入呼吸作用，海洋生态系统会变得更加异养（López-Urrutia et al.，2006；O'Gorman et al.，2012；Regaudie-de-Gioux and Duarte，2012；Kvale et al.，2015），这一观点已经得到大量的研究验证。由于光合作用的活化能低于呼吸作用，升温导致异养呼吸升高的速度是初级生产力升高速度的 2 倍，当温度上升 4℃，初级生产力增加 20%，异养代谢会增加 43%（Harris et al.，2006）。在地中海西北部 Blanes 湾的升温实验中，随着温度升高细菌的生长效率没有改变，但是细菌的呼吸率和生长率都有增加，这意味着升温增加了细菌对碳的需求（Vázquez-Domínguez et al.，2007）。在波罗的海春季水华期间，围隔培养发现随温度升高细菌生产力比初级生产力增加得更快（Hoppe et al.，2008），水温升高 2~6℃，浮游生态系统吸收 CO_2 的能力降低 31%（Wohlers et al.，2009）；在波罗的海北部，升温 3℃后细菌的群落呼吸速率平均提高 2.6 倍（Panigrahi et al.，2013）。

热带和寒带海区细菌的呼吸作用对升温的反应不同。在热带潟湖，细菌的呼吸率随温度的升高而降低，在 38℃培养条件下的耗氧率比 23℃时低 25%（Pires et al.，2014）。在寒带，升温对细菌呼吸的促进作用有季节性，有研究在北冰洋海区进行了 4 个季节的升温实验，发现在春季和夏季，升温显著促进了细菌的呼吸作用，而秋、冬季的升温则没有明显影响（Vaquer-Sunyer et al.，2010）。

随水温升高，自养过程在混合营养生物代谢中的比例可能出现降低，这是生态系统变得更加异养的另一个表现。例如，在 13~21℃的水温培养条件下，淡水混合营养棕鞭藻 *Ochromonas* sp.自养生长的贡献为 6%，在 33℃时则降低至 2%（Wilken et al.，2013）。目前自然海区浮游混合营养生物的异养情况尚未见报道。

六、变暖对细菌丰度的影响

MTE 理论预测随着温度升高，生物的代谢速率会增加，所以在饵料总量不变的情况下，高温时生态系统的承载力（carrying capacity）将变小（Savage et al.，2004；Anderson-Texeira and Vitousek，2012）。这意味着相同的资源供应仅可以支持较少的具有更高代谢率的生物,海洋细菌的丰度和生物量将随着温度的升高而降低。

根据MTE,若以6℃温度差和生物异养过程的活化能为0.65eV进行粗略估计,细菌丰度预计会减少58%(Morán et al.,2015),但实际研究的结果与这一预测并不完全一致。例如,根据MTE,若地中海西北部温度升高2℃,细菌的生物量将会降低18.6%。该海区目前的升温速度约为0.032℃/a,按上述预测则细菌丰度大约每年降低3%(Sarmento et al.,2010)。在地中海西北部的Blanes湾的实际调查中,已观测到2003~2010年细菌丰度有下降趋势,每年丰度约降低10%,是理论预测值的3倍(Sarmento et al.,2010)。在大西洋东北部的比斯开湾,2002~2011年春季平均温度升高了1℃,平均0.086℃/a,但细菌的丰度不仅没有降低,反而每年增加3%,细菌个体体积每年缩小1%(Morán et al.,2015)。

由于纬度不同,全球不同海区的水温有较大差异,比对不同温度海区的生物量变化情况可作为预测全球变暖影响的方法之一,称为空间换时间法(space-for-time substitution)。Li等(2004)统计了全球不同海区的资料,发现在-2~15℃条件下,海洋细菌的丰度会随水温升高而增加,当温度高于15℃后,细菌丰度反而会随水温升高而降低。这一结果意味着MTE理论关于升温与细菌生物量变化趋势的预测仅适用于水温较高的海区。

许多海水升温培养实验的结果与Li等(2004)的结论较为接近。例如,有研究在北冰洋不同海区分别进行升温培养实验,从巴伦支海采集的海水(-1.19℃)升温到5.5℃过程中,异养生物包括细菌、鞭毛虫、纤毛虫的丰度都显著增加;从Isfjorden峡湾采集的海水(受到大西洋水影响,温度6.2℃)进行升温实验,异养生物丰度的增加就不明显,表明不同温度海水中的生物对升温的反应不同(Lara et al.,2013)。在比斯开湾的海水初始温度为12.7~21.2℃,大多数情况下升温3℃培养后细菌的生物量保持不变,个别情况下还出现了细菌生物量升高的情况(Huete-Stauffer et al.,2015)。还有研究预测在年平均表层水温小于26℃的海域,温度升高1℃将造成异养细菌生物量增加(Morán et al.,2017)。这些实验和理论推测结果与MTE关于升温与细菌生物量变化趋势的预测并不相符,表明MTE也有一定的局限性。

七、变暖对浮游病毒的影响

病毒是海洋生态系统中数量最多的生命形式,可能会感染海水中每一种生物体。海洋病毒最主要的宿主为细菌、蓝细菌、藻类、原生生物等微食物网生物,病毒和这些宿主生物也组成了海洋中绝大多数生物量,在海洋生物地球化学循环中发挥着重要作用。尽管已有大量研究对CO_2浓度增加和全球变暖对海洋生态系统的潜在影响进行了评估，但其中对病毒丰度和多样性影响的研究还很有限。已有的研究认为，全球变暖可能会影响所有的海洋生态系统组分，其中病毒不仅会受到全球变暖影响，还是气候变化的潜在参与者。

气候变化通过多种直接和间接机制影响海洋病毒（Danovaro et al., 2011）。气候变化对病毒的直接影响主要是升温影响病毒的失活速率。间接影响包括：①影响宿主群落组成、新陈代谢和生长效率，从而对其侵染病毒丰度、种类组成和病毒-宿主相互作用产生影响；②影响病毒的生命策略，即溶原性感染与裂解性感染转换。

通过升温培养实验，有研究发现无论是实验室纯培养病毒（de Paepe and Taddei, 2006）还是自然水体中的浮游病毒（Wei et al., 2018），温度升高都可导致病毒的失活速率增加。不同病毒亚群对升温的反应也不一致，与高荧光病毒亚群相比，升温条件下低荧光病毒亚群失活速率的增加幅度更大（Wei et al., 2018）。海水中存在多种胞外蛋白酶（extracellular protease），可以破坏由蛋白质组成的病毒衣壳（viral capsid），使病毒失活。在一定温度范围内，升温造成胞外蛋白酶活性增加，因而提高了病毒的失活速率。此外，温度还可能影响病毒衣壳蛋白或脂膜的稳定性，从而影响病毒活性（Mojica and Brussaard, 2014）。病毒失活速率升高意味着宿主感染率的降低，病毒失活也会向水体释放大量的营养元素，可能对微食物网的物质和能量流动产生重要影响。

由于病毒复制和生命周期与宿主新陈代谢密切相关，温度升高可能会影响病毒-宿主相互作用，如随着原核生物生长速率的增加，病毒的裂解周期（lytic cycle）缩短，裂解量增大（Proctor et al., 1993；Hadas et al., 1997），从而导致更高的病毒生产速率和丰度。已有的观测结果发现，在水温较低的高纬度海区（–2~14℃），病毒生产速率随温度升高而提高，当水温超过14℃时，病毒生产速率反而随温度继续升高而下降。各不同纬度海区的病毒丰度与温度之间均分别存在正相关关系，其中，温带大洋的相关性最为显著，温度升高4℃病毒丰度即翻倍（Danovaro et al., 2011）。但若将各不同纬度海区的数据汇总分析，则总体上病毒丰度与温度之间呈负相关，这可能是纬度变化导致海区光照和营养条件的差异，影响了宿主的生长，进而通过级联效应影响了病毒的侵染和复制速率。

海洋变暖将引起水体层化加强，抑制了大洋上层营养盐的垂直交换，营养条件变化将对病毒的生命策略产生重要影响。温和病毒在侵染宿主后，可以根据环境不同选择进入溶原途径或者裂解途径，称为溶原决定（lysogenic decision）。营养盐浓度是溶原决定的重要影响因素（Williamson and Paul, 2004；Long et al., 2008；Payet and Suttle, 2013），当营养盐浓度较低，且病毒/宿主丰度值较高的时候，海洋病毒倾向于进入溶原途径（Wilson and Mann, 1997）。从富营养的密西西比河羽流到寡营养的墨西哥湾，随营养盐浓度降低，噬藻体和噬菌体的溶原性都出现提高（Long et al.，2008）。磷酸盐可能会诱导原病毒从宿主染色体上脱离，进入裂解循环，对已被病毒感染的聚球藻，在磷酸盐限制条件下绝大多数噬藻体都处于溶原状态，只有9.3%的聚球藻宿主被裂解，而当磷酸盐浓度较高时，100%的聚球藻细胞被病毒裂解，表明此时裂解途径占绝对优势（Wilson et al.，1996）。随全球变暖，近海和河口的水文环境可能变得更利于病毒的裂解途径，而开阔大洋上层营养盐限制则更利于病毒的溶原途径。这种层化加强、营养盐限制导致的病毒的生命策略改变可能首先在高纬度极地海区发生。更高比例的细菌和蓝细菌等宿主与病毒共存而不是被裂解，意味着原核生物在海洋营养盐再生过程中的作用降低，微食物网病毒回路的重要性下降，有机物向更高营养水平的传递增加（Danovaro et al.，2011）。

第四节　温度升高对微食物网生物地理分布的影响

生物地理分布和多样性是生态学研究的重要内容。在微食物网的各生物类群中，砂壳纤毛虫可以根据壳的形态进行种类鉴定，易于开展生物地理分布研究，而且纤毛虫对环境变化响应迅速，适宜环境能快速繁殖并建立种群，不同水团中种类组成差异显著，可作为水团指示生物。无壳纤毛虫形态分类需要进行蛋白银染色等操作，在生态分布研究中很难广泛使用。微食物网中的其他生物类群缺乏明显的形态分类特征，难以开展传统的生物地理分布研究。

砂壳纤毛虫的生物地理分布格局与中型浮游动物一致，主要分为大洋类群和近岸类群，其中北半球的大洋类群分为北极类群、亚极地类群、亚极地和亚热带之间的过渡带类群、亚热带环流中心类群和赤道类群等。北极类群主要是盾形笛杯虫（*Ptychocylis urnula*），位于北太平洋和白令海的亚极地类群有钝笛杯虫（*Ptychocylis obtusa*）、网纹虫（*Parafavella* spp.）、寒冷类铃虫（*Codonellopsis frigida*）、挪威棘口虫（*Acanthostomella norvegica*）和一种未定种号角虫（*Salpingella* sp.1）等（Wang et al.，2019）。北冰洋是对全球变暖最敏感的区域之一，总体变化趋势是入流水量增大、海水升温、海冰融化。在北冰洋-太平洋扇区，太平洋入流水是连通白令海（亚极地类群）和北冰洋（北极类群）浮游生物的桥梁。北太平洋的亚北极砂壳纤毛虫随着太平洋入流水进入北冰洋，随着温度降低和环境的改变逐渐消亡。对比2016年和2019年白令海至北冰洋相似海区的纤毛虫群落，可发现浮游纤毛虫的粒级正在趋向于小型化，部分砂壳纤毛虫种类分布区北移，表明该海域中微型浮游动物正在经历着剧烈的北方化进程。近年来随着全球变暖，从北太平洋进入北冰洋的入流水量增加，促使亚北极砂壳纤毛虫向北移动的距离越来越远，其中号角虫（*Salpingella* sp.1）已经到达波弗特环流的中心，将来能否在北冰洋建立常年存在的种群是全球是否变暖的重要标志（Wang et al.，2022）。

在自养微食物网生物中，原绿球藻主要分布在全球大洋 40°S～40°N 的温暖水域，如印度洋和西太平洋寡营养环流中，在近岸和中高纬度海区（水温 15℃以下）几乎不存在。目前全球海洋中约有 (2.9±0.1)×10^{27} 个原绿球藻（Flombaum et al., 2013）。有模型研究显示，随着全球变暖，到 21 世纪末期全球海洋原绿球藻的丰度将增加 29%，分布区域也会极向扩大（Flombaum et al., 2013）。

全球变暖带来的浮游生物的分布区发生极向移动，也将导致各不同纬度海区的生物多样性发生变化，其中北冰洋的生物多样性增加将最为明显。模型研究预测在北冰洋海区微食物网生物中，光合自养细菌（主要为聚球藻）的多样性增加最为显著，微型浮游动物的多样性增加幅度次之，异养细菌的多样性增幅最小（Ibarbalz et al., 2019）。

第五节 全球变暖对微食物网重要性的潜在影响

根据上述对微食物网重要性的介绍，我们可以大致预测全球变暖的影响。在全球变暖背景下，海水温度升高，水体层化加剧，从底层混合到表层的营养盐减少，全球海洋总体的寡营养程度增加，自养浮游生物的群体粒径变小，表现在小粒级自养浮游生物的生物量占比升高，或自养浮游生物群落平均（或中值）粒径变小。这意味着微食物网占主导的海区将扩大，优势度加强。

已有证据证明，随全球变暖海洋自养浮游生物的群体粒径正在变小。例如，北冰洋微微型浮游植物的丰度升高，而其他粒级浮游植物的丰度在降低（Li et al., 2009），聚球藻将会进入北冰洋海区（Paulsen et al., 2016）。2006～2019 年，德国 Wadden 海大多数浮游植物种类的细胞体积变小，群落的加权平均粒级下降（Hillebrand et al., 2022）。2003～2010 年，大西洋经向断面（Atlantic meridional transect, AMT）赤道及赤道以北海区，小粒级（pico 级和 nano 级）的浮游植物增加，大粒级（micro 级）浮游植物减少，但总叶绿素浓度没有明显变化（Agirbas et al., 2015）。根据叶绿素遥感资料估算浮游植物的粒径中值的研究发现，1998～2007 年在亚热带海区叶绿素粒径中值从约 1.4μm 下降为约 1.3μm（Polovina and Woodworth, 2012）。这些结果表明随全球变暖微食物网的重要性正在增加。

参 考 文 献

Aberle N, Bauer B, Lewandowska A, et al. 2012. Warming induces shifts in microzooplankton phenology and reduces time-lags between phytoplankton and protozoan production. Marine Biology, 159(11): 2441-2453.

Aberle N, Malzahn A M, Lewandowska A M, et al. 2015. Some like it hot: the protozooplankton-copepod link in a warming ocean. Marine Ecology Progress Series, 519:

103-113.

Agirbas E, Martinez-Vicente V, Brewin R J W, et al. 2015. Temporal changes in total and size-fractioned chlorophyll-a in surface waters of three provinces in the Atlantic Ocean (September to November) between 2003 and 2010. Journal of Marine Systems, 150: 56-65.

Allen A P, Gillooly J F, Brown J H. 2005. Linking the global carbon cycle to individual metabolism. Functional Ecology, 19(2): 202-213.

Anderson-Texeira K J, Vitousek P M. 2012. Ecosystems//Sibly R M, Brown J H, Kodric-Brown A. Metabolic Ecology: a Scaling Approach. UK: John Wiley & Sons: 99-111.

Apple J K, del Giorgi P A, Kemp W M. 2006. Temperature regulation of bacterial production, respiration, and growth efficiency in a temperate salt-marsh estuary. Aquatic Microbial Ecology, 43(3): 243-254.

Arrieta J M, Mayol E, Hansman R L, et al. 2015. Dilution limits dissolved organic carbon utilization in the deep ocean. Science, 348(6232): 331-333.

Azam F, Smith D C, Hollibaugh J T. 1991. The role of the microbial loop in Antarctic pelagic ecosystems. Polar Research, 10(1): 239-244.

Bird D, Karl D. 1999. Uncoupling of bacteria and phytoplankton during the austral spring bloom in Gerlache Strait, Antarctic Peninsula. Aquatic Microbial Ecology, 19(1): 13-27.

Boltzmann L. 1966. Further studies on the thermal equilibrium of gas molecules//Brush S G, Harr D T. Kinetic Theory. Oxford: Pergamon: 88-175.

Brody S, Procter R C, Ashworth U S. 1934. Growth and development with special reference to domestic animals. XXXIV, Basal metabolism, endogenous nitrogen, creatinine and neutral sulphur excretions as functions of body weight. University of Missouri Agricultural Experimental Station Residential Bulletin, 220: 1-40.

Brown J H, Gillooly J F, Allen A P, et al. 2004. Toward a metabolic theory of ecology. Ecology, 85(7): 1771-1789.

Calbet A, Landry M R. 2004. Phytoplankton growth, microzooplankton grazing, and carbon cycling in marine systems. Limnology and Oceanography, 49(1): 51-57.

Calbet A, Saiz E, Almeda R, et al. 2011. Low microzooplankton grazing rates in the Arctic Ocean during a *Phaeocystis pouchetii* bloom(Summer 2007): fact or artifact of the dilution technique? Journal of Plankton Research, 33(5): 687-701.

Calbet A, Sazhin A F, Nejstgaard J C, et al. 2014. Future climate scenarios for a coastal productive planktonic food web resulting in microplankton phenology changes and decreased trophic transfer efficiency. PLoS One, 9(4): e94388.

Caron D A, Dennett M R, Lonsdale D J, et al. 2000. Microzooplankton herbivory in the Ross Sea, Antarctica. Deep Sea Research Part II: Topical Studies in Oceanography, 47(15-16): 3249-3272.

Chen B, Landry M R, Huang B, et al. 2012. Does warming enhance the effect of microzooplankton grazing on marine phytoplankton in the ocean? Limnology and Oceanography, 57(2): 519-526.

Danovaro R, Corinaldesi C, Dell'Anno A, et al. 2011. Marine viruses and global climate change. FEMS Microbiology Reviews, 35(6): 993-1034.

de Paepe M, Taddei F. 2006. Viruses' life history: towards a mechanistic basis of a trade-off between survival and reproduction among phages. PLoS Biology, 4(7): e193.

del Giorgio P A, Cole J J. 1998. Bacterial growth efficiency in natural aquatic systems. Annual Review of Ecology and Systematics, 29(1): 503-541.

Fedosov M V, Ermachenko I A. 1969. On primary production in the Northwest Atlantic. Special

Publication of the International Commission for the Northwest Atlantic Fisheries, 7: 87-93.

Flombaum P, Gallegos J L, Gordillo R A, et al. 2013. Present and future global distributions of the marine Cyanobacteria *Prochlorococcus* and *Synechococcus*. Proceedings of the National Academy of Sciences of the United States of America, 110(24): 9824-9829.

Franzè G, Lavrentyev P J. 2014. Microzooplankton growth rates examined across a temperature gradient in the Barents Sea. PLoS One, 9(1): e86429.

Fuhrman J A, Azam F. 1980. Bacterioplankton secondary production estimates for coastal waters of British Columbia, Antarctica, and California. Applied and Environmental Microbiology, 39(6): 1085-1095.

Garzio L M, Steinberg D K, Erickson M, et al. 2013. Microzooplankton grazing along the Western Antarctic Peninsula. Aquatic Microbial Ecology, 70(3): 215-232.

Gillooly J F, Brown J H, West G B, et al. 2001. Effects of size and temperature on metabolic rate. Science, 293(5538): 2248-2251.

Hadas H, Einav M, Fishov I, et al. 1997. Bacteriophage T4 development depends on the physiology of its host *Escherichia coli*. Microbiology, 143(1): 179-185.

Hanson R, Lowery H, Shafer D, et al. 1983. Microbes in Antarctic waters of the Drake Passage: vertical patterns of substrate uptake, productivity and biomass in January 1980. Polar Biology, 2(3): 179-188.

Harris L A, Duarte C M, Nixon S W. 2006. Allometric laws and prediction in estuarine and coastal ecology. Estuaries and Coasts, 29(2): 340-344.

Hillebrand H, di Carvalho A J, Dajka J C, et al. 2022. Temporal declines in Wadden Sea phytoplankton cell volumes observed within and across species. Limnology and Oceanography, 67(2): 468-481.

Hodson R E, Azam F, Carlucci A, et al. 1981. Microbial uptake of dissolved organic matter in McMurdo Sound, Antarctica. Marine Biology, 61(2-3): 89-94.

Hoppe H G, Breithaupt P, Walther K, et al. 2008. Climate warming in winter affects the coupling between phytoplankton and bacteria during the spring bloom: a mesocosm study. Aquatic Microbial Ecology, 51: 105-115.

Huete-Stauffer T M, Arandia-Gorostidi N, Díaz-Pérez L, et al. 2015. Temperature dependences of growth rates and carrying capacities of marine bacteria depart from metabolic theoretical predictions. FEMS Microbiology Ecology, 91(10): fiv111.

Ibarbalz F M, Henry N, Brandão M C, et al. 2019. Global trends in marine plankton diversity across kingdoms of life. Cell, 179(5): 1084-1097.

Jiao N, Legendre L, Robinson C, et al. 2015. Comment on "dilution limits dissolved organic carbon utilization in the deep ocean". Science, 350(6267): 1483.

Kirchman D L, Moran X A, Ducklow H. 2009. Microbial growth in the polar oceans-role of temperature and potential impact of climate change. Nature Reviews Microbiology, 7(6): 451-459.

Kleiber M. 1932. Body size and metabolism. Hilgardia, 6: 315-353.

Kleiber M. 1947. Body size and metabolic rate. Physiological Reviews, 27(4): 511-541.

Kvale K F, Meissner K J, Keller D P. 2015. Potential increasing dominance of heterotrophy in the global ocean. Environmental Research Letters, 10(7): 074009.

Landry M R, Brown S L, Selph K E, et al. 2001. Initiation of the spring phytoplankton increase in the Antarctic Polar Front Zone at 170°W. Journal of Geophysical Research: Oceans, 106(C7):

13903-13915.

Landry M R, Calbet A. 2004. Microzooplankton production in the oceans. ICES Journal of Marine Science, 61(4): 501-507.

Lara E, Arrieta J M, Garcia-Zarandona I, et al. 2013. Experimental evaluation of the warming effect on viral, bacterial and protistan communities in two contrasting Arctic systems. Aquatic Microbial Ecology, 70(1): 17-32.

Legrand C, Fridolfsson E, Bertos-Fortis M, et al. 2015. Interannual variability of phyto-bacterioplankton biomass and production in coastal and offshore waters of the Baltic Sea. Ambio, 44(Suppl 3): 427-438.

Levinsen H, Nielsen T G, Hansen B W. 1999. Plankton community structure and carbon cycling on the western coast of Greenland during the stratified summer situation. II. Heterotrophic dinoflagellates and ciliates. Aquatic Microbial Ecology, 16(3): 217-232.

Li W K, McLaughlin F A, Lovejoy C, et al. 2009. Smallest algae thrive as the Arctic Ocean freshens. Science, 326(5952): 539.

Li W K W, Head E J H, Harrison W G. 2004. Macroecological limits of heterotrophic bacterial abundance in the ocean. Deep Sea Research Part I: Oceanographic Research Papers, 51(11): 1529-1540.

Lignell R. 1993. Fate of a phytoplankton spring bloom: sedimentation and carbon flow in the planktonic food web in the northern Baltic. Marine Ecology Progress Series, 94: 239-252.

Long A, McDaniel L D, Mobberley J, et al. 2008. Comparison of lysogeny(prophage induction) in heterotrophic bacterial and *Synechococcus* populations in the Gulf of Mexico and Mississippi river plume. The ISME Journal, 2(2): 132-144.

López-Urrutia Á. 2008. Comment: the metabolic theory of ecology and algal bloom formation. Limnology and Oceanography, 53(5): 2046-2047.

López-Urrutia A, Martin E S, Harris R P, et al. 2006. Scaling the metabolic balance of the oceans. Proceedings of the National Academy of Sciences of the United States of America, 103(23): 8739-8744.

Marrasé C, Lim E L, Caron D A. 1992. Seasonal and daily changes in bacterivory in a coastal plankton community. Marine Ecology Progress Series, 82(3): 281-289.

Mojica K D, Brussaard C P. 2014. Factors affecting virus dynamics and microbial host-virus interactions in marine environments. FEMS Microbiology Ecology, 89(3): 495-515.

Morán X A, Alonso-Saez L, Nogueira E, et al. 2015. More, smaller bacteria in response to ocean's warming? Proceedings of the Royal Society B: Biological Sciences, 282(1810): 20150371.

Morán X A G, Gasol J M, Pernice M C, et al. 2017. Temperature regulation of marine heterotrophic prokaryotes increases latitudinally as a breach between bottom-up and top-down controls. Global Change Biology, 23(9): 3956-3964.

O'Gorman E J, Pichler D E, Adams G, et al. 2012. Impacts of warming on the structure and functioning of aquatic communities. Advances in Ecological Research, 47: 81-176.

Panigrahi S, Nydahl A, Anton P, et al. 2013. Strong seasonal effect of moderate experimental warming on plankton respiration in a temperate estuarine plankton community. Estuarine, Coastal and Shelf Science, 135: 269-279.

Paulsen M L, Doré H, Garczarek L, et al. 2016. *Synechococcus* in the Atlantic Gateway to the Arctic Ocean. Frontiers in Marine Science, 3: 191.

Payet J P, Suttle C A. 2013. To kill or not to kill: the balance between lytic and lysogenic viral infection is driven by trophic status. Limnology and Oceanography, 58(2): 465-474.

Pearce I, Davidson A T, Thomson P G, et al. 2010. Marine microbial ecology off East Antarctica(30°～80°E): rates of bacterial and phytoplankton growth and grazing by heterotrophic protists. Deep Sea Research Part II: Topical Studies in Oceanography, 57(9): 849-862.

Pearce I, Davidson A T, Thomson P G, et al. 2011. Marine microbial ecology in the sub-Antarctic Zone: rates of bacterial and phytoplankton growth and grazing by heterotrophic protists. Deep Sea Research Part II: Topical Studies in Oceanography, 58(21): 2248-2259.

Pires A P F, Guariento R D, Laque T, et al. 2014. The negative effects of temperature increase on bacterial respiration are independent of changes in community composition. Environmental Microbiology Reports, 6(2): 131-135.

Polovina J J, Woodworth P A. 2012. Declines in phytoplankton cell size in the subtropical oceans estimated from satellite remotely-sensed temperature and chlorophyll, 1998～2007. Deep Sea Research Part II: Topical Studies in Oceanography, 77-80: 82-88.

Pomeroy L R, Deibel D. 1986. Temperature regulation of bacterial activity during the spring bloom in Newfoundland coastal waters. Science, 233(4761): 359-361.

Pomeroy L R, Macko S A, Ostrom P H, et al. 1990. The microbial food web in Arctic seawater: concentration of dissolved free amino-acids and bacterial abundance and activity in the Arctic-Ocean and in Resolute Passage. Marine Ecology Progress Series, 61(1-2): 31-40.

Pomeroy L R, Wiebe W J. 1988. Energetics of microbial food webs. Hydrobiologia, 159(1): 7-18.

Pomeroy L R, Wiebe W J. 2001. Temperature and substrates as interactive limiting factors for marine heterotrophic bacteria. Aquatic Microbial Ecology, 23(2): 187-204.

Pomeroy L R, Wiebe W J, Deibel D, et al. 1991. Bacterial responses to temperature and substrate concentration during the Newfoundland spring bloom. Marine Ecology Progress Series, 75: 143-159.

Pomeroy L, Wiebe W. 1993. Energy sources for microbial food webs. Marine Microbial Food Webs, 7(1): 101-118.

Proctor L M, Okubo A, Fuhrman J A. 1993. Calibrating estimates of phage-induced mortality in marine bacteria: ultrastructural studies of marine bacteriophage development from one-step growth experiments. Microbial Ecology, 25(2): 161-182.

Regaudie-de-Gioux A, Duarte C M. 2012. Temperature dependence of planktonic metabolism in the ocean. Global Biogeochemical Cycles, 26(1): GB1015.

Rivkin R B, Anderson M R, Lajzerowicz C. 1996. Microbial processes in cold oceans. I. Relationship between temperature and bacterial growth rate. Aquatic Microbial Ecology, 10: 243-254.

Rivkin R B, Legendre L. 2001. Biogenic carbon cycling in the upper ocean: effects of microbial respiration. Science, 291(5512): 2398-2400.

Rose J M, Caron D A. 2007. Does low temperature constrain the growth rates of heterotrophic protists? Evidence and implications for algal blooms in cold waters. Limnology and Oceanography, 52(2): 886-895.

Rose J M, Feng Y, Gobler C J, et al. 2009. Effects of increased pCO_2 and temperature on the North Atlantic spring bloom. II. Microzooplankton abundance and grazing. Marine Ecology Progress Series, 388: 27-40.

Rose J M, Fitzpatrick E, Wang A, et al. 2013. Low temperature constrains growth rates but not short-term ingestion rates of Antarctic ciliates. Polar Biology, 36(5): 645-659.

Safi K A, Griffiths F B, Hall J A. 2007. Microzooplankton composition, biomass and grazing rates along the WOCE SR3 line between Tasmania and Antarctica. Deep Sea Research Part I: Oceanographic Research Papers, 54(7): 1025-1041.

Sarmento H, Montoya J M, Vazquez-Dominguez E, et al. 2010. Warming effects on marine microbial food web processes: how far can we go when it comes to predictions? Philosophical Transactions of the Royal Society B, 365(1549): 2137-2149.

Savage V M, Gillooly J F, Brown J H, et al. 2004. Effects of body size and temperature on population growth. The American Naturalist, 163(3): 429-441.

Sherr E B, Sherr B F, Hartz A J. 2009. Microzooplankton grazing impact in the Western Arctic Ocean. Deep Sea Research Part II: Topical Studies in Oceanography, 56(17): 1264-1273.

Sommer U, Adrian R, Bauer B, et al. 2012. The response of temperate aquatic ecosystems to global warming: novel insights from a multidisciplinary project. Marine Biology, 159(11): 2367-2377.

Sorokin Y I. 1971. On the role of bacteria in the productivity of tropical oceanic waters. Internationale Revue der Gesamten Hydrobiologie und Hydrographie, 56(1): 1-48.

Stoecker D K, Nejstgaard J C, Madhusoodhanan R, et al. 2015. Underestimation of microzooplankton grazing in dilution experiments due to inhibition of phytoplankton growth. Limnology and Oceanography, 60(4): 1426-1438.

Vaqué D, Guadayol Ò, Peters F, et al. 2009. Differential response of grazing and bacterial heterotrophic production to experimental warming in Antarctic waters. Aquatic Microbial Ecology, 54(1): 101-112.

Vaquer-Sunyer R, Duarte C M, Santiago R, et al. 2010. Experimental evaluation of planktonic respiration response to warming in the European Arctic Sector. Polar Biology, 33(12): 1661-1671.

Vázquez-Domínguez E, Vaqué D, Gasol J M. 2007. Ocean warming enhances respiration and carbon demand of coastal microbial plankton. Global Change Biology, 13(7): 1327-1334.

Vázquez-Domínguez E, Vaqué D, Gasol J M. 2012. Temperature effects on the heterotrophic bacteria, heterotrophic nanoflagellates, and microbial top predators of the NW Mediterranean. Aquatic Microbial Ecology, 67(2): 107-121.

Verity P G, Wassmann P, Frischer M E, et al. 2002. Grazing of phytoplankton by microzooplankton in the Barents Sea during early summer. Journal of Marine Systems, 38(1-2): 109-123.

Wang C, Wang X, Xu Z, et al. 2022. Planktonic tintinnid community structure variations in different water masses of the Arctic Basin. Frontiers in Marine Science, 8: 775653.

Wang C, Xu Z, Liu C, et al. 2019. Vertical distribution of oceanic tintinnid(Ciliophora: Tintinnida) assemblages from the Bering Sea to Arctic Ocean through Bering Strait. Polar Biology, 42(11): 2105-2117.

Wei W, Zhang R, Peng L, et al. 2018. Effects of temperature and photosynthetically active radiation on virioplankton decay in the western Pacific Ocean. Scientific Reports, 8(1): 1525.

Wilken S, Huisman J, Naus-Wiezer S, et al. 2013. Mixotrophic organisms become more heterotrophic with rising temperature. Ecology Letters, 16(2): 225-233.

Williamson S, Paul J H. 2004. Nutrient stimulation of lytic phage production in bacterial populations of the Gulf of Mexico. Aquatic Microbial Ecology, 36(1): 9-17.

Wilson W H, Carr N G, Mann N H. 1996. The effect of phosphate status on the kinetics of cyanophage infection in the oceanic cyanobacterium *Synechococcus* sp. WH7803. Journal of Phycology, 32(4): 506-516.

Wilson W H, Mann N H. 1997. Lysogenic and lytic viral production in marine microbial communities. Aquatic Microbial Ecology, 13(1): 95-100.

Wohlers J, Engel A, Zöllner E, et al. 2009. Changes in biogenic carbon flow in response to sea surface warming. Proceedings of the National Academy of Sciences of the United States of America, 106(17): 7067-7072.

Yang E J, Jiang Y, Lee S. 2016. Microzooplankton herbivory and community structure in the Amundsen Sea, Antarctica. Deep Sea Research Part II: Topical Studies in Oceanography, 123: 58-68.